Nano-photonics for Advanced Networks

Kiyoshi Asakawa · Yoshimasa Sugimoto ·
Shigeru Nakamura

Nano-photonics for Advanced Networks

Springer

Kiyoshi Asakawa
University of Tsukuba
Tsukuba, Japan

Yoshimasa Sugimoto
National Institute for Materials Science
Tsukuba, Japan

Shigeru Nakamura
NEC Corporation
Tokyo, Japan

ISBN 978-3-031-80636-0 ISBN 978-3-031-80637-7 (eBook)
https://doi.org/10.1007/978-3-031-80637-7

This Springer imprint is published by the registered company Springer Nature Switzerland AG
The registered company address is: Gewerbestrasse 11, 6330 Cham, Switzerland

Preface

The terminology "nano-photonics" has been defined by the National Academy of Science of the USA as "the science and engineering of light interactions that take place on wavelength and sub-wavelength scales where the physical, chemical, or structural nature of natural or artificial nano-structured matter controls the interactions". Since the 1990s, novel nano-scale photonic structures and materials such as quantum dots, photonic crystals, surface plasmons, and metamaterials, all qualifying as nano-photonics under the aforementioned definition, have emerged in a wide-ranging science and technology of optoelectronics.

Quantum dots have primarily been developed for nano-scale lasers and single photon sources, while metamaterials have been explored for implementing unique light control techniques made possible by metallic and negative refractive-index nano-structures/-materials.

Photonic crystals—nano-scale periodic structures with high refractive-index contrast—have attracted much attention to their ultimate light-confinement and dynamic-control behaviors. Typical applications include extremely miniaturized high-power surface-emitting lasers and nano-photonic signal processing devices.

Recently, silicon photonics with high refractive-index contrast waveguides on silicon on insulator (SOI) has emerged as a new nano-photonics family. Owing to its small footprint, low cost, low power consumption, and high-speed performance, all made available by complementary metal–oxide–semiconductor (CMOS)-compatible process technologies, silicon photonics has provided many potentially key solutions to problems of ever-increasing network demands such as high power efficiency, wide bandwidth, and high transmission speed, strongly triggered by the explosively expanding traffics in the telecom/data-com worlds. Typical applications include compact optical interconnect transceivers, innovative multi-cast photonic switches in the photonic core network, optical phased array light detection and ranging (LiDAR), photonic neuro-computing accelerators, etc.

Surface plasmon—a composite electromagnetic wave extremely confined at a metallic/dielectric interface—is now under urgent development for a novel possibility: a compact waveguide platform capable of integration with nano-electronics, thus exhibiting diffraction-limit free small footprint, wide bandwidth, and high transmission speed performances. Typical applications include 500-GHz plasmonic modulator for short optical link interconnect transceivers. This approach paves the way for the surface plasmon-based hybrid nano-optoelectronics.

By putting the main purpose of this book on the applications in advanced photonic networks, quantum dots, and metamaterials are not included because of their inherently different application areas: This book is composed of three nano-photonic structures and performances; silicon photonics, photonic crystal, and surface plasmon.

This book is composed of seven chapters: Chaps. 1–3 review fundamentals of silicon photonics, photonic crystal, and surface plasmon. Chapters 4–7 describe the applications of these three topics to advanced photonic networks. In particular, Chap. 4 is directed to silicon photonic waveguides for network transceivers and switches, while Chap. 5 is directed to silicon LiDAR and optical neural network.

Applications in this book are intended for both specialists in photonics and newcomers. In addition, photonic network applications are specifically written to be beneficial not only to photonic device engineers but also to photonic network system engineers who wish to study further the expanding challenges posed by the explosive increase in photonic network traffics and the latest solutions to them. This book contains a wide range of nascent applications enabled by the aforementioned three nano-photonics; in the end, note that these nascent applications will definitely inspire nano-photonic industries and engineers to hit on the killer applications for which they are searching continuously.

Tsukuba, Japan — Kiyoshi Asakawa
Tsukuba, Japan — Yoshimasa Sugimoto
Tokyo, Japan — Shigeru Nakamura
June 2024

Acknowledgements

The book—nano-photonics—has been based on the lecture notes made by one of the authors, K. Asakawa, for an intensive topical lecture in the Indian Institute of Information, Technology, Design and Manufacturing (IIITDM) at Jabalpur, India, in 2017–2020. The author would like to acknowledge the support of Prof. V. Dinesh Kumar to give me a motivation of publishing the book to spread the contemporary topics in the nano-photonics throughout the world.

March 2024 Kiyoshi Asakawa

Contents

1 **Survey of Dielectric Photonic Waveguides** 1
1.1 Introduction 1
1.2 Mode Propagation and Dispersion Relation 2
1.2.1 Ray-Optics Approach 2
1.2.2 Electromagnetic-Wave Approach 5
1.2.3 Phase and Group Velocities 7
1.3 Waveguide Configurations and Materials 9
1.3.1 Two-/Three-Dimensional Waveguide Structures 9
1.3.2 Materials and Refractive Indices 10
1.4 Core Waveguide Structures 10
1.4.1 Interferometric Waveguide 11
1.4.2 Directional Coupler 13
1.4.3 Optical Resonator 14
1.4.4 Arrayed Waveguide Grating 17
1.4.5 Polarization Beam Splitter and Rotator 18
1.4.6 Fiber-to-Chip Light Coupler 21
1.5 Summary 23
References 24

2 **Survey of Basic Photonic Crystal Waveguides** 25
2.1 Introduction 25
2.2 Type of Photonic Crystal 26
2.2.1 One-Dimensional to Three-Dimensional Structures 26
2.2.2 Lattice Configuration 28
2.3 Light in Photonic Crystal 28
2.3.1 Light in a Uniform Medium 28
2.3.2 Photonic Band in a Periodic Medium 29
2.3.3 Photonic Band Gap 30

2.4 Defect Mode in Photonic Crystal 31
2.4.1 Type of Defect 31
2.4.2 Defect Mode in Band Gap 32
2.5 Two-Dimensional Photonic Crystal Slab 32
2.5.1 Modes in Photonic Band Gap 32
2.5.2 Type of Photonic Crystal Slab 34
2.6 Pioneering Works on Photonic Crystal Slab Waveguide 35
2.6.1 Theoretical Work on Bend Waveguide 35
2.6.2 Experimental Work on Low-Loss Waveguide 36
2.7 Peculiar Phenomena in Photonic Crystal 37
2.7.1 Super-prism Effect in Three-Dimensional Photonic Crystal 38
2.7.2 High-*Q* Nano-cavity in Two-Dimensional Photonic Crystal 40
2.7.3 Slow-Light Effect in Photonic Crystal Slab Waveguide 42
2.7.4 Band-Edge Emission in Two-Dimensional Photonic Crystal 43
2.8 Summary 44
References 45

3 Survey of Surface Plasmon Waveguides 47
3.1 Introduction 47
3.2 Metal Optics of Plasmon-Polariton 48
3.2.1 Drude Model 48
3.2.2 Dielectric Function of Metal 49
3.2.3 Dispersion Relation of Plasmon 51
3.3 Surface Plasmon 52
3.3.1 Dispersion Relation of Surface Plasmon 52
3.3.2 Physical Properties of Surface Plasmon 54
3.4 Excitation of Surface Plasmon 57
3.4.1 Configurations of SP Excitation 57
3.4.2 Kretschmann Configuration 59
3.5 Localized Surface Plasmon 60
3.5.1 Localized Surface Plasmon Resonance 60
3.5.2 Metal Nano-particle Optics 60
3.5.3 Metal Nano-antenna Optics 63
3.6 Summary 66
3.7 Appendix 66
References 69

4 Applications of Silicon Photonic Waveguides (I) Network Transceivers and Switches 71
4.1 Introduction 71
4.2 Progress of Silicon Photonic Platform 73
4.2.1 CMOS-Compatible Silicon Photonic Platform 73

4.2.2 Advanced Laser-on-Silicon Platform 78
4.2.3 Forecast of Silicon Photonics 81
4.3 Silicon Photonic Interconnect for Data Center and Short Optical Links 82
4.3.1 Front Runners in Silicon Photonic Transceiver Research 82
4.3.2 Monolithically Integrated Silicon Photonic Transceiver 85
4.3.3 Ultra-compact Silicon Photonic Transceiver 88
4.4 Silicon Photonic Switch for Core Network and Data Center 93
4.4.1 Architecture of CDC–ROADM 94
4.4.2 Multi-cast Silicon Photonic Switch for CDC–ROADM 97
4.4.3 Densely Integrated Silicon Photonic Switch 101
4.4.4 Silicon Photonic Switch for Data Center 103
4.5 Summary 106
References 107

5 Applications of Silicon Photonic Waveguides (II) LiDAR and Optical Neural Network 111
5.1 Introduction 111
5.2 Si Photonic LiDAR for Remote Sensing 112
5.2.1 Integrated Optical Beam Steering Arrays 112
5.2.2 FMCW–LiDAR with High-Performance OPA 119
5.2.3 LiDAR with Focal Plane Arrays 124
5.2.4 LiDAR with Slow-Light Grating and Prism Lens 126
5.3 Optical Neural Network for Deep Learning 129
5.3.1 Photonic Accelerator for "Beyond CMOS" Platform 129
5.3.2 Programmable Nano-photonic Processor for On-Chip ONN 132
5.3.3 Complex-Valued High-Accuracy ONN 134
5.3.4 New Model for Nonlinear Mapping-Based Classification 138
5.3.5 Space-Efficient Integrated Diffractive ONN 142
5.3.6 Ever-Growing Si Photonic Neural Networks 145
5.4 Summary 146
References 147

6 Applications of Photonic Crystal Waveguides 151
6.1 Introduction 151
6.2 Slow-Light Photonic Crystal Device 153
6.2.1 Potential Application of Slow-Light Devices 153
6.2.2 Dispersion Control of Slow-Light Waveguide 154
6.2.3 Demonstration of Slow-Light Device 159

6.3 Ultra-fast All-Optical Photonic Crystal Switch 162
6.3.1 Quantum Dot-Combined Analogue Photonic Crystal Switch 162
6.3.2 Flip-Flop-Based Digital Photonic Crystal Switch 165
6.4 Nano-cavity-Based Photonic Crystal DRAM 170
6.4.1 Principle of Optical Bi-stability in Fabry–Perot Cavity 170
6.4.2 Bi-stable Switch in Si High-Q Nano-cavity 171
6.4.3 Bi-stable Switch in InGaAsP/InP BH Laser Nano-cavity 174
6.4.4 Large-Scale All-Optical DRAM with PC Nano-cavity 176
6.5 Extreme High-Power Photonic Crystal Surface-Emitting Laser 179
6.5.1 Photonic Crystal Surface-Emitting Laser: PCSEL 180
6.5.2 Polarization Control by Unit Shape 182
6.5.3 Room-Temperature Continuous-Wave PCSEL 183
6.5.4 Watt-Class RT CW–PCSEL 183
6.5.5 Double-Hole Lattice PCSEL for 10 W-Class Lasing 185
6.5.6 Dually Modulated PCSEL for Smart Sensing 188
6.5.7 Application of PCSEL to LiDAR Sensing Mobility 190
6.6 Summary 192
References 194

7 Applications of Surface Plasmon Waveguides 197
7.1 Introduction 197
7.2 Development of Plasmonic Waveguides 199
7.2.1 Metal-Based Plasmonic Waveguide 199
7.2.2 Hybrid Plasmonic Waveguide 202
7.3 High-Speed and Broadband Plasmonic Devices 204
7.3.1 High-Efficiency Plasmonic Mode Converter 204
7.3.2 High-Speed and Broadband Plasmonic Modulator 206
7.3.3 High-Speed and Broadband Plasmonic Detector 208
7.4 Application to Microwave Plasmonics in Wireless Communication 208
7.4.1 Plasmonic Nano-antenna 209
7.4.2 Optical Interconnect with Plasmonic Device Array 209
7.4.3 500 GHz Plasmonic Modulator for Sub-THz Photonics 211
7.5 Summary 213
References 213

Index 217

Abbreviations

2-D	Two-dimensional
3-D	Three-dimensional
5G	Fifth generation
ADM	Add/drop multiplexer
AI	Artificial intelligence
ANN	Artificial neural network
APN	All-photonic network
ATR	Attenuated total reflection
AWG	Arrayed waveguide grating
BCB	Benzo-cyclobutene
BER	Bit error rate
BFO	Bacterial foraging optimization
BH-PC	Buried hetero-structure photonic crystal
BOX	Buried oxide
BPSK	Binary phase shift keying
BS	Beam splitter
CCD	Charge coupled device
CDC	Colorless, directionless, and contention-less
CMOS	Complementary metal–oxide–semiconductor
CNN	Convolutional neural network
CP	Control pulse
CW	Continuous wave
DBR	Distributed Bragg reflector
DC	Directional coupler
Demux	De-multiplexer
DH	Double-hole lattice
DH-PCSEL	Double-hole lattice photonic crystal surface-emitting laser
DMMC	Diagonal matrix multiplication core
DM-PCSEL	Dually modulated photonic crystal surface-emitting laser
DQPSK	Differential quadrature phases-shift keying
DWDM	Dense wavelength division multiplex

EB	Electron beam
EM	Electromagnetic
EO	Electro-optic
EO effect	Electro-optic Pockels effect
ER	Extinction ratio
FDTD	Finite-difference-time-domain
FE	Field enhancement
FFP	Far field pattern
FMCW	Frequency-modulated continuous wave
FPA	Focal plane array
FPGA	Field programmable gate array
FPR	Fabry–Perot resonator
FPG	Free-propagation region
FSR	Free spectral range
FWHM	Full-width at half maximum
GC	Grating coupler
GF	GlobalFoundry
GVD	Group velocity dispersion
GW	Grating waveguide
HBM	High-bandwidth memory
HCG	High index-contrast grating
HDLPW	Hybrid dielectric-loaded plasmonic waveguide
HIMI	Hybrid insulator-metal-insulator
HMC	Hybrid memory cube
HMIM	Hybrid metal-insulator-metal
IDNN	Integrated diffractive neural network
IF	Intermediate frequency
III-V	III-V compound semiconductor
IL	Insertion loss
IMI	Insulator–metal–insulator
INT	Inter-leaver block
IoT	Internet of Things
IOWN	Innovative optical and wireless network
IR	Infra-red
LCOS	Liquid crystal on silicon
LD	Laser diode
LGA	Land grid array
LiDAR	Light detection and ranging
LO	Local oscillator
LRSP	Long-range surface plasmon
LSI	Large-scale integration
LSP	Localized surface plasmon
LSPCW	Lattice-shifted photonic crystal waveguide
LSPR	Localized surface plasmon resonance
LVLD	Low group-velocity and low group-velocity dispersion

MBE	Molecular beam epitaxy
MCM	Multi-chip-module
MCS	Multicast switch
MEMS	Micro-electromechanical system
MIM	Metal–insulator–metal
ML	Machine learning
MMF	Multi-mode fiber
MMI	Multi-mode interferometer
MMW	Millimeter-wave
MOSFET	Metal–oxide–semiconductor field-effect transistor
MOVPE	Metal organic vaper phase epitaxy
MQW	Multiple quantum-well
MRM	Micro-ring modulators
MRR	Micro-ring resonator
MSE	Mean squared error
MSM	Metal–semiconductor–metal
Mux	Multiplexer
MVM	Matrix-vector multiplication
MZI	Mach–Zehnder interferometer
MZM	Mach–Zehnder modulator
NIR	Near infra-red
NRZ	Non-return-to-zero
NW	Nano-wire
OCS	Optical communication system.
ODFT	Optical discrete Fourier transform
OEICs	Optoelectronic integrated circuit
OIDFT	Optical inverse discrete Fourier transform
OIU	Optical interference unit
ONC	Optical neural chip
ONL	Optical nonlinearity
ONN	Optical neural network
ONU	Optical network unit
ONLU	Optical nonlinearity unit
OPA	Optical phased array
o-RAM	Optical random access memory
PB	Photonic band
PBG	Photonic band gap
PBS	Polarization beam splitter
PC	Photonic crystal
PCCW	Photonic crystal coupled waveguide
PCSEL	Photonic crystal surface-emitting laser
PD	Photodetector
PDL	Polarization-dependent loss
PIC	Photonic integrated circuit
PILOSS	Path-independent-loss

PL	Propagation length
PLC	Planar lightwave circuit
PLL	Phase-locked loop
PNP	Programmable nano-photonic processor
PON	Passive optical network
PP	Plasmon polariton
PPC	Projection-based photonic classifier
PR	Polarization rotator
PhS	Phase shifter
PS	Polarization splitter
PSP	Propagating surface plasmon
PSR	Polarization splitter-rotator
PSTM	Photon scanning tunneling microscope
PW	Plasmonic waveguide
QD	Quantum dot
QPSK	Quadrature phase shift keying
QW	Quantum well
RAM	Random access memory
RAU	Remote antenna unit
ROADM	Reconfigurable optical add-drop multiplexing
RoF	Radio-over-fiber
RT	Room temperature
Rx	Receiver
S&S	Switch and select
SDM	Space-division multiplexing
SEM	Scanning electron microscope
SMF	Single-mode fiber
SOA	Semiconductor optical amplifier
SoC	System-on-chip
SOI	Silicon-on-insulator
SgP	Signal pulse
SP	Surface plasmon
SPGC	Single polarization grating couplers
SPM	Self-phase modulation
SPR	Surface plasmon resonance
SSC	Spot size converter
SVM	Support vector machine
SWG	Sub-wavelength grating
T/R	Transmitter/receiver
TDM	Time domain multiplexing
TE	Transverse electric
TIR	Total internal reflection
TM	Transverse magnetic
TO	Thermo-optic
ToP	Topology optimization

ToF	Time-of-flight
TPA	Transponder aggregator
TwPA	Two-photon absorption
TSV	Through silicon via
TWDM	Time and wavelength division multiplexing
Tx	Transmitter
UV	Ultra-violet
VCSEL	Vertical-cavity surface-emitting laser
VGC	Vertical grating coupler
VMM	Vector-matrix multiplication
VOA	Variable optical attenuators
WDM	Wavelength-division multiplex
WSS	Wavelength selective switch
WWW	World Wide Web
XOR	Exclusive OR
YI	Young's interferometer

Chapter 1
Survey of Dielectric Photonic Waveguides

Abstract Nano-photonic waveguides are categorized into three types: silicon (Si) photonics, photonic crystals, and surface plasmons. Si photonics is based on integrated high refractive-index contrast Si-on-insulator (SOI) waveguides with nanoscale widths and thicknesses called nano-wires (NWs). This chapter surveys fundamentals of Si photonics based on the conventional dielectric optical waveguides, that is, mode theories, geometries, materials, and device configurations: First, guided mode theories and dispersion properties are described based on ray-optics and electromagnetic-wave approaches. Second, two- to three-dimensional waveguide structures and materials are surveyed. Finally, configurations and functions of typical passive waveguides as core elements of Si photonic platform are surveyed. They include interferometric waveguides, directional couplers, optical resonators, arrayed waveguide gratings, polarization beam splitters/rotators, and fiber-to-chip light couplers. Applications of Si photonic waveguides such as hybrid lasers, detectors, modulators/switches are described in Chap. 4 (optical transceiver and switch) and Chap. 5 (LiDAR and optical neural network).

1.1 Introduction

A free-space light wave is approximated by a plane wave with a constant frequency ω, an infinite parallel-plane wave front vertical to a wave vector k, and a constant peak-to-peak amplitude, whereas a guided light wave is an optical wave propagating along a dielectric waveguide with total internal reflection (TIR) repeated between core/clad interfaces within the waveguide, like single- or multi-mode in an optical fiber. Here, on the premise that this book mainly focuses on planar optical waveguide devices/circuits, we consider a slab waveguide and investigate characteristics of the guided mode such as propagation constant, guided mode condition, intensity distributions, etc. Next section begins with a ray-optics approach: a classical geometric approach with a good approximation on direct ray tracing along the waveguide, omitting detailed waveform characteristics. This approach is followed by the more strict approach: an electromagnetic-wave approach based on Maxwell's equations [1–3].

K. Asakawa et al., *Nano-photonics for Advanced Networks*,
https://doi.org/10.1007/978-3-031-80637-7_1

1.2 Mode Propagation and Dispersion Relation

1.2.1 Ray-Optics Approach

Guided Mode

Figure 1.1a shows a schematic of a slab waveguide with a core region of refractive index n_1 and clad region of index n_2, where $n_1 > n_2$. A light wave propagates along the waveguide with an intensity mainly confined in the core region. Figure 1.1b is a cross-section of the slab waveguide based on ray-optics description in which a light beam propagates with TIR repeated at the core/clad interface. When a wave vector (or propagation constant) in vacuum is $\boldsymbol{k}_o$, a wave vector $\boldsymbol{k}$ in the n_1 medium is $\boldsymbol{k} = \boldsymbol{k}_0 n_1$. Then, the longitudinal $\boldsymbol{k}$-component $\boldsymbol{\beta}$ and transverse $\boldsymbol{k}$-component $\boldsymbol{\gamma}$ are expressed as:

$$\boldsymbol{\beta} = \boldsymbol{k}_0 n_1 \cos\theta, \quad \boldsymbol{\gamma} = \boldsymbol{k}_0 n_1 \sin\theta, \tag{1.1}$$

where θ is an angle between $\boldsymbol{k}$ and $\boldsymbol{\beta}$, as shown in Fig. 1.1b, and $0 < \theta < \theta_c$. Here, θ_c is a critical angle that gives $\cos\theta_c = n_2/n_1$ and determines the condition that TIR disappears and a leaky mode appears due to penetration of the light beam into the clad region. In this sense, θ_c is a critical parameter that allows a dispersion relation characterized by the discrete guided modes, as described later. Note that confinement

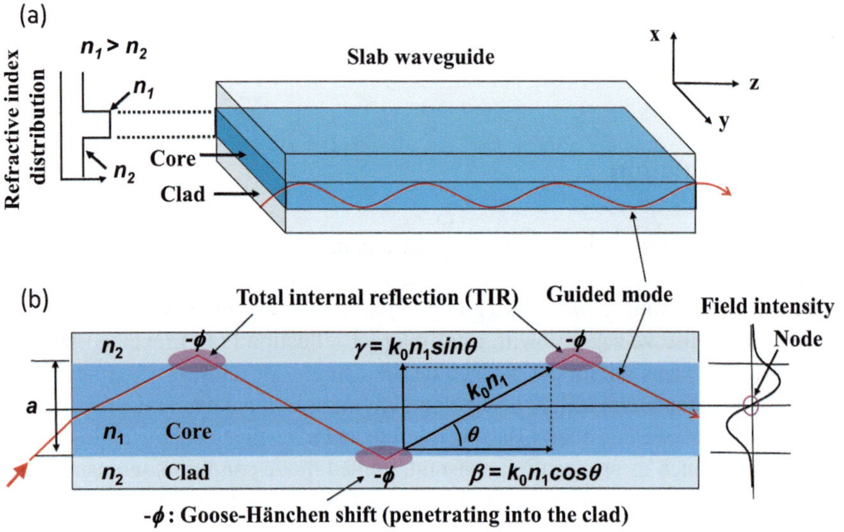

Fig. 1.1 **a** Schematic of a slab waveguide with core and clad layers, where their refractive indices are n_1 and n_2, respectively. **b** Cross-section of the slab waveguide based on ray-optics description where the light beam propagates with TIRs repeated at the core/clad interface

of the guided mode in the core region without leakage enforces the transverse wave vector γ to satisfy the following equation in a phase shift $\Delta\Psi$ as a result of the resonance within a waveguide cavity:

$$\Delta\Psi = 2a\boldsymbol{\gamma} + 2\phi = 2\pi N \quad (N : \text{integer}\, 0, 1, 2, \ldots). \tag{1.2}$$

This is called a mode equation based on the ray-optics approach, from which N-dependent standing waves arise within the core thickness a. Here, ϕ is called the Goose–Hänchen shift: To get insight into the TIR phenomenon in the vicinity of the core/clad interface, fields of the guided light homogeneously penetrate into the clad region, which causes an extra phase shift ϕ for the propagating guided light accompanied by the TIR. When θ is much smaller than θ_c, it nearly equals $-\pi$. For simplicity, a description of the phase shift ϕ including the waveguide parameters is omitted here.

In Eq. (1.2), an example of the field intensity profile for $N = 1$ is plotted at the right-side panel in Fig. 1.1b, where the first-order standing wave is depicted within the thickness a. In addition, a number of nodes (one in this case) is indicated. As shown on the right side, N equals the node number in the standing wave profile. In Fig. 1.2, transverse field distributions of the light propagating along the waveguide for three lowest-order modes $N = 0$, 1, and 2 are schematically displayed in color. See also the right-side panels in Fig. 1.2, which shows different-order standing wave profiles depending on N.

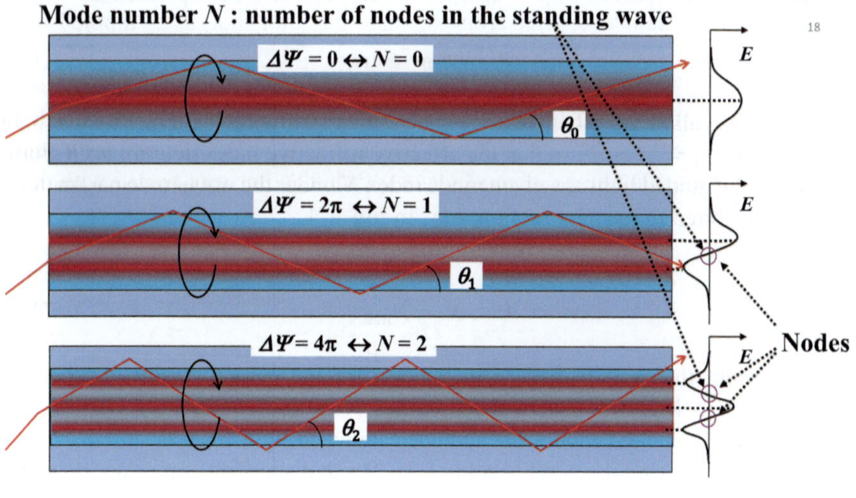

Fig. 1.2 Schematics of thickness-direction distributions of the light intensity along the waveguide for the three lowest-order modes $N = 0$, 1, and 2. Intensity profiles corresponding to the color display are also shown in standing wave profiles at the right edge panels. Courtesy of Prof. H. Yamada

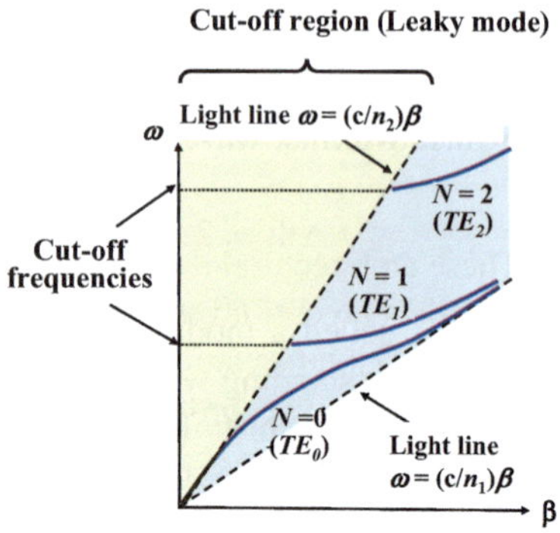

Fig. 1.3 Schematic of a $\omega - \beta$ dispersion relation for optical guided modes propagating along the slab waveguide with core (index: n_1) and clad (n_2) structures. Curves in blue show the lowest-order three modes $N = 0$, 1, and 2. TE: transverse electric

Mode Dispersion

An electromagnetic field distribution of a discrete guided mode in the waveguide is characterized by the mode number N in Eq. (1.2), while its fundamental features are briefly viewed by a dispersion relationship between a frequency ω and a longitudinal wave number β. As shown in Fig. 1.1b, N is a discrete number, while θ is limited by $0 < \theta < \theta_c$, where θ_c is a critical angle for the TIR effect as described before. The above inequality on θ can be changed into the following:

$$k_0 n_1 \cos\theta_c = k_0 n_2 < \beta < k_0 n_1 \quad (\cos\theta_c = n_2/n_1). \tag{1.3}$$

Since β can be replaced by $\beta = k_0 n_{\text{eff}}$, Eq. (1.3) can also be expressed as:

$$n_2 < n_{\text{eff}} < n_1, \tag{1.4}$$

where n_{eff} is called an effective refractive index. Considering Eq. (1.1), n_{eff} is expressed as $n_{\text{eff}} = n_1 \cos\theta$, so that the effective refractive index determines a phase velocity of the guided light at a given mode index N under the waveguide parameters. It is also exhibited in the next section, e.g., in Fig. 1.4.

Back to Eq. (1.2), an integer N is limited to $N_{\max}$:

$$0 < N < N_{\max}. \tag{1.5}$$

As a result, the $\omega - \beta$ dispersion relation for the guided modes is schematically shown in Fig. 1.3, where three modes $N = 0$, 1, and 2 are depicted as examples. These guided modes are included between light lines $\omega = (c/n_2)\beta$ and $\omega = (c/n_1)\beta$, reflecting the condition in (1.3). These dispersion curves are theoretically derived in principle from the following equations: Maxwell's equations, wave equations, assumed harmonic time-dependent electric field functions, and boundary conditions at the core/clad interface in the dielectric slab waveguide. Mathematical treatments with these equations are omitted here. Refer to [2, 3] for detail.

Frequencies for $\beta = k_0 n_2$ are called cut-off frequencies, which allow the guided modes to vary into the leaky modes extending to the cut-off region (in yellow) in Fig. 1.3.

1.2.2 Electromagnetic-Wave Approach

Extension of Maxwell's Equations

Mode dispersions of the guided waves in a symmetric slab waveguide are derived by a ray-optics approach, as shown in Fig. 1.3. The result shows that light lines in the core and clad regions control the zone allowed for the guided modes, while cut-off frequencies clarify the boundary between the guided modes and leaky modes except for the lowest-order mode.

Yet, for more practical design and simulation techniques of waveguide devices, an electromagnetic-wave approach based on Maxwell's equations provides a more detailed formulation of the guided wave. Based on this approach, we shortly overview the theoretical treatment which explicitly relates waveguide parameters to propagation characteristics of the guided waves with several normalized wave numbers. Now we adopt an asymmetric planar dielectric slab waveguide, as shown in Fig. 1.4a, assuming wider applications such as Si photonics under an air/waveguide/substrate system. Under this structure, light propagates in the z-direction with fields confined in the x-direction and extending infinitely in the y-direction. A refractive index of

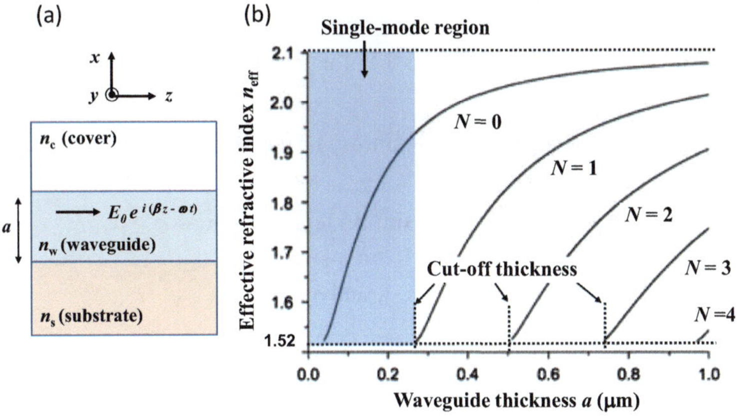

Fig. 1.4 **a** Schematic cross-section of an asymmetric dielectric slab waveguide with refractive indices of n_w, n_c, and n_s for cover (upper clad), waveguide (core), and substrate (lower clad) regions, respectively. a is a waveguide thickness. **b** Calculation of the TE-mode diagram for the asymmetric dielectric planar waveguide with refractive indices of the substrate, $n_s = 1.52$; waveguide, $n_w = 2.1$; cover, $n_c = 1.333$; at wavelength, $\lambda = 675$ nm. The single-mode region and cut-off thicknesses are indicated

the waveguide (core) region is n_w, whereas those of cover (upper clad) and substrate (lower clad) regions are n_c and n_s, respectively: $n_{c,s}$ for simplicity. A waveguide thickness is *a,* while the material is assumed to be non-magnetic ($\mu_0 = 1$), current- and charge density-free ($J = \rho = 0$) isotropic dielectrics. Moreover, we consider only a TE mode. Then only E_y, H_x, and H_z are nonzero components, and $E_z = E_x = H_y = 0$. Here, a solution from Maxwell's equations is assumed to be a harmonic time-dependent waveform: $A(t) = A\ e^{i\omega t}$, with an angular frequency $\omega = 2\pi c/\lambda$. In the planar waveguide, solutions from Maxwell's equations can be described as $A\ (x, y, z) = A_m\ (x, y)\ e^{-i\beta mz}$, where $\beta_m = (2\pi/\lambda_0)\ n_w \cos\theta$ is a propagation constant and *m* is a mode index. Hereafter, *m* is omitted for simplicity.

From these conditions, curl equations in Maxwell's equations are:

$$H_x = -\frac{\beta}{\omega}E_y, \quad H_z = \frac{i}{\omega}\frac{\partial E_y}{\partial x}, \quad \frac{\partial H_z}{\partial x} - i\beta H_x = -i\omega m_w^2 E_y. \tag{1.6}$$

The wave equation for the TE mode is:

$$\frac{\partial^2 E_y}{\partial x^2} + \left\{[n(x)k_0]^2 - \beta^2\right\}E_y = 0, \tag{1.7}$$

where k_0 is a wave number in vacuum ($k_0 = 2\pi/\lambda_0 = \omega/c$), while $n(x)$ is a refractive-index distribution in the waveguide cross-section: $n(x) = n_w$, n_c, and n_s in the waveguide, cover region, and substrate region, respectively. For boundary conditions, tangential components of E_y and H_z are continuous. Guided modes are bounded by the condition: $k_0\ n_{c,s} \le \beta \le k_0\ n_w$. Inserting this boundary condition into Eqs. (1.6) and (1.7) yields for TE mode:

$$E_y = \mathrm{A}e^{-\gamma x} \quad \text{for } x \ge 0 \tag{1.8}$$

$$E_y = A\left(\cos(\delta x) - \frac{\gamma}{\delta}\sin(\delta x)\right) \quad \text{for } 0 \ge x \ge -a, \tag{1.9}$$

$$E_y = \mathrm{A}\left(\cos(\delta x) - \frac{\gamma}{\delta}\sin(\delta x)\right)e^{\alpha(x+h)} \quad \text{for } -a \ge x, \tag{1.10}$$

where *A* is an arbitrary constant, and α, γ, and δ are:

$$\alpha = \sqrt{\beta^2 - k_0^2 n_s^2}, \tag{1.11}$$

$$\gamma = \sqrt{\beta^2 - k_0^2 n_c^2} \tag{1.12}$$

$$\delta = \sqrt{k_0^2 n_w^2 - \beta^2}. \tag{1.13}$$

Tangential components E_ys should be continuous at each interface between the media. These conditions provide a set of linear equations having nontrivial solutions if the determinant of the coefficient becomes zero, whereby the following eigenvalue is obtained:

$$\tan(\delta a - m\pi) = \frac{2\gamma\delta}{\delta^2 - \gamma\alpha}. \tag{1.14}$$

Since $\beta = n_{\text{eff}}\, k$, as shown earlier, Eq. (1.14) provides an important relationship between the effective refractive index n_{eff} and waveguide thickness a.

TE-Mode Diagram

The eigenvalue equation (1.14) can be solved either graphically or numerically. Figure 1.4b shows calculated mode diagrams on an effective refractive index n_{eff} as a function of the waveguide thickness a for the TE mode. Parameters in the used asymmetric dielectric planar waveguide are: $n_{\text{s}} = 1.52$, $n_{\text{w}} = 2.1$, $n_{\text{c}} = 1.33$ at $\lambda_0 =$ 675 nm. The result shows the single-mode region and cut-off thicknesses for a given waveguide structure. When an n_{eff} comes closer to the n_{w}, a transverse field profile of the guided wave is confined stronger into the waveguide, while for an n_{eff} closer to the n_{s}, transverse field profile is extended wider in the substrate. These data function as important guidelines for designing the waveguide devices such as passive/active photonic network devices.

1.2.3 *Phase and Group Velocities*

This section surveys wave-propagation-related fundamental phenomena: a phase velocity and group velocity in the light wave. As a fundamental technology in optical

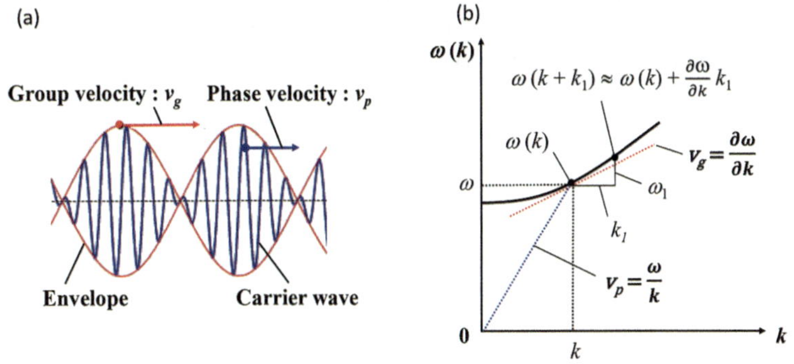

Fig. 1.5 **a** Schematic of an amplitude-modulated waveform showing the phase velocity v_{p} and group velocity v_{g}. **b** $\omega(k)$ point on the dispersion curve showing phase velocity v_{p} (dotted blue line) and group velocity v_{g} (dotted red line)

communication, a carrier light wave essentially needs to be amplitude- (phase-, frequency-) modulated for transmitting signal information during propagating in the medium, like audio signals in a radio wave and video signals in a TV wave. Here, an amplitude-modulated light wave function in a simplified form is illustrated in Fig. 1.5a: An amplitude-modulated profile is called an envelope. Then, a propagation velocity of the carrier wave is called a phase velocity v_p, whereas a propagation velocity of the envelope is called a group velocity v_g. When an amplitude profile $g(x, t)$ of the carrier wave (in blue) $\cos(kx - \omega t)$ is modulated into an envelope profile (in red) $g_0 \sin(k_1 x - \omega_1 t)$, then the total wave function $f(x, t)$ is presented as:

$$f(x,t) = g_0 \sin(k_1 x - \omega_1 t)\cos(kx - \omega t), \quad \text{where } k_1 \ll k, \omega_1 \ll \omega. \tag{1.15}$$

This function involves a wave number component, $k \pm k_1$, and a frequency component, $\omega \pm \omega_1$. As stated above, a carrier wave propagates at a phase velocity $v_p = \omega/k$, while a modulated wave, i.e., an envelope, propagates at a group velocity $v_g = \omega_1/k_1$, much slower than the ω/k. As illustrated in Fig. 1.5b, formulation of the group velocity v_g is generalized using the following Taylor expansion around the point (ω, k) on the dispersion curve $\omega(k)$ as:

$$\omega(k + k_1) \approx \omega(k) + \frac{\partial \omega}{\partial \mathrm{k}} k_1. \tag{1.16}$$

Since the second term is nothing but determining the group velocity, v_g is generalized in the $\omega(k)$ dispersion curve as:

$$v_g = \frac{\omega_1}{k_1} = \frac{\left(\frac{\partial \omega}{\partial \mathrm{k}}\right)k_1}{k_1} = \frac{\partial \omega}{\partial \mathrm{k}}. \tag{1.17}$$

In this way, the group velocity can be newly defined as $v_g = \partial\omega/\partial k$.

Typically, the phase velocity v_p, illustrated in the tilted dotted line in blue in Fig. 1.5b, is larger than the group velocity v_g, similarly illustrated in the tilted dotted line in red. The reason is verified by noticing the two arbitrary points, (ω, k) and $(\omega + \omega_1, k + k_1)$ on the dispersion curve as:

$$v_g = \frac{\partial \omega}{\partial \mathrm{k}} = \frac{\partial \mathrm{k} v_p}{\partial \mathrm{k}} = \frac{\partial \mathrm{k}}{\partial \mathrm{k}} v_p + k \frac{\partial v_p}{\partial k} = v_p - \lambda \frac{\partial v_p}{\partial \lambda} (k = 2\pi/\lambda). \tag{1.18}$$

When the last term $\lambda\, (\partial v_p/\partial \lambda) > 0$, Eq. (1.18) means that $v_p > v_g$; the phase velocity is proved to be larger than the group velocity. However, this result depends on the properties of the medium: The medium in which $v_p = v_g$ is called a non-dispersive medium, while the medium in which $v_g < v_p$ is called a normal-dispersive medium. The medium in which $v_g > v_p$ is called an anomalous-dispersive medium. Note that the dispersion is not the property of the waves themselves but the property of the medium in which a wave travels. As a particular example of an artificially designed or structured waveguide, a photonic crystal waveguide provides a low (or even zero) group velocity device called a slow-light device. Refer to Chaps. 2 and 6 for detail.

1.3 Waveguide Configurations and Materials

1.3.1 Two-/Three-Dimensional Waveguide Structures

In an integrated optical device/circuit, three-dimensional (3-D) waveguides are widely used, while two-dimensional (2-D) waveguides are used limitedly for sufficiently wide waveguides where in-plane expansion of the beam due to diffraction can be ignored. Figure 1.6 shows schematic cross-sections of 2-D and 3-D waveguides. The region in dark red shows a core region with high refractive index and light energy confined.

2-D Waveguide

- *Slab waveguide*: A high-index planar layer is operated as a 2-D wave-guiding core region on the low-index substrate. Propagating light is confined in the vertical direction only, while expanding infinitely in the lateral direction.

3-D waveguide

- *Buried waveguide*: A high-index wave-guiding core region is buried in a surrounding low-index clad region. The core region often has a rectangular shape.
- *Ridge waveguide*: A high-index wave-guiding core region with a strip shape is placed on top of its low-index planar substrate as a clad structure. Since three sides are surrounded by low-index air, optical confinement is strong in the ridge waveguide. Si waveguide platform (high-index region) fabricated on the SiO_2 layer (low-index region) is used mostly with this configuration for wide Si photonic applications because Si-on-insulator (SOI) complementary metal–oxide–semiconductor (CMOS) process technologies are available without serious difficulty.
- *Strip-loaded waveguide*: A dielectric strip with a low index n_{strip} is formed on a planar waveguide with a high index n_{planar} ($n_{\mathrm{strip}} < n_{\mathrm{planar}}$). Since an effective refractive index of the n_{planar} region under the loading strip is larger than that under the air, the waveguide core is located at the n_{planar} region under the strip.

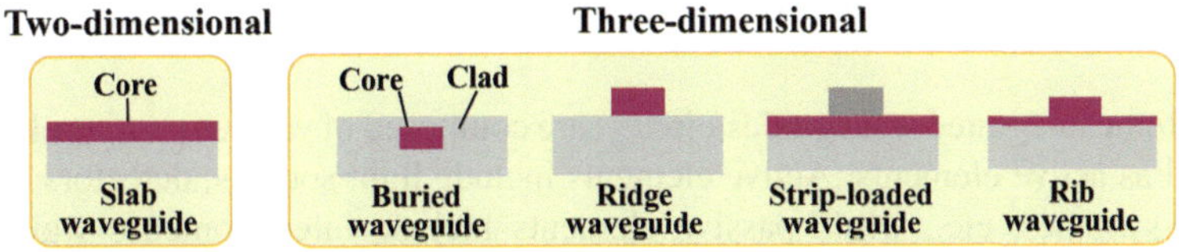

Fig. 1.6 Schematic of cross-sectional 2-D and 3-D waveguide configurations

Geometry of the optical confinement is given both by the thickness of the n_{planar} layer and width of the loading strip.

- *Rib waveguide*: It is similar to the ridge waveguide, but the rib is formed by an etching process on the substrate surface and becomes part of the wave-guiding core. Simple material preparation is an advantage of this type of waveguide.

Usually, these 3-D waveguides are called rectangular waveguides, although their forms are normally not exactly rectangular. Besides these 3-D waveguides, there is a diffused waveguide formed by the diffusion of dopants with high refractive indices in a substrate. A typical example includes $LiNbO_3$ waveguide with a titanium (Ti) diffused core. The diffused profile is not exactly defined depending on the thermal diffusion condition.

1.3.2 Materials and Refractive Indices

Typical waveguide materials and refractive indices are shown in Table 1.1. They are classified into two groups: a visible region and near infra-red (NIR) region, depending on whether the wavelength to be used is transparent or absorbed in the given materials. In Si photonics for applications to optical communication systems, a pair of Si and SiO_2 is mostly used as high-index wave-guiding core and low-index clad regions, respectively. This is because Si has many features in the NIR region such as high transparency, high index contrast with SiO_2, and compatibility with SOI CMOS processing.

Table 1.1 Typical waveguide materials and refractive indices

Wavelength	Core (n_1)	Clad (n_2)
Visible (~ 600 nm)	Si_3N_4 (2.0) TiO_2 (2.2) SU-8 (1.6)	SiO_2 (1.45) Water (1.33) Air (1.0)
Near infrared (1.3–1.5μm)	Si (3.5) Ge (4.0)	SiO_2 (1.45) Air (1.0)

1.4 Core Waveguide Structures

Si photonic integrated waveguides/circuits are composed of various passive elements as well as active elements. Active elements include light sources, detectors, modulators, switches, etc., while passive elements include interferometric waveguide (Mach–Zehnder interferometer, Young's interferometer, and multi-mode interferometer), directional coupler, optical resonator (Fabry–Perot resonator, micro-ring resonator), arrayed waveguide grating, polarization beam splitter and rotator, fiber-to-chip light coupler, inverse taper coupler, and grating coupler. In this section, we take up such passive elements only, focusing on their structures, principles, and functions, and active elements are described in Chap. 4 as a part of applications of Si photonic waveguides.

1.4.1 Interferometric Waveguide

Mach–Zehnder interferometer

Figure 1.7a shows a schematic configuration of the Mach–Zehnder interferometric (MZI) waveguide. An incident light beam is divided into two waveguide arms (Arm 1 and Arm 2) and combined to one at the output port. Mode fields E_1 in Arm 1, E_2 in Arm 2, and output intensity I at the output port are expressed using wave numbers (propagation constants) β_1 and β_2 in two arms as:

$$E_1 = c_1 \mathrm{e}^{-i\beta_1 L}, \quad E_2 = c_2 \mathrm{e}^{-\mathrm{i}\beta_2 \mathrm{L}}, \tag{1.19}$$

$$I = |E_1 + E_2|^2 = c_1^2 + c_2^2 + 2c_1 c_2 \cos(\Delta\varphi), \tag{1.20}$$

where L is a length of the arm and $\Delta\varphi$ is a phase difference between two arms as: $\Delta\varphi = (\beta_1 - \beta_2)\,L$. Note that, since β_1 and β_2 depend on refractive indices n_1, n_2, and geometries of the waveguides in real devices, $\Delta\varphi$ is also varied by these waveguide parameters after two optical beams interfere to each other, propagating through the two arms. An output intensity I varies with $\Delta\, n_{\mathrm{eff}}$, as shown in Fig. 1.7b, where n_{eff} is an effective refractive index: $n_{\mathrm{eff}} = \beta/k_0$, as defined in Eq. (1.4). For applications to optical modulators and switches, electrodes are arranged asymmetrically above or beside either waveguide to control the n_{eff} via thermo-optic or other physical effects. Practical applications of MZIs are seen most frequently in Chaps. 4–7.

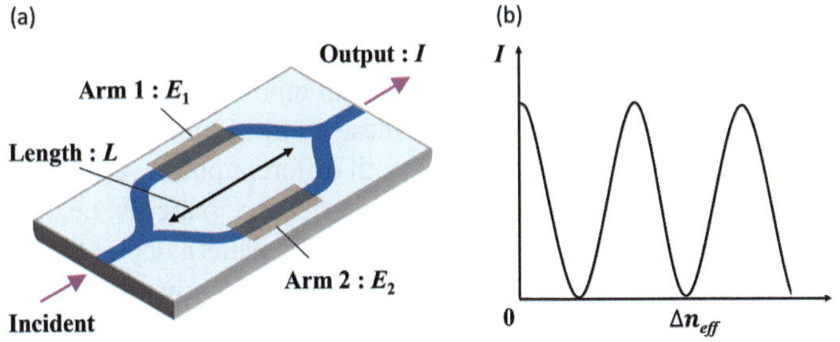

Fig. 1.7 **a** Schematic of a Mach–Zehnder interferometer. **b** An example of output intensity variation as a function of n_{eff}

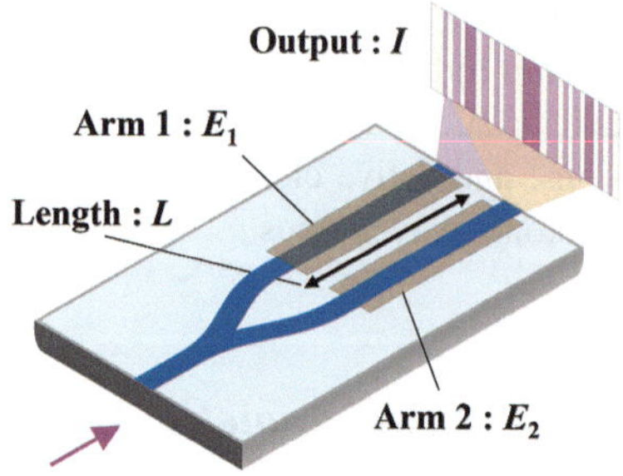

Fig. 1.8 Schematic of Young's interferometer waveguide. Two parallel beams are radiated into air at the edge of the output ports and projected on the screen, producing an interference diffraction pattern on it

Young's Interferometer

Young's interferometer (YI), as shown in Fig. 1.8, is another important interferometric optical waveguide. An incident light beam is divided into two channel waveguides (Arm 1 and Arm 2) like the MZI. Yet, different from the MZI, two parallel beams propagating through the channel waveguides are radiated at the edge into air and projected on a screen, often on a charge-coupled device (CCD) detection camera via a cylindrical lens, where an interference diffraction pattern is generated on the screen as an output signal. Due to the 3-D detection geometry, this type of interferometer is not compatible to a planar optical integrated circuit. Nevertheless, the YI has long been used for optoelectronic applications like the MZI, including exceptional application such as a biophotonic sensor [4].

Multi-mode Interferometer

A multi-mode interferometer (MMI) is widely used for optical beam splitters and combiners in a planar integrated optics owing to low optical losses, wide optical bandwidths, small footprints, and low sensitivities to fabrication tolerances. Figure 1.9a shows a bird's eye view of the MMI, where a wide-width multi-mode waveguide is coupled to an input single-mode waveguide port and doubled output ports. In general, the MMI adopts $m \times n$ input/output ports, while the figure shows the 1×2 port configuration. The multi-mode waveguide regime allows a finite number of eigenmodes, here indicating the lowest four modes for instance: $\varphi\ (x, y)$ (φ_{00}, φ_{10}, φ_{20}, φ_{30}) (showing standing waves), as shown on the left in Fig. 1.9a. When an input light impinges into the multi-mode waveguide, an input field is matched with mode fields of the multi-mode waveguide. Since these components are eigenmodes, they propagate independently from one another with different phase velocities and intervene mutually. As a result, they form unique interference patterns, i.e., self-images, as shown by an example on the right in Fig. 1.9a. The pattern depends on the propagation length of light: In Fig. 1.9a, the length is arranged so that the double peaks in the field distribution nearest to the output end of the MMI match the double output ports. These phenomena constitute a basis for the operation of MMI structures [5].

Figure 1.9b shows a schematic of an application to an incident beam splitter and output beam combiner of the MZI modulator. In this application, two ports of the

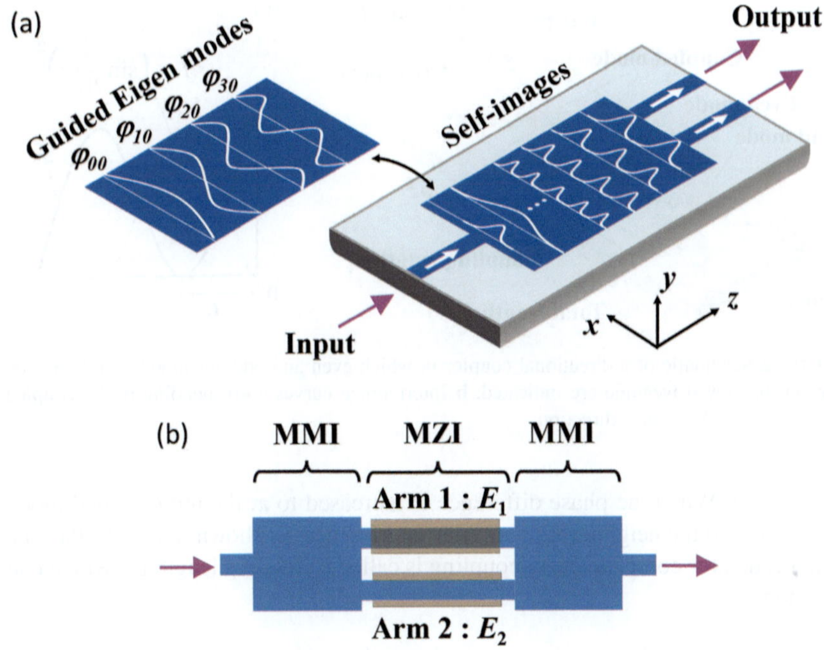

Fig. 1.9 **a** Schematic of the MMI with waveforms of the lowest four eigenmodes (left panel) and self-images (right panel), each as an example, within the MMI region. **b** Application of the MMI to an input beam splitter and output beam combiner coupled to the MZI modulator

MMI are directly connected to two arms of the MZI. For careful design of the joint region between the MMI and MZI, the input and output waveguides are often tapered to reduce the coupling loss due to the modal mismatch between them.

1.4.2 Directional Coupler

A directional coupler (DC) is composed of two parallel waveguides closely arranged to each other and input/output ports connected to them, as shown in Fig. 1.10a. Fundamental behaviors of the light beam in the parallel waveguides are given by the coupled mode theory [6], where the parallel waveguides can be considered as a new "coupled waveguide" rather than two independent waveguides. When input light is launched in one of the input ports, an even mode and odd mode (black thin lines in the figure), both eigenmodes in the new coupled waveguide, are excited simultaneously and coupled to each other. As a result, interference between them forms a newly synthesized guided mode called a coupled mode (red thick lines in the figure). Since the even and odd modes have different wave numbers (propagation constants) β_{even} and β_{odd}, the coupled mode propagates with the phase difference involving a term of

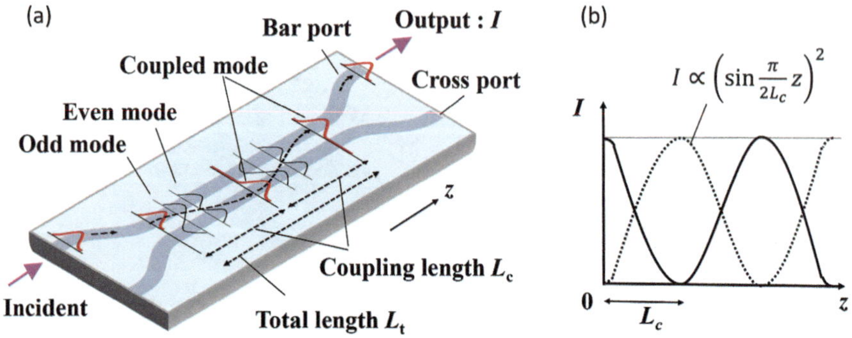

Fig. 1.10 **a** Schematic of a directional coupler in which even and odd eigenmodes and coupled modes in the new waveguide are indicated. **b** Interference curves corresponding to the coupled mode intensities along the *z*-direction

$|\beta_{even} - \beta_{odd}|$. When the phase difference is increased to π, the input-guided mode is transferred to the neighboring port after interference, as shown in Fig. 1.10a. The length needed for complete cross-coupling is called a coupling length L_c, which can be written as:

$$L_c(\lambda) = 2\pi/\left[\beta(\lambda)_{even} - \beta(\lambda)_{odd}\right] = \lambda/\left[n(\lambda)_{eff}^{even} - n(\lambda)_{eff}^{odd}\right], \quad (1.21)$$

where $n\ (\lambda)_{eff}{}^{even}$ and $n\ (\lambda)_{eff}{}^{odd}$ are effective refractive indices in the even and odd guided modes, respectively. When the total length L_t of the DC waveguide is odd times the coupling length L_c, incident light is transferred to the neighboring port, while, when the L_t is even times the coupling length L_c, the light is coupled back to the original channel waveguide: the bar port. Figure 1.10b shows interference curves corresponding to the coupled mode intensities along the z-direction. The DC waveguide is promising for a power splitter in the Si photonic integrated circuit, as shown in Chap. 4 (Si photonic multi-port switches, etc.), or for a TE-TM polarization divider.

1.4.3 Optical Resonator

Optical resonators are widely used in optical engineering fields for generic laser equipment and photonic integrated circuit elements in optical communication networks. In this section, two types of optical resonators are introduced: a traditional Fabry–Perot resonator (FPR) providing a fundamental principle of an optical resonance and a modern micro-ring resonator (MRR) with capability of compact photonic resonator/filter elements.

Fabry–Perot Resonator

A Fabry–Perot resonator (FPR) consists of two mirrors facing each other, as shown schematically in Fig. 1.11a [7], where a reflectance and transmittance of the front mirror are r_f and t_f, while those of the end mirror are r_e and t_e, respectively. Since mirrors are partially transmitting, a light wave incident into the resonator through the input mirror propagates back and forth between two mirrors many times. This process induces constructive and destructive interferences in the light wave. Constructive interference amplifies the electric field of light in the resonator, causing the standing wave under the resonance condition described in the later Eq. (1.24). At this condition, all of the incident beams are transmitted through the resonator. On the other hand, destructive interference out of the resonance condition (i.e., off-resonance condition) enforces the light wave with high reflection and low transmission, thus weakening its field. At the resonance condition, the resonator transmits the incident beam only at separate frequencies, leading to resonator modes. These modes are derived from a plane wave that propagates back and forth in the resonator with an amplitude of Ee^{-ikz}. Assuming the cavity length L between the mirrors, a multiply reflected and transmitted light E_t is:

$$\begin{aligned} E_t &= E_i\left(t_f t_e e^{-i\phi/2} + t_f r_e r_f t_e e^{-3i\phi/2} + t_f r_e^2 r_f^2 t_e e^{-5i\phi/2} + \cdots\right. \\ &= E_i \frac{t_f t_e e^{-i\phi/2}}{1 - r_e r_f e^{-i\phi}}, \end{aligned} \tag{1.22}$$

where ϕ is a round-trip phase variation in the cavity: $\phi = 2L\omega/c = 4\pi L\nu/c$. An optical power P_t transmitted through the resonator dependent on the frequency of light is revealed as:

$$P_t = |E_t|^2 = \frac{(t_f t_e)^2}{(1 - r_e r_f)^2 + 4 r_e r_f \sin^2\left(\frac{\phi}{2}\right)} |E_i|^2. \tag{1.23}$$

The resonance condition is:

$$\phi_m = 2\pi m \text{ or } 2L = m\lambda \quad (\text{m: integer}). \tag{1.24}$$

At resonance, a series of resonance modes appear in the frequency spectrum, as shown in Fig. 1.11b, where a frequency separation between resonances is named free spectral range (FSR), $\Delta\nu_{FSR}$, as:

$$\Delta\nu_{FSR} = \nu_m - \nu_{m-1} = \frac{c}{4\pi L}(\phi_m - \phi_{m-1}) = \frac{c}{2L}. \tag{1.25}$$

Performance of the FPR strongly depends on the reflectivity of the mirror. High-reflectivity mirror produces a narrow (sharp) linewidth $\Delta\nu$ of the resonance mode, while low-reflectivity mirror brings a broad linewidth. The linewidth $\Delta\nu$ is given by the ratio of the frequency separation $\Delta\nu_{FSR}$ (FSR) versus finesse F, as:

$$\Delta\nu\frac{\Delta\nu_{\mathrm{FSR}}}{F}. \tag{1.26}$$

The FPR with a high finesse achieves a sharp transmission peak, which enables clear discrimination of the neighboring transmission peaks owing to the enhanced resolution of the FPR. These performance parameters are applicable to the FPR-based photonic components in general, including a MRR, as described next.

Micro-ring Resonator

As a wavelength-selective device, a Si waveguide-based MRR plays an important role for its large FSR (several tens of nm) owing to a small bend radius (less than several μm) and ring with high contrast in refractive index between Si and Si-oxide. Figure 1.12a schematically shows an add-drop MRR. It connects the ring and two adjacent bus waveguides via co-directional evanescent coupling. Refer to Fig. 1.10a for the principle of the co-directional evanescent coupling. The MRR serves as a compact spectral filter for applications to a wavelength-division multiplex (WDM) optical communication system.

Figure 1.12a derives a basic spectral property of a MRR filter using several parameters (r_j, k_j, and a, where $j = 1, 2$): When a phase shift of light propagating through a ring waveguide is an integer times 2π, the light interferes positively and the resonance state occurs in the cavity. In the coupling region with the ring and two adjacent bus waveguides, r_j is a self-coupling coefficient, whereas k_j is a cross-coupling coefficient ($j = 1, 2$). For the lossless coupling, ${r_j}^2 + {k_j}^2 = 1$. Furthermore, a is a transmission through a single-pass. Next, transmissions from an input port to a pass port and drop port are described as T_{pass} and T_{drop}, respectively. Here, they involve structural parameters, r_j, k_j and a. Additionally, an incident field is assumed to be partly transmitted to a drop port via resonance in the cavity. Figure 1.12b shows a schematic of transmission spectra, T_{pass} and T_{drop}, at the pass port and drop port, respectively, where $\phi = \beta L$ is a phase shift through a single-pass, L is a length through a round-trip of the ring, and β is a propagation constant of the circulating mode. See [8] for mathematical derivation of the T_{pass} and T_{drop}. FSR, defined as a wavelength range between two resonances in Fig. 1.11, is also given for the add-drop MRR filter as:

$$\mathrm{FSR} = \frac{\lambda^2}{n_{\mathrm{g}}L} \tag{1.27}$$

Here n_{g} is a group index, defined by taking into account the dispersion of the waveguide, as:

$$n_{\mathrm{g}} = n_{\mathrm{eff}} - \lambda\frac{\mathrm{d}n_{\mathrm{eff}}}{\mathrm{d}\lambda}, \tag{1.28}$$

where n_{eff} means an effective refractive index in the waveguide.

The add-drop multiplexer is one of the main components in WDM for separating the required channel to be dropped from a set of channels that pass through. The

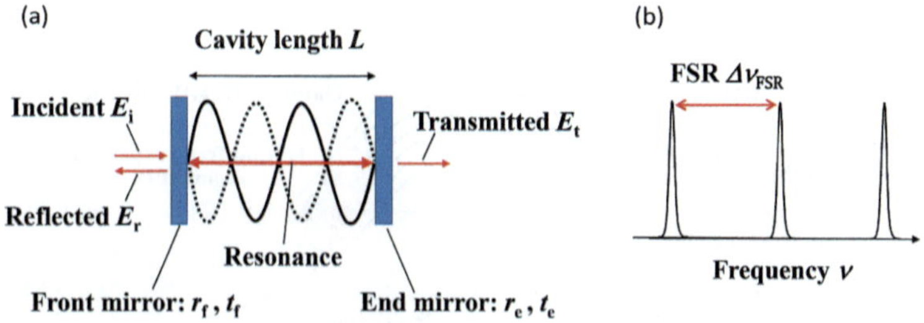

Fig. 1.11 **a** Schematic of an optical resonance in the FPR. **b** Series of resonance peaks and the FSR showing a peak space

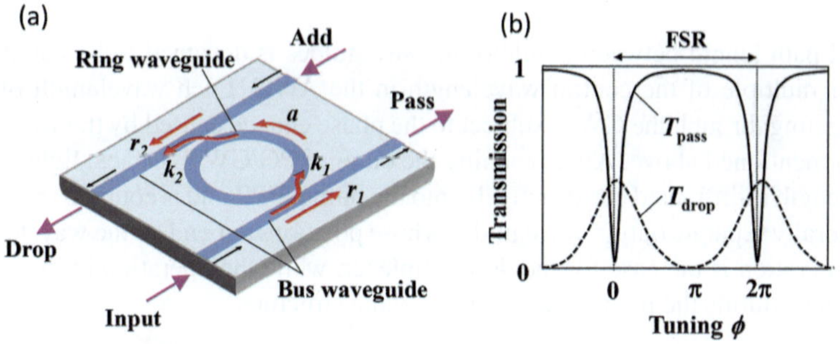

Fig. 1.12 **a** Schematic of add-drop micro-ring resonator. **b** Transmission spectra T_{pass} and T_{drop} at the pass port and drop port, respectively

add-drop MRR filter composed of Si photonic platform exhibits a low loss and wide FSR. For instance, it is used in the reconfigurable optical add-drop multiplexing (ROADM) system in Sect. 4.4.3.

1.4.4 Arrayed Waveguide Grating

An arrayed waveguide grating (AWG) is inevitable in the WDM (wavelength-division multiplexing) system. The AWG in a multiplexer is needed for multiplexing optical signals with different wavelengths into an optical waveguide at the end of the transmission line, while that in a de-multiplexer is needed for de-multiplexing individual optical signals with different wavelengths at the end of the receiving line [9]. Figure 1.13 shows a configuration of the AWG. The central grating waveguide (GW) designed as a phased array is attached by two identical free-propagation regions (FPGs) at input/output FPG/GW interfaces [20]. Optical signals with several wavelengths in the input waveguide are emitted into the input FPG, reach the input FPG/GW interfaces with a Gaussian distribution, and impinge into the GW. The grating is composed of multiple waveguides with different lengths. Difference of an

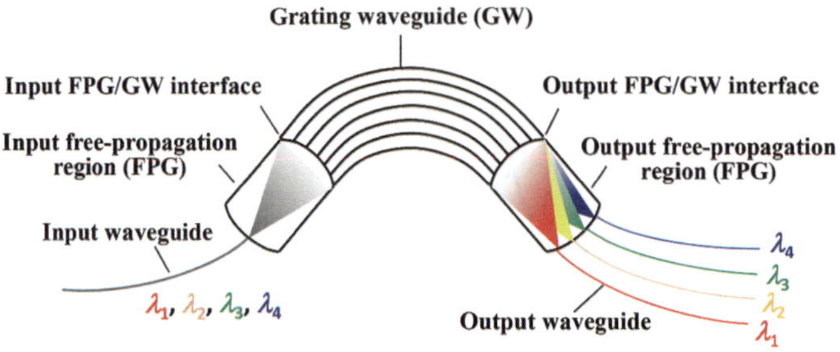

Fig. 1.13 Schematic of the AWG composed of input/output free-propagation regions (FPRs) and grating waveguide (GW). Adapted with permission from Ref. [20], © 2020 OEA

optical path length between neighboring waveguides is designed to be equal to an integer multiple of the central wavelength in the AWG. Each wavelength of light propagating through the GW is subject to the phase change caused by the waveguide length mentioned above. After reaching the output FPG/GW interface, light in each wavelength is diffracted constructively into the output FPR and is coupled into one of the laterally separate output waveguides whose positions depend on the wavelengths. This operation is the case for the de-multiplexer, while the operation in the reverse direction exhibits the multiplexer with the same structure.

1.4.5 Polarization Beam Splitter and Rotator

Polarization Diversity System

In general, several dielectric materials have refractive indices depending on the polarization direction of light, a property called an optical birefringence. In integrated photonic circuits, photonic functional devices often suffer from unfavorable effects based on such birefringence, namely polarization-dependent effects in dispersion, loss, wavelength characteristics, etc. A polarization diversity system equipped with polarization splitters/rotators is a versatile way to mitigate such polarization-dependent problems for designing integrated photonic circuits.

Recently, an attractive polarization diversity system was proposed [10]. Figure 1.14 shows a schematic of the polarization diversity system with two identical polarization splitters (PS1 and PS2) and polarization rotators (PR1 and PR2) besides other functional devices (e.g., filters). When incident TE and TM modes impinge into an input port, they are separated by the PS1: The TM and TE modes propagate through different roots: Route 1 and Route 2, as shown in the lower panel in Fig. 1.14. In Route 1, the TM mode is rotated 90° by the PR1 into the TE mode and reaches the PS2 through the functional device (filter). In Route 2, the TE mode propagates to the PR2 through the functional device and is rotated 90° by the PR2 into the TM wave and then reaches the PS2. This system has two features: One is that the functional device

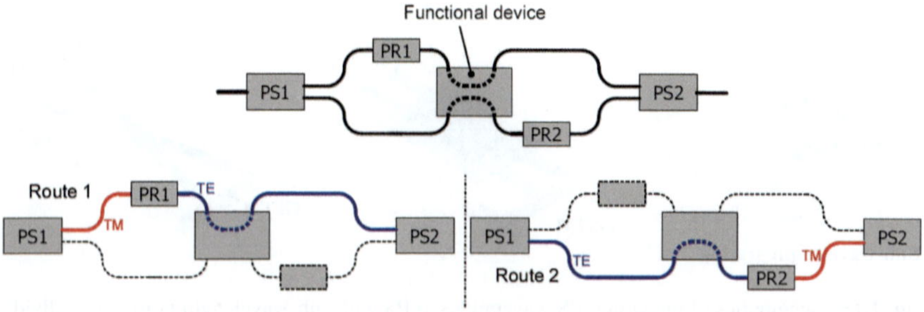

Fig. 1.14 Schematic of a polarization diversity system with two identical polarization splitters (PS1 and PS2) and polarization rotators (PR1 and PR2) besides other functional devices (e.g., filters). Reprinted with permission from Ref. [10], © 2008 The Optical Society (OSA)

is designed for the TE mode only and not needed to be designed for both modes. The other is that the total propagation lengths for the TM (in red) and TE (in blue) modes are the same between two routes. Consequently, polarization compensation is made between the two separate routes, so that input light with different polarization does not influence badly on the performance of functional devices. As an exceptional case without the need of such a polarization diversity system, a polarization-independent MZI modulator on a SOI substrate for a multi-cast Si photonic switch platform was proposed. Refer to Sect. 4.4.2 for detail. Typical PSs and PRs with their geometries and performances are introduced in the later sections.

Polarization Splitter

An integrated polarization splitter (PS) is a key optical interconnect component to use both TE and TM modes on the same chip. Various PS waveguides have been reported to date [11]. Typical integrated PSs are introduced, as follows:

- *PS with directional coupler* [10]: In the directional coupler (DC) (Sect. 1.4.2), the coupling length L_c—the length needed for complete cross-coupling—is clearly different between the TE and TM modes; L_c for the TM is smaller than L_c for the TE. Use of this characteristic in the DC is effective for the PS. In Fig. 1.14, the DC was used for PSs (PS1 and PS2) in the integrated Si photonic polarization diversity circuit [10].
- *PS with sub-wavelength grating* [12]: The 2-D grating, as shown in Fig. 1.15a, is capable of functioning efficient vertical-to-surface light coupling and TE-to-TM polarization splitting simultaneously. The geometry of the 2-D grating coupler enables the TE- and TM-combined incident light to be radiated independently into 90°-different in-plain directions.
- *PS with MMI waveguide* [13]: The MMI-based PS, as shown in Fig. 1.15 b, has been widely investigated similar to the DC-based PS. The DC-based PS needs precise control of a narrow coupling gap between the parallel waveguides, while the MMI-based PS has advantages of a compact footprint and looser fabrication tolerant.

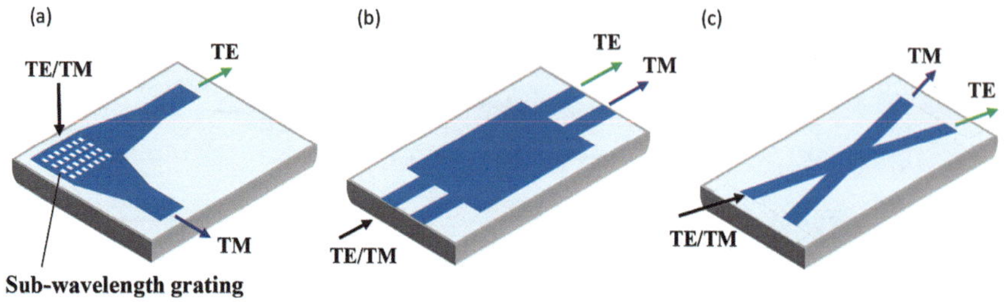

Fig. 1.15 Schematics of integrated PS waveguides. **a** PS with sub-wavelength grating that divides orthogonal polarizations into two different output ports. **b** PS with a MMI waveguide. **c** Ultra-compact PS with crossed waveguides

- *Ultra-compact PS with crossed waveguide* [14]: Different from the MMI-based PS with a long MMI length, the PS with flare-designed crossed waveguides, as shown in Fig. 1.15c, achieves only one beat-length of the splitter. A fabricated 0.94 μm long 2 × 2 crossed waveguide succeeded in a splitting ratio > 15 dB over 50 nm bandwidth.

Polarization Rotator

An integrated PR (polarization rotator) is another key device as well as the PS for the polarization diversity system. Here, typical integrated PRs reported to date have been introduced, as follows:

- *PR with double-core structure* [10]: A PR with a unique double-core structure was developed. The PR was connected to input and output Si single-mode waveguides. In the double-core region, a smaller (200-nm) Si square core is buried in the second larger (840-nm) SiO_xN_y square core (refractive index n_{second} of 1.60). In the second core, the Si square core is shifted so that edges of both side and bottom overlap edges of the second core. PR is mainly determined by n_{second} and sizes of both cores. For the 35-μm rotator length, a TM input mode was sufficiently suppressed and a TE mode was transmitted with 1-dB loss only. See [10] for detail.
- *PR with mode-evolution* [15]: Another type of PR capable with a SOI technology was proposed, as shown in Fig. 1.16. A Si input waveguide with 400 nm in height and 200 nm in width is connected to the middle section with a tapered Si waveguide, which continues up to the output section with 400-nm width and 200-nm height. The whole Si structure is surrounded by SiO_2. Lower panels show calculated optical mode distributions of a quasi-TM polarization at the input waveguide region, while quasi-TE polarizations at the transition and output waveguide regions. Reverse polarization rotation from the quasi-TE to quasi-TM mode is available also with the same device.

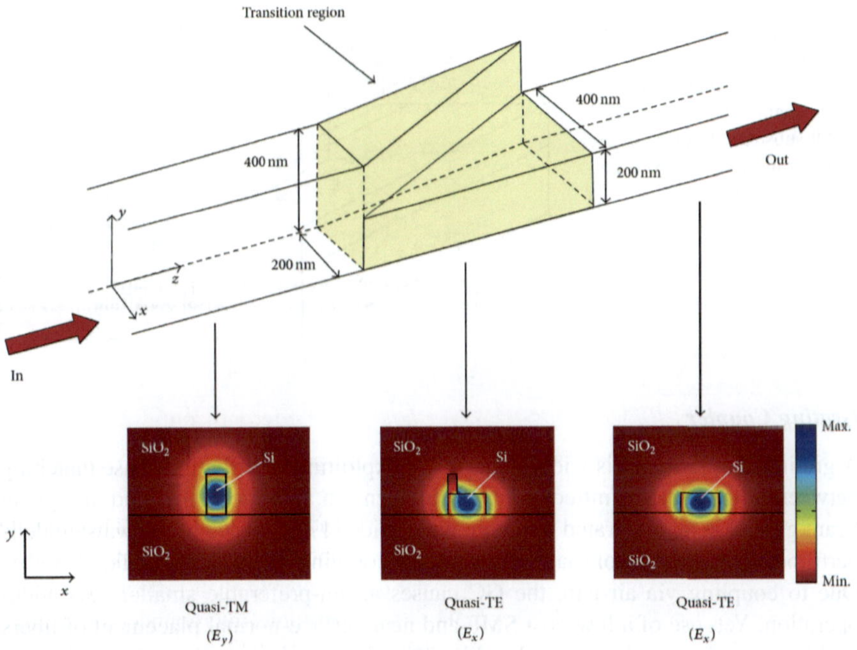

Fig. 1.16 Schematics of PR configurations. PR is based on a mode-evolution. Each lower part shows simulated spatial distributions of optical modes in different portions of the structure. Reprinted with permission from Ref. [15], © 2014 Hindawi Publishing Corporation

1.4.6 Fiber-to-Chip Light Coupler

Inverse Taper Coupler

For conventional butt-coupling between an optical fiber and a channel waveguide, a taper coupler with a SiO_2 or polymer material was proposed: The coupler provides approximate matching of the optical beam profiles at fiber/coupler and coupler/channel-waveguide interfaces via an adiabatic expansion/reduction of the beam profiles. Yet, it does not provide suitable coupling due to deficient matching of effective refractive indices at each interface. Instead, a new inverse taper coupler covered with a rectangular high-index polymer waveguide was proposed, as shown in Fig. 1.17 [16]. The inverse taper region (as long as ~ hundred μm) is an end portion with a Si single-mode waveguide (0.3-μm square) fabricated on an SOI wafer, while the rectangular high-index waveguide region (3- × 3-μm square) acts effective butt-coupling with the single-mode fiber (SMF). The new coupler provides good coupling efficiencies both in optical mode profiles and effective refractive indices between the Si waveguide and SMF. Demonstration resulted in an acceptable 1-dB conversion loss for a 200-μm long taper with a 60-nm wide tip [16].

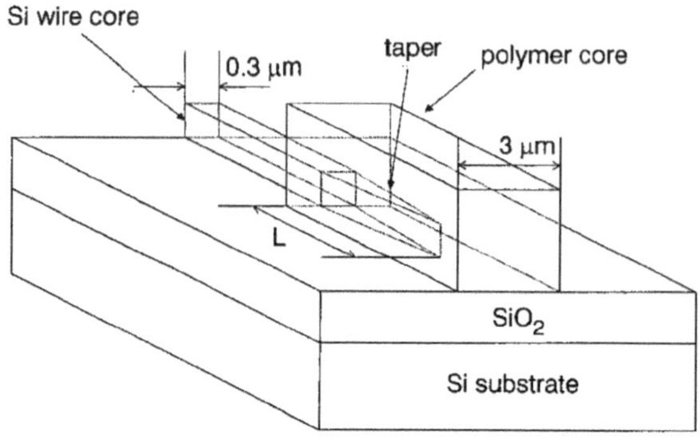

Fig. 1.17 Schematic of an inverse taper coupler. An end portion of the Si taper waveguide is connected to the Si channel waveguide on a SOI substrate. Reprinted with permission from Ref. [16], © 2002 IET

Grating Coupler

A grating coupler (GC) is widely known for exploiting an effective phase matching between a light beam emitted from and incident on an optical fiber and an optical beam guided in an integrated photonic waveguide [17]. Surface of the waveguide is perturbed with a grating of shallow grooves for coupling between fiber/guided modes. Due to coupling via air-gap, the GC causes an un-preferable smaller bandwidth operation. Yet, use of a low-cost SMF and near surface-normal placement of fibers enabled to probe any device on the chip. This situation is not achievable with other methods like butt-coupling mentioned above. In the following, advanced GCs with large coupling efficiencies, shallow-etched and fully etched GCs, are introduced.

- *Shallow-etched GC* [18]: Fig. 1.18a shows a newly proposed shallow-etched SOI GC overlaid with an over-layer (poly-Si) to improve coupling between the fiber and waveguide modes [18]: A high coupling efficiency over 70% between the fiber mode and guided wave mode was theoretically estimated. The GC is defined first by locally depositing a poly-Si layer (thickness of ~ 150 nm) on top of the un-etched Si waveguide layer (thickness of ~ 220 nm). After that, grating trenches are etched by ~ 200 nm from the overlaid top surface. If the vertical grating structure is close to symmetry, a coupling efficiency is at most 50% because the diffracted radiation occurs identically in top and bottom cladding. In contrast, the grating structure with the poly-Si over-layer presents large vertical asymmetry. In addition, by optimizing other structure parameters for further increasing vertical asymmetry, the resultant coupling efficiency has proved to be as high as 78% at a wavelength of 1.55 μm and an optical 3 dB bandwidth of about 85 nm. Device fabrication compatible with standard CMOS processing technology is another feature of this GC.
- *Fully etched GC* [19]: Fig. 1.18b shows a schematic of a newly proposed GC with fully etched slots optimized in coupling between an optical fiber and SOI slab waveguide. This type of coupler has the most important feature of capability to produce in one lithography step together with other SOI components [19]. In the conventional fully etched GC, strong Fresnel reflection at an air-Si interface is

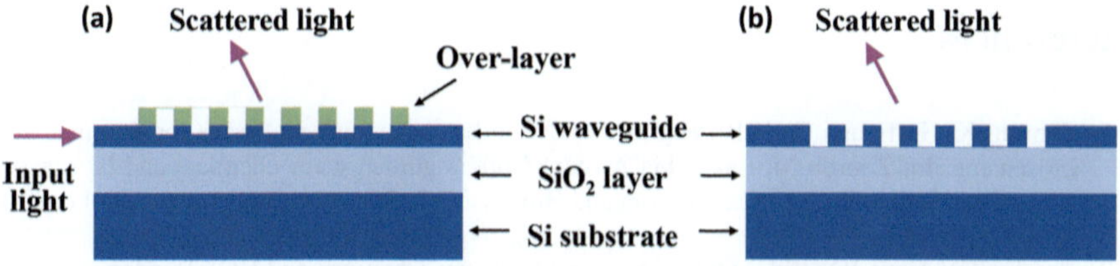

Fig. 1.18 Schematic of new types of GCs. **a** Shallow-etched GC. **b** Fully etched GC

a drawback, while, in the shallow-etched GC mentioned above, need of the additional lithography step after the step of overlaying the un-patterned Si waveguide area is a significant drawback. In contrast, the proposed GC mitigates several drawbacks including an antireflection at the SiO_2-layer/Si substrate interface by optimizing grating parameters such as grating pitch, filling factor of the grating array, slot width, SiO_2-layer thickness, etc. Constructive interference is used to suppress radiation into the substrate for an optimal power coupling into the fiber mode. Thus, a theoretical maximal coupling efficiency of 49% has proved to be available with a 3 dB bandwidth of 35 nm.

Recently, a further new type of the through-etched GC has been gaining popularity: sub-wavelength GC (SWG) utilizing sub-wavelength artificial material. Topics on the SWG are taken up in Sect. 4.2.1 for important technologies in the Si photonic platform. Refer to [21–23] for detail.

1.5 Summary

Basic mode theories, fundamental structures, and primary characteristics of optical waveguides were surveyed in general. A dispersion relation $\omega - \beta$ for optical guided modes was derived through a ray-optics approach, while mode diagrams on the effective refractive index vs waveguide parameters were derived through an electromagnetic-wave approach. Typical passive waveguide devices represented by MZIs and MRRs are most applicable to optical modulators and switches, while MMIs, directional couplers, and AWGs are widely used for optical beam dividers and frequency discriminators. Performances of these waveguides were explained by showing various conventional and improved structures. Configuration and an operational principle of polarization control waveguides such as polarization splitters and rotators, inevitable for polarization-independent systems in the telecom networks, were described in detail. Recent development of fiber-to-chip light couplers including inverse taper coupler and grating coupler was also discussed for implementing optical integration of Si platform chips in the optical fiber networks. For applications of Si photonic waveguides including active waveguides such as modulators and switches, refer to Chap. 4 (Si optical transceivers in the data center and switches in the network core) and Chap. 5 (Si LiDAR and optical neural network).

References

1. Schmitt K, Hoffmann C (2010) High-refractive-index waveguide platforms for chemical and biosensing. In: Zourob M, Lakhtakia A (eds) Optical guided-wave chemical and biosensors I. Springer Series on Chemical Sensors and Biosensors 7. Springer, Berlin. https://doi.org/10.1007/978-3-540-88242-8_2
2. Snyder AW, Love JD (1983) Optical waveguide theory. Chapman and Hall, London
3. Kogelnik H (1988) Theory of optical waveguides. Springer, New York
4. Mulder HKP, Ymeti A, Subramaniam V, Kanger JS (2012) Size-selective detection in integrated optical interferometric biosensors. Opt Express 20:20934–20950
5. Blahut M, Kasprzak D (2004) Multimode interference structures—properties and applications. Optica Applicata XXXIV:573–587
6. Yariv A (1973) Coupled-mode theory for guided-wave optics. IEEE J Quantum Electron 9:919–933
7. Vaughan M (1989) The Fabry-Perot interferometer: history, theory, practice and applications. Series in optics and optoelectronics. Taylor & Francis, Boca Raton. ISBN: 9780852741382
8. Bogaerts W, Heyn PD, Vaerenbergh TV, Vos KD, Selvaraja SK, Claes T, Dumon P, Bienstman P, Thourhout DV, Baets R (2012) Silicon micro-ring resonators. Laser Photonics Rev 6:47–73
9. Elrofai SE (2015) Review paper of array waveguide grating (AWG). Int Res J Eng Technol (IRJET) 02:1010–1029
10. Fukuda H, Yamada K, Tsuchizawa T, Watanabe T, Shinojima H, Itabashi S (2008) Silicon photonic circuit with polarization diversity. Opt Express 16:4872–4880
11. Subbaraman H, Xu X, Hosseini A, Zhang X, Zhang Y, Kwong D, Chen RT (2015) Recent advances in silicon-based passive and active optical interconnects. Opt Express 23:2487–2511
12. Taillaert D, Chong H, Borel P, Frandsen L, De La Rue R, Baets R (2003) A compact two-dimensional grating coupler used as a polarization splitter. IEEE Photon Technol Lett 15:1249–1251
13. Kiyat I, Aydinli A, Dagli N (2005) A compact silicon-on-insulator polarization splitter. IEEE Photon Technol Lett 17:100–102
14. Hosseini A, Rahimi S, Xu X, Kwong D, Covey J, Chen RT (2011) Ultra-compact and fabrication-tolerant integrated polarization splitter. Opt Lett 36:4047–4049
15. Troia B, De Leonardis F, Lanzafame M, Muciaccia T, Grasso G, Giannoccaro G, Campanella CE, Passaro VMN (2014) Design and optimization of polarization splitting and rotating devices in silicon-on-insulator technology. Adv OptoElectron 2014:16. Article ID 490405. https://doi.org/10.1155/2014/490405
16. Shoji T, Tsuchizawa T, Watanabe T, Yamada K, Morita H (2002) Low loss mode size converter from 0.3 μm square Si wire waveguides to single mode fibers. Electron Lett 38:1669–1670
17. Yariv A, Nakamura M (1977) Periodic structures for integrated optics. IEEE J Quantum Electron 13:233–253
18. Roelkens G, Van Thourhout D, Baets R (2006) High efficiency Silicon-on-Insulator grating coupler based on a poly-Silicon overlay. Opt Express 14:11622–11630
19. Schmid B, Petrov A, Eich M (2009) Optimized grating coupler with fully etched slots. Opt Express 17:11066–11076
20. Asakawa K, Sugimoto Y, Nakamura S (2020) Silicon photonics for telecom and data-com applications. Opto-Electron Adv 3:200011-1–200011-26
21. Wood M, Sun P, Reano RM (2012) Compact cantilever couplers for low-loss fiber coupling to silicon photonic integrated circuits. Opt Express 20:164–172
22. Cheben P, Xu D-X, Janz S, Densmore A (2006) Subwavelength waveguide grating for mode conversion and light coupling in integrated optics. Opt Express 14:4695-4702
23. Lalanne P, Hugonin J-P (1998) High-order effective-medium theory of subwavelength gratings in classical mounting: application to volume holograms. J Opt Soc Am A 15:1843–1851

Chapter 2
Survey of Basic Photonic Crystal Waveguides

Abstract Photonic crystals (PCs)—nano-scale periodic air-hole lattice structures with a high refractive-index contrast—have attracted much attention to ultimate light-confinement and dynamic-control behaviors. This chapter surveys fundamentals of unique PC concepts such as lattice geometry, specific band structure, and peculiar optical phenomenon: The first topic is index contrast-based 1-D to 3-D PC structures and dispersion relations with band diagrams and full band gaps strongly dependent of lattice configuration. The second topic is significances of point/line defects specific to the PC and defect modes in the band gap. These concepts straightforwardly generate nano-scale cavities and waveguide devices. In particular, 2-D PC slab waveguides and their unique features such as mode confinement in the slab, light cone, and leaky mode are indispensable design guidelines for constructing an integrated PC platform. As the third topic, initial pioneering works on the theoretical and experimental demonstration of 2-D PC waveguides are described. The final topic is peculiar phenomena of the PC leading to unique applications, a super-prism effect based on the extraordinary refraction, an extremely high-*Q* nano-cavity, a slow-light effect resulting from the low group velocity, and a coherent surface emission at a Γ-point as an origin of the 2-D PC surface-emitting laser.

2.1 Introduction

A dielectric 1-D periodic structure such as multiple-layer dielectric mirror has been widely used in optics since the early twentieth century. It is considered to be the 1-D PC structure. Yet, it is useful only for a limited purpose as a light control method [1]. In 1987, Yablonovitch [2] and John [3] reported the new concept of the PC: the structure for light control means independently. E. Yablinovitch theoretically showed existence of a photonic band gap (PBG), where any PC mode does not exist in any direction. In turn, S. John stated that the PC provides a new phenomenon represented by light localization. These two papers gave a great impact on the field of optoelectronic science and technology. In advance of these papers, K. Otaka reported calculated results of the band structure in the 3-D PC in 1979 [4]. Nevertheless, he

K. Asakawa et al., *Nano-photonics for Advanced Networks*,
https://doi.org/10.1007/978-3-031-80637-7_2

did not mention the existence of the full PBG. After these events, many works in this field started: The main parts of seven to eight years after 1987 were spent mainly for theoretically searching new PBG-possessing dielectric crystals and their band calculations. Since the middle of the 1990s, fabrication of 2-D and 3-D PCs has been developed as well as investigation of many unique properties of the PCs. This chapter surveys fundamental aspects of the PC such as unique concept, lattice geometry, specific band structure, and peculiar optical phenomenon. A variety of applications of the PC waveguide and cavity to nano-scale photonic devices in the advanced photonic networks are shown in Chap. 6.

2.2 Type of Photonic Crystal

2.2.1 One-Dimensional to Three-Dimensional Structures

The PC is defined as a crystal in which a dielectric constant (ε) or refractive index (n) periodically varies spatially to form a lattice in the specific direction. Early theoretical work on the PC has been described in the book "Photonic Crystals" authored by Joannopoulos et al. [5]. In the 1-D, 2-D, and 3-D PCs, variation is caused along one, two, and three directions, respectively, as shown in Fig. 2.1. In Fig. 2.1a, the 1-D PC is well known as a multi-layer dielectric film. Light propagates along the direction perpendicular to the film, where A and B present materials with different dielectric constant (ε) or refractive index (n), while a is a periodicity in the 1-D PC. In the 2-D PC in Fig. 2.1b, vertical dielectric rods in light blue and light brown are alternately arranged to form a 2-D lattice in a horizontal plane, namely a 2-D PC plane. In actual materials, each rod is replaced by an air-hole or pillar, and the 2-D structure is included in a rather thin slab whose thickness is on the order of light wavelength, so that light propagates along any direction in the 2-D PC slab. Similarly in a 3-D PC in Fig. 2.1c, each box in light blue or light brown is replaced by an air-sphere, dielectric-sphere, etc. Actually reported PC structures are shown later. The 3-D PC is also constructed by an accumulation of cross-stacked thin bars/rods. Light propagates along any direction in the 3-D PC. Here, we consider states of light inside the PC, i.e., eigenmodes of light with a vector $\boldsymbol{k}$ parallel to this perpendicular direction. Then these eigenmodes in the PC are quite different in uniform 1-D, 2-D, and 3-D dielectrics.

So far, a variety of 1-D to 3-D PC structures have been proposed. Figure 2.2a–c show a family of typical 1-D to 3-D PCs. All the structures except for those in light blue/brown involve fine structures close to more realistic PCs. Among all the PC families, an air-hole type 2-D PC slab at the right end often called the "air-bridge" is used most often for integrated PC device applications. This is because the light is vertically confined in the slab owing to high refractive-index contrast between air and dielectric and propagates through the slab without loss under the 2-D PC band structure. In addition, such a 2-D PC slab allows for planar micro-/nano-scale fabrication technologies.

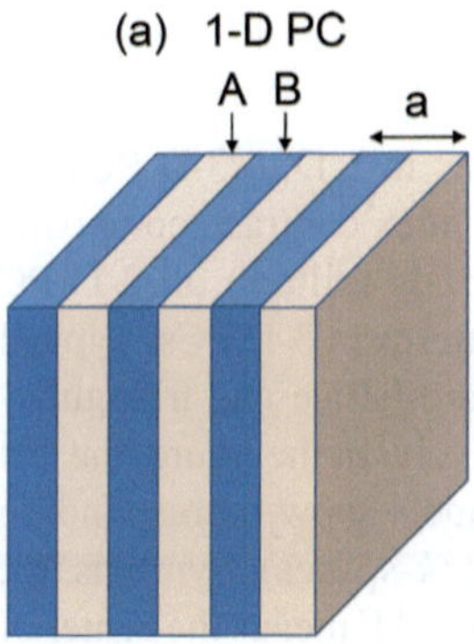

Fig. 2.1 Periodic high index contrast distributions in 1-D PC (**a**), 2-D PC (**b**), and 3-D PC (**c**)

In Chap. 6, a variety of applications to advanced photonic network devices have been introduced (Sects. 6.2–6.5). They include a slow-light device, ultra-fast all-optical switching device, nano-cavity-based optical random access memory (o-RAM), and extremely high-power 2-D PC surface-emitting laser. Practical applications of 3-D PCs have challenges in manipulating light propagation in a specific region of the 3-D structure as well as complicated fabrication of the desired 3-D PC structures. Although 3-D volume-based control of light propagation using a nano-rod accumulated 3-D PC has ever been demonstrated, this book omits such 3-D PC applications.

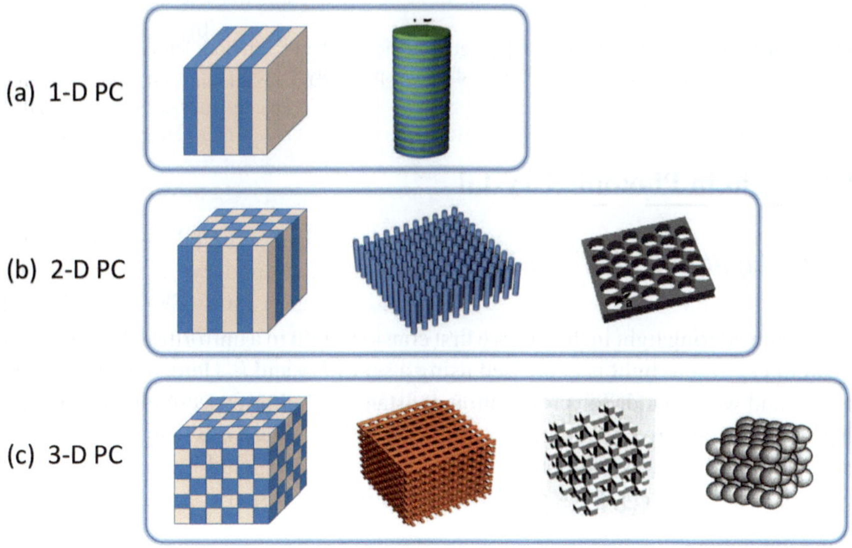

Fig. 2.2 Schematic structures of typical 1-D to 3-D PC family

2.2.2 *Lattice Configuration*

As shown in Figs. 2.1 and 2.2, periodic lattice components in 1-D to 3-D PCs are composed of a variety of geometrical elements with high index contrast, composite thin films in 1-D PCs, air-holes in 2-D PC slabs, nano-rods (pillars) on 2-D PC substrates, and composite layers with in-plane periodic patterns in 3-D PCs. Typical lattice configurations in 2-D PCs are represented by square-lattice and triangular-lattice configurations, as shown in Fig. 2.3a and b, respectively. In the figure, the left panels show real-space configurations, while the right panels show corresponding 2-D Brillouin zones (BZs) presented by reciprocal lattices (***k***-spaces). Symbols, M, X, K, Γ, etc. are given to singular points in the BZs. The symbol Γ means the center of the BZ. In a simple cube, M and X mean the centers of an edge and face, respectively, while, in a hexagonal, K means a middle of an edge joining two rectangular faces and M means a center of a rectangular face. These symbols are used also in other lattices.

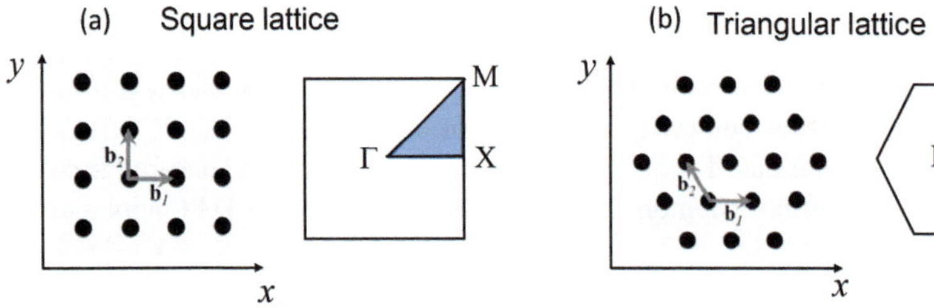

Fig. 2.3 **a** Square lattice of the PC. **b** Triangular lattice of the PC. In both lattices, left and right panels show lattice configurations and the first BZ, respectively

2.3 Light in Photonic Crystal

2.3.1 *Light in a Uniform Medium*

Before considering light in the PC, we first consider light in a uniform and transparent medium [1], where light is expressed using a set of $h\omega$ and $\boldsymbol{k}$. Here, $h\omega$ and $\boldsymbol{k}$ mean energy and wave vector (or propagation constant) of light, respectively. In vacuum, their relations are given simply by $\omega = ck$ ($k = |\boldsymbol{k}|$), where c is a light velocity in vacuum. In addition, light has another state: polarizations including electric field ($\boldsymbol{E}$) and magnetic field ($\boldsymbol{H}$). Each polarization has its own dispersion relation between $h\omega$ and k.

2.3.2 Photonic Band in a Periodic Medium

Now we consider light in the PC [1]. In the PC, eigenstates of light largely differ from those in a homogeneous medium; $\varepsilon(\boldsymbol{r})$ varies periodically with $\boldsymbol{r}$. Here, we suppose a dielectric material, i.e., the magnetic permeability $\mu(\boldsymbol{r}) = \mu_0$, where μ_0 is a permeability in vacuum. Now, we consider the wave equation for $\boldsymbol{H}(\boldsymbol{r})$ given by Maxwell's equations, as:

$$\nabla \times \left[\varepsilon(\boldsymbol{r})^{-1}(\nabla \times \boldsymbol{H}(\boldsymbol{r}))\right] = \mu_0\omega^2\boldsymbol{H}(\boldsymbol{r}), \tag{2.1}$$

where $\boldsymbol{H}(\boldsymbol{r}, t)$ and $\boldsymbol{E}(\boldsymbol{r}, t)$ are supposed to be harmonic modes as:

$$\boldsymbol{H}(\boldsymbol{r}, t) = \boldsymbol{H}(\boldsymbol{r})\exp(\mathrm{i}\omega t) \quad \text{and} \quad \boldsymbol{E}(\boldsymbol{r}, t) = \boldsymbol{E}(\boldsymbol{r})\exp(\mathrm{i}\omega t). \tag{2.2}$$

By solving the eigenvalue equation regarding the $\boldsymbol{H}(\boldsymbol{r})$, $\boldsymbol{E}(\boldsymbol{r})$ is given by the following relation:

$$\boldsymbol{E}(\boldsymbol{r}) = (\mathrm{i}\omega)^{-1}\left[\varepsilon(\boldsymbol{r})^{-1}(\nabla \times \boldsymbol{H}(\boldsymbol{r}))\right]. \tag{2.3}$$

Generally, resultant eigenenergy $h\omega(\boldsymbol{k})$, i.e., a dispersion relation between $h\omega$ and $\boldsymbol{k}$, is depicted by specific curves, i.e., photonic bands (PBs). The dispersion curve is often depicted in the reduced BZ (the first BZ) like a familiar electronic band diagram in semiconductors such as Si and GaAs. Figure 2.4 represents a typical photonic band diagram with a state density in the 2-D square lattice with dielectric circular rods for E polarization [6]. The ordinate axis shows a normalized eigenfrequency, $\omega a/2\pi c$, while the abscissa axis shows a wave vector represented by the singular points (Γ, M, X), as shown in Fig. 2.3. ω, a, and c indicate an angular frequency, lattice constant, and vacuum light velocity, respectively. The 2-D PC used in the calculation has proved to possess three PBGs at around $\omega a/2\pi c = 0.25$, 0.45, and 0.65. Symbols, M, X, and Γ specific to the BZ are explained in Fig. 2.3.

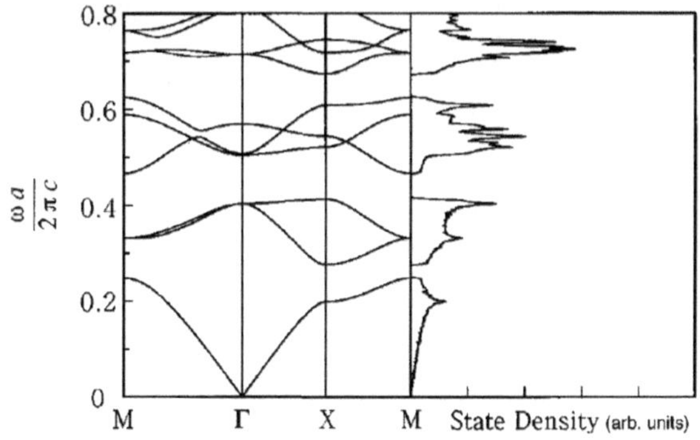

Fig. 2.4 Example of photonic band diagram and state density for E-polarized light in the square lattice with circular rods. Reprinted with permission from Ref. [6], © 1997 The American Physical Society

2.3.3 Photonic Band Gap

Many types of realistic PB structures in 2-D and 3-D PCs have been reported to date [5]. Typical examples of 2-D PB structures include an air-hole triangular lattice and a nano-rod square lattice. It is well known that only the 2-D PB structure for an air-hole triangular lattice exhibits a 2-D PBG irrespective of polarization if the refractive-index contrast in a lattice is large. In Ref [1], an example of the PBG is shown for the 2-D triangular-lattice air-hole PC structure. The PBG lies in the full BZ, so that this state is called a full PBG. It is noted that the full PBG does not appear in the square-lattice 2-D PC. Later, it is found that, if particular defects (such as point and line defects. See Sect. 2.4) are inserted in the entire periodic PC structure, PBs specific to the defect appear within the full PBG and exhibit practically important roles of waveguide operations. Figure 2.5 shows an example of a specific full PBG (yellow zone) in a particular 2-D PC structure with a hexagonal array of circular columns where each column is connected to its nearest neighbors by thin rectangular rods [7]. The result indicates maximum large full PBGs for both E and H polarizations caused by optimizing the radius of the circular columns and the width of the connecting rods. This is an example of possible flexible design of the PC against both polarizations.

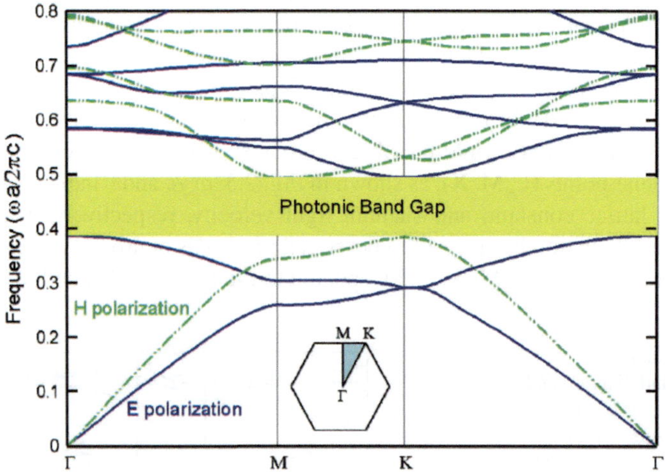

Fig. 2.5 Example of large full PBGs in 2-D PC structure with hexagonal lattice of circular columns each connected to its nearest neighbors by thin rectangular rods. Reprinted with permission from Ref. [7], © 2003 The American Physical Society

2.4 Defect Mode in Photonic Crystal

2.4.1 Type of Defect

Thus far, a completely periodic PC lattice without a singularity, namely a bulk PC, has been considered. Here, we introduce a lattice singularity "defect" that means a disorder or irregularity brought into a particular lattice point, line, or plane in the bulk PC. Such a defect breaks a symmetry that the PC originally reserved. Introduction of the defect is almost essential for confining or guiding light in some particular points or regions, thus constructing nano-scale cavities or waveguides for several kinds of PC applications. Now, we consider typical realistic lattice geometry of the PC defects based on the 2-D bulk PC, as shown in Fig. 2.2b, for instance. In the 2-D PC pillar structure, as shown in the center PC in Fig. 2.2b, when a diameter or shape of the cut-end of the pillar in any single line is changed from the surrounding media, it breaks the symmetry of the initial 2-D bulk PC and serves as a 2-D line defect. Figure 2.6a shows a missing-line defect of the 2-D PC pillar array as a particular case with the pillar diameter $d = 0$ along the line-defect structure. It serves as a 2-D PC waveguide. We consider another 2-D PC, a 2-D PC slab (such as air-bridge), as shown in the right PC in Fig. 2.2b, where air-hole arrays are arranged in the 2-D slab. The colored circle shows the top-view air-hole. If the diameter or shape of the air-hole in any single line is changed from the surrounding media, it breaks the symmetry of the initial 2-D PC slab and serves as a 2-D line defect. Figure 2.6b shows a missing-line defect of the 2-D air-hole PC slab as a particular case with the diameter $d = 0$ in the line-defect structure. It also serves as the 2-D PC waveguide. In addition, such a missing air-hole line defect exhibits strong vertical confinement of light, depending on the index contrast of the slab itself. Since the fabrication of any line shape (e.g., bend in the figure) is easily allowed with the current dry etching technology, such a missing air-hole defect is mostly applicable to a variety of PC waveguide devices. Section 2.6 in this chapter describes simulated and measured characteristics in the

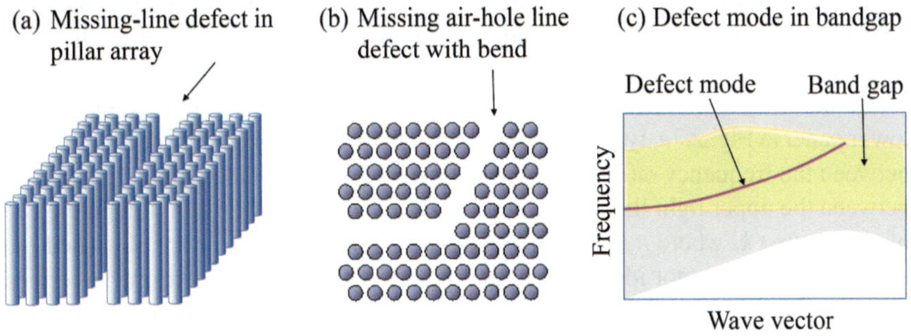

Fig. 2.6 Schematics of PC defects and defect mode. **a** Missing-line defect in 2-D PC pillar array. **b** Missing air-hole line defect with a bend in the 2-D PC air-hole slab. **c** Defect mode in the PBG in the band diagram

2-D PC waveguides. Chapter 6 also describes many sophisticated 2-D PC applications (slow-light, all-optical switch, and o-RAM applications). The introduction of such a 2-D defect is similarly applicable to 1-D and 3-D PCs. Note that such a defect induces a defect-based eigenmode, i.e., defect mode in a PBG of the PC band, as shown by the red curve in Fig. 2.6c. Performance of light propagating through the defect waveguide is controlled by the defect mode in the PBG. Detailed characteristics of the defect mode are discussed in the next (Sect. 2.4.2).

2.4.2 Defect Mode in Band Gap

Introduction of defects into the bulk PC produces an eigenmode called a defect mode in the PBG [1]. A point-defect mode in the bandgap, i.e., defect mode composed of point defects, is similar to an impurity mode in the semiconductor. When this mode is excited, light is localized around the defect. A line-defect mode, i.e., defect mode composed of line defects, has a specific dispersion relation, where the mode frequency ω varies with $\boldsymbol{k}$ in the band gap. Figure 2.6c shows a schematic line-defect mode (in red) in the 2-D PC with a square lattice of dielectric pillars in the air. Note that a single missing-line defect induces a single defect mode in the band gap, while doubled missing lines in the same 2-D PC induce doubled defect modes in the same bandgap.

2.5 Two-Dimensional Photonic Crystal Slab

2.5.1 Modes in Photonic Band Gap

Guided Modes Out of the Light Cone

Here, we focus on light modes in the 2-D PC slab with a thickness comparable to the wavelength of light [1]. To begin with, guided modes in the homogeneous dielectric slab are surveyed. The upper panel in Fig. 2.7a shows a uniform slab waveguide with a core index $n_2 = 2.67$, clad index $n_1 = 1$ (air), and thickness d, for instance. In the lower panel in Fig. 2.7a, typical guided modes are presented in the dispersion relation between the frequency $\omega(\pi c/\mathrm{d})$ and wave vector $k_{//}\,(\pi/d)$. Guided modes are located between the upper light line in vacuum of $\omega = ck$ and the lower light line in the core of $\omega = (c/n_2)\,k$, where n_2 is a refractive index in the core. They are classified into a TE or TM mode (not identified in the figure). These guided mode curves appear similarly in Fig. 1.3. Now, we consider the 2-D PC slab with a square lattice of air-holes indicated at the upper panel (cross-sectional air-bridge PC waveguide) in Fig. 2.7a [1]: As shown in the lower panel, the stop band (bandgap zone) emerges, as indicated by Bloch folding lines in Fig. 2.7b, while each band in Fig. 2.7a is folded into the reduced BZ. Moreover, mode mixing between the TE and TM modes occurs and any guided modes in the PC slab are no longer TE (TM) modes but are called TE (TM)-like modes. Regarding 2-D PC device applications, most practical photonic devices use guided mode waveguides mentioned above in the 2-D PC slab with any defect lattice structures. Refer to the practical photonic devices for detail in Chap. 6, along with an exceptional application using no guided mode but leaky mode introduced next.

Light Cone and Leaky Mode

As shown in Fig. 2.7b, guided modes under the light line in vacuum in the 2-D PC slab are extended to extended continuous modes in air above the vacuum light line, i.e., within the light cone (gray zone). Note that any mode within the light cone can be coupled to modes in air: These modes are called leaky modes or radiation modes. In Sect. 6.5, we introduce an important application of the 2-D PC slab functioning at the Γ-point within the light cone, an excellent PC surface-emitting laser (PCSEL). Note that the PCSEL uses a 2-D slab with no defect points but uniform or modified PC lattice points only. It functions as lasing in the vertical direction on the slab surface (the reason for "surface-emitting laser") owing to diffractive coupling between the

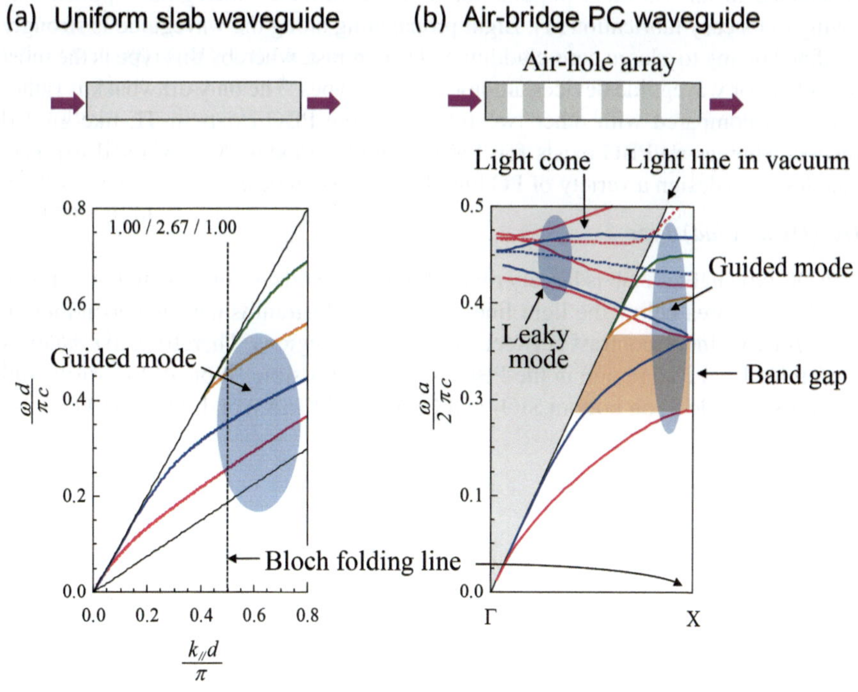

Fig. 2.7 Degenerate expression of band diagrams in the 2-D slabs. **a** Dispersion diagram in uniform (homogeneous dielectric) slab waveguide. **b** Dispersion diagram in the air-bridge PC slab waveguide. The frequency ω and the wave vector $k_{//}$ in (**a**) are given in units of $\pi c/d$ and π/d, respectively, while those in (**b**) are given in units of $2\pi c/a$ and π/a, respectively, where a is an air-bridge PC slab thickness

guiding modes in the 2-D slab and the vertical radiation mode at the Γ-point in the light cone. The PCSEL should also be noted because of outstanding performances in high-power single-mode emission with extremely small beam divergences. Refer to Sect. 6.5 for more detail.

2.5.2 Type of Photonic Crystal Slab

The 2-D PC slab device with air-hole lattices is relatively easy to fabricate as compared with the 3-D PC device because current semiconductor micro-/nano-fabrication technologies are available. Here, taking the 2-D PC core with an air-hole array into account, the following three sorts of PC slabs have been reported to date [1]: AB (air-bridge) type, OC (oxide- or low-index-cladding) type, and SC (semiconductor- or hetero-structure-cladding) type, as shown in Fig. 2.8.

AB (Air-Bridge) Type

The AB type in Fig. 2.8a using dielectric slabs with line-defect waveguides—suspended in air—is most popular for high-performance waveguide applications owing to its easy fabrication [1]. Light propagating along the waveguide is strongly confined owing to a large core-cladding index contrast, whereby this type is the most attractive for waveguide devices among the three types. The only drawback is rather fragile as compared with other two types. A large PBG exists in TE-like guided modes, whereas no PBG exists for TM-like guided modes. A large band gap also enables us to design a variety of PC line-defect waveguides.

OC (Oxide Clad) Type

The OC type in Fig. 2.8b is largely robust for applications to compare to the AB type [1]. Yet, the area below the light line in the band diagram is narrow depending on the refractive-index contrast between core and clad regions. Therefore, low index is preferable for a clad region in the PBG as large as possible. Typical core slabs with air-holes include Si on bottom SiO_2 and GaAs (or AlGaAs) on bottom Al_xO_y.

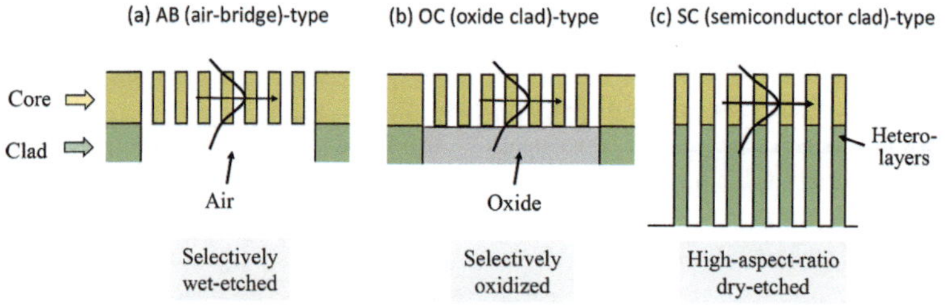

Fig. 2.8 Schematics of three types of 2-D PC slab waveguides with air-holes. **a** AB-type wet-etched selectively. **b** OC-type oxidized selectively. **c** SC-type dry-etched in high aspect ratio

SC (Semiconductor Clad) Type

The SC type in Fig. 2.8c is the most attractive for robust air-hole slab from the viewpoint of fabrication [1]: Air-hole structures in the core extend to the bottom cladding substrate to form high-aspect-ratio air-holes, as shown in the figure. Yet, no guided mode exists outside the light cone and all modes are leaky modes. However, the SC type is useful for applications depending on the performance in radiation loss [8].

2.6 Pioneering Works on Photonic Crystal Slab Waveguide

So far, 2-D and 3-D PCs were extensively studied for practical applications in the fields of ultimate radiation control devices and optical signal processing devices. A new concept of light guiding using PBGs was put forward, in place of the light guiding using a traditional refractive-index gap [9, 10]. From then on, applications of the PC device have focused on small-scale integration of photonic active and passive elements for the coming optical communication. In this section, typical theoretical and experimental pioneering works on the 2-D PC waveguides performed in an early stage of the PC development are selectively introduced, as follows.

2.6.1 Theoretical Work on Bend Waveguide

Theoretical transmission characteristics of the PC waveguides started from the middle of the 1990s. For instance, Fig. 2.9 shows theoretically simulated reflection coefficients and electric field distribution using a 2-D PC bend waveguide composed of dielectric rods with a square lattice of a lattice periodicity a [9]. In Fig. 2.9a, a missing row of rods is distributed in a bent waveguide. In more detail, note that single rod at the corner of the inner square lattice is shifted to the inside of the missing corner in L shape for mitigating steepness of the abrupt bent corner. Figure 2.9b shows calculated reflection coefficients at the bent vs normalized frequency by 1-D scattering model (solid line in red) and numerical simulations (thick-dotted line in black). Refractive index of rods and their radius were assumed as 3.4 and $0.18a$, respectively. The result indicates almost zero reflection at frequencies around 0.35, resulting in the presumed 98% for maximum transmission. In addition, average high transmission of > 95% was presumed also for a wide frequency range. Note that the maximum value of 98% is extremely higher than 30% for similar conventional bend. As shown by the electric field distribution for $\omega = 0.353 \times 2\pi c/a$ in Fig. 2.9c, light wave is completely confined inside the waveguide and propagates smoothly around the abrupt bend with a bend curvature radius on the order of light wavelength.

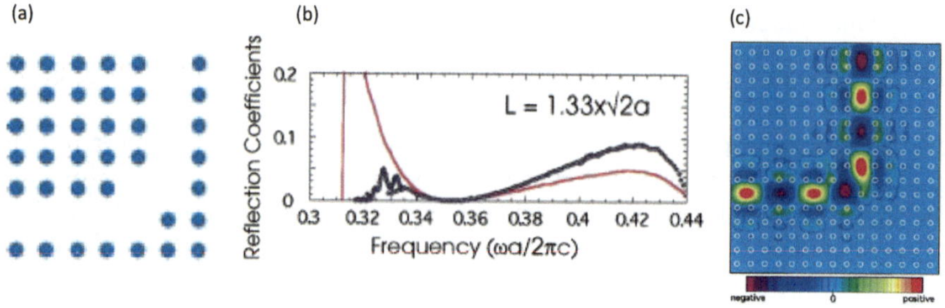

Fig. 2.9 Theoretical transmission characteristics of light in 2-D PC bend waveguides. **a** Square-lattice distribution of dielectric rods with missing row of rods, constructing a bend waveguide. **b** Calculated reflection coefficients as a function of frequency for 1-D scattering model (solid line in red) and the numerical simulations (diamonds in black). **c** Electric field distribution of the 2-D PC bend waveguides. Reprinted with permission from Ref. [9], © 1996 APS

Another numerical analyses of T-shaped 2-D PC waveguide branches with a square lattice of dielectric rods were reported [11]. It resulted in high transmission coefficients of more than 49.5% when a dielectric constant ε of the rod and curvature radius of the T-shape were 11.56 and $0.2a$, respectively, where a was a lattice constant. It was verified that electric fields were split almost equally at the T-shaped branch.

2.6.2 Experimental Work on Low-Loss Waveguide

Along with the theoretical works mentioned above, experimental works on fabrication of 2-D PC components were undertaken with straight/bend waveguides, Y-splitters, directional couplers (DCs), etc., needed for implementing typical integrated 2-D PC circuits. Our group (book authors) realized dry-etched smooth 2-D PC straight waveguides, 60°-bends, and Y-splitters using an air-bridge type PC on the GaAs substrate [12]. Figure 2.10a indicates a schematic of a triangular-lattice waveguide with a line defect composed of a single row of missing air-holes in the Γ–K direction (refer to Fig. 2.3b). As shown by a SEM photograph in Fig. 2.10b, a GaAs PC slab with an air-bridge type was fabricated by electron-beam lithography, Cl_2-gas reactive-ion-beam etching, and selective wet-etching. Figure 2.10c shows simulated transmission spectra of two defect modes (odd and even) in the bandgap for the 2-D PC slab. For calculation, a 3-D finite-difference-time-domain (FDTD) method was used with a slab thickness d of 250 nm, lattice constant a of 350 nm, and air-hole radius R of 220 nm. Figure 2.10d shows transmittance spectra of the sample shown above, where a thick-dotted line and a thin solid line correspond to the experimental and simulated results, respectively. 60°-bends and Y-splitters were fabricated as well. Resultant propagation loss was 1.5 ± 0.5 dB/mm for the straight waveguide and below 1 dB for the bent waveguide. Bandwidths were also found to

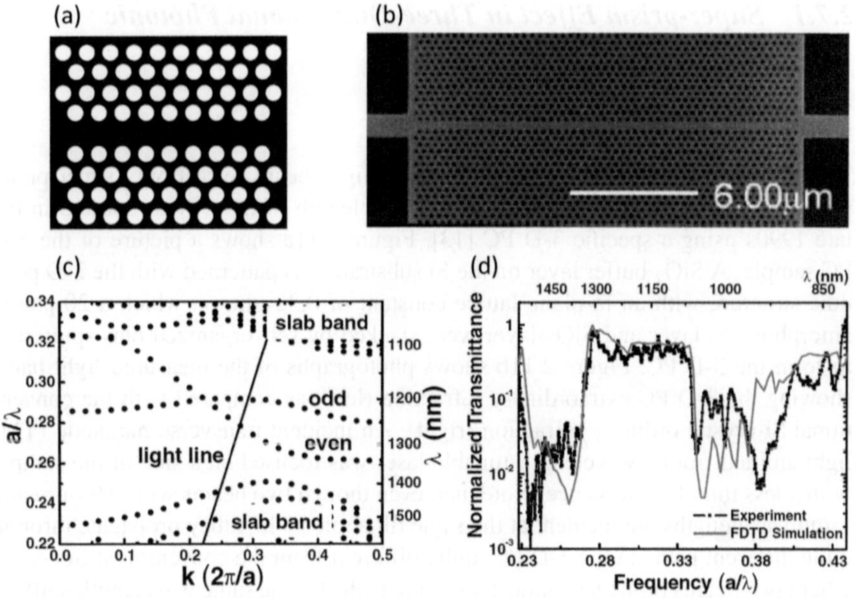

Fig. 2.10 **a** Schematic triangle lattice pattern of a GaAs PC slab straight waveguide with a single-line defect. **b** SEM photograph of the demonstrated air-bridge structured straight PC waveguide. **c** Calculated band diagram in the *Γ*–*K* direction. **d** Measured (thick line) and calculated (thin line) transmittance spectra. Reprinted with permission from Ref. [12], © 2004 The Japan Society of Applied Physics

be wider than 40 nm. Measurement of the Y-splitter also suggested the capability to split light almost equally with more than 85% in total transmittance. These results suggest large capability of integrated low-loss 2-D PC circuits.

2.7 Peculiar Phenomena in Photonic Crystal

Together with theoretical research on novel characteristics of 2-D to 3-D PCs, several pioneering works on finding peculiar phenomena suggesting new applications of PCs have been reported in the late 1990s–early 2000s. This section selects four topics: super-prism phenomenon as extraordinary refraction of the PC, high-Q nano-cavity in the 2-D PC, slow light in the PC waveguide, and band-edge emission at Γ-point in 2-D PC, as follows:

2.7.1 Super-prism Effect in Three-Dimensional Photonic Crystal

Discovery of extraordinary refraction

The first pioneering work on the surprising light beam refraction: "super-prism phenomenon", extremely sensitive to the wavelength of light, was reported in the late 1990s using a specific 3-D PC [13]. Figure 2.11a shows a picture of the 3-D PC sample: A SiO_2 buffer layer on the Si substrate was patterned with the 2-D periodic structure with an in-plane lattice constant of 0.32 μm, on which a 20-paired amorphous Si layer and SiO_2 layer were stacked by self-organized bias sputtering to form the 3-D PC. Figure 2.11b shows photographs of the measured light paths showing the 3-D PC extraordinary refraction (left), as compared with the conventional Si-crystal ordinary refraction (right). An incident transverse magnetic (TM) light emitted from a wavelength-tunable laser was focused on a side of the sample with a less than 1° divergence. Note that, even though two beams with 0.99-μm and 1-μm wavelengths are incident at the same tilt angle of 15°, they propagate through quite different paths in the 3-D PC, quite different from the conventional Si crystal where two beams propagate almost the same paths for the same wavelength shift. In the 3-D PC, angles for two beams are separated by about 50° for the wavelength shift of only 0.01 μm (1%) around 1 μm. Detailed investigation on changing the incident

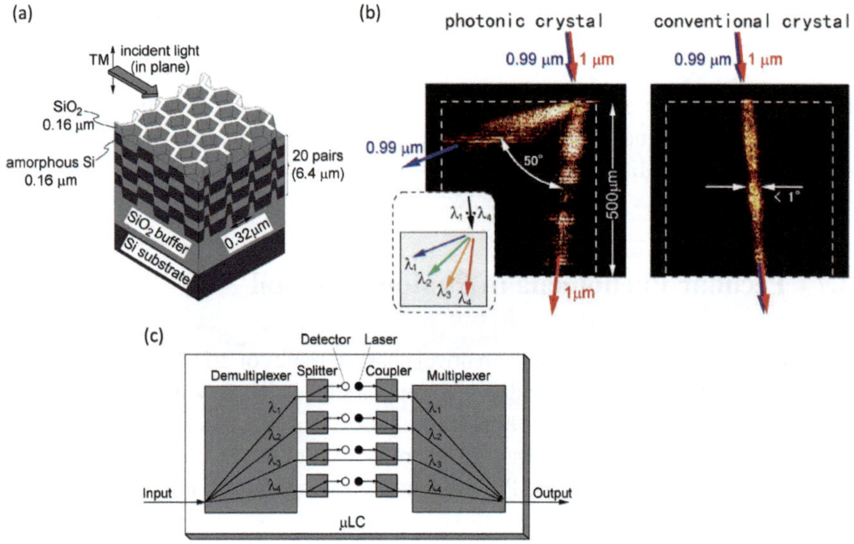

Fig. 2.11 **a** Schematic of the 3-D PC self-formed on a patterned Si substrate. **b** Micrographs of light beams propagating through a 3-D PC (left panel) and a Si wafer (right panel) when beams at different wavelengths, 0.99 and 1.0 μm, are incident with a tilting angle of 15°. **c** Schematic of a PC-based integrated WDM-ADM involving a (de-)multiplexer, splitter, and coupler. Reprinted with the permission from Ref. [13], © 1999 AIP

angle of 15° has proved that the maximum angular dispersion reaches 50° with only a 1% change of the wavelength. The super-prism is promising for the practical (de-) multiplexers in the WDM communication, as shown in Fig. 2.11c, where an add/drop multiplexer (ADM) for WDM is proposed. An estimated device scale is on the order of ~ 200 μm: two orders smaller than the conventional device such as arrayed waveguide grating (AWG) with a few centimeters in size.

Mechanism of Super-prism Effect

The super-prism phenomenon was theoretically explained by a 3-D PC-based extraordinary refraction model in an anisotropic ***k***-space, as shown in Fig. 2.12 [14]. Figure 2.12a shows a measured laser-beam propagation angle θ_p versus incident angle θ_i, where $\Omega = a/\lambda$ is a normalized frequency given by a lattice constant a of 0.4 μm and 0.32 μm for wavelength $\lambda = 0.956$ μm. Solid curves and dotted plots

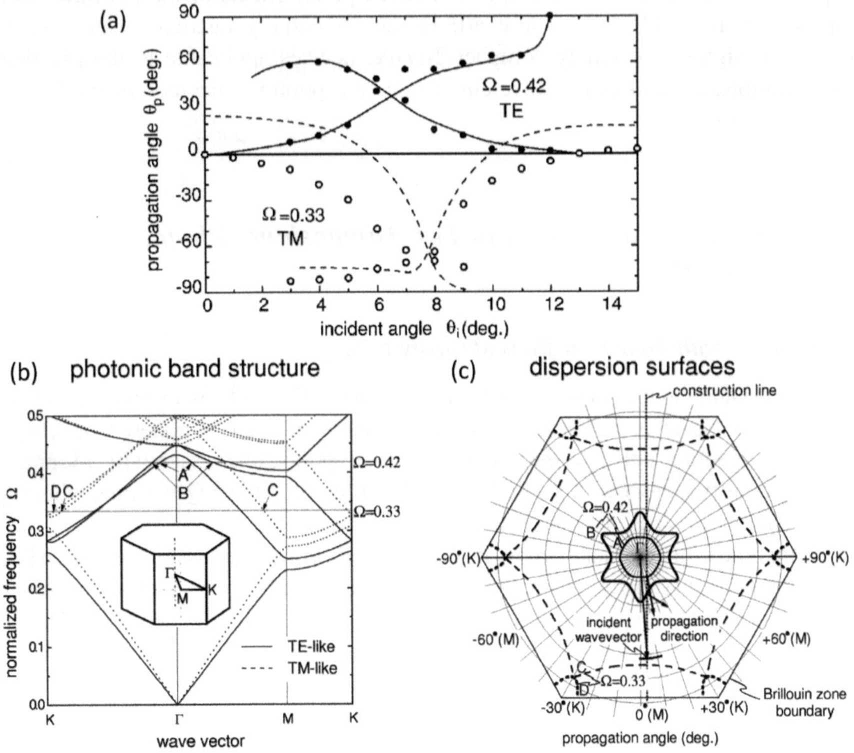

Fig. 2.12 **a** Measured propagation angle θ_p versus incident angle θ_i. Solid and dashed line curves correspond to the calculated bands. **b** Simulated TE-like mode (solid line) and TM-like mode (dotted line) curves in the photonic band structure. **c** Simulated constant-frequency dispersion surfaces. Incident wave vector is emitted from Γ-point, while propagation directions are indicated by vectors normal to the dispersion surface. Reprinted with permission from Ref. [14], © 1998 APS

reveal calculated and measured results, respectively. The result shows that propagation angles largely depend on the incident angle below 12°. Moreover, these curves exhibit unique branches with opposite angular dependence. Such dispersion characteristics cannot be explained by conventional optics in crystal, but need an extraordinary refraction model in an anisotropic ***k***-space. Figure 2.12b and c shows simulated photonic band structures and corresponding dispersion surfaces, respectively, for the same 3-D PC as used in the experiment using a plain-wave expansion method. In these curves, A and B ($\Omega = 0.42$) are caused by TE-like mode bands, while C and D ($\Omega = 0.33$) by TM-like mode bands. The result shows extreme anisotropy particularly in the bands: B, C, and D. Considering that a propagation direction is normal to the dispersive surface at a propagation wave vector, good agreement between calculated and measured data means that characteristic angular dependence is explained well using the starfish-shaped dispersion surface in Fig. 2.12c.

After the proposal of the "super-prism", more advanced theoretical as well as experimental analyses on the dispersion-based prism effects in the PC have been reported in Refs. [15–18]. These articles are essentially instructive and important for designing extremely compact 2-D PC devices applicable to the practical (de-)multiplexers and other wavelength-selective photonic devices in the WDM communication.

2.7.2 High-Q Nano-cavity in Two-Dimensional Photonic Crystal

Nano-cavity with Tuned Air Rods at Cavity Edge

PC-based nano-cavities with a small group of point/line defects are promising for applications based on the strong optical confinement such as coherent electron-photon interaction and nonlinear optics. These applications essentially require a high Q (quality factor) with a small V (modal volume), a large Q/V value, where Q means energy loss per cycle vs energy saved in the cavity.

To this end, a high-Q nano-cavity in a 2-D PC slab with three missing air rods in line is theoretically investigated by pursuing the optimum configuration of triangle lattice air-rod parameters, as shown in Fig. 2.13a [19]. Here, a slab thickness and air-rod radius are $0.6a$ and $0.29a$, respectively, while a lattice constant a is 0.42 μm. In the in-plane direction, an optical energy is confined by the Bragg reflection due to a periodic air-rod lattice, while in the vertical direction, an optical energy is confined within the slab based on the total internal reflection (TIR). Figure 2.13b shows a shape of an optical electric field in the cavity with a length of $L = 2.5\lambda$, where λ is a wavelength of light. Meanwhile, Fig. 2.13c shows a Fourier-transformed electric field. Note that large components (indicated by "leaky region") remain inside the leaky region (smaller than $2\pi/\lambda$) as one of the side lobes: It means increase of radiation loss, that is, decrease of Q-value in the cavity. In contrast, Fig. 2.13d shows an improved cavity structure where air rods at both edges of the cavity are shifted by $0.15a$, as indicated

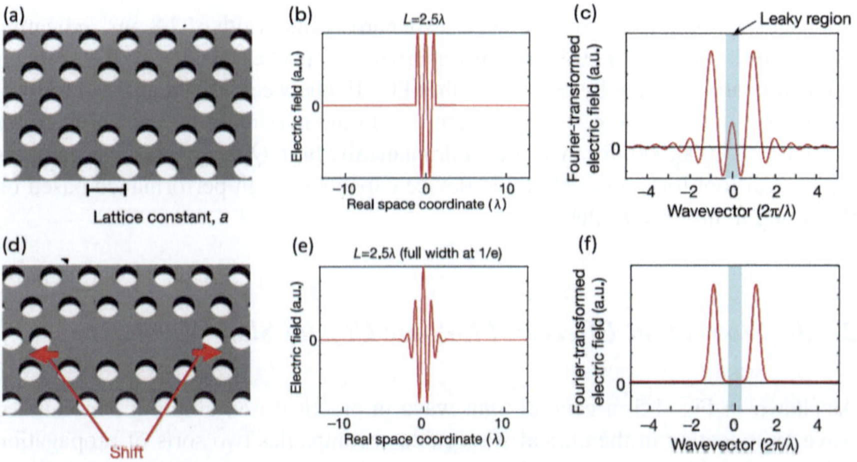

Fig. 2.13 2-D PC nano-cavities with a triangular lattice of air rods. **a** Initial cavity with three missing air rods. **b** and **c** Electric field shape in the cavity, and Fourier-transformed one, respectively. A blue area means the leaky region. **d** Cavity improved by shifting air rods at both edges of the cavity in the arrow directions. **e** and **f** Improved electric field shape and Fourier-transformed one, respectively. Reprinted with permission from Ref. [19], © 2003 Springer Nature

by arrows in red. Figure 2.13e and f shows resultant electric field with a Gaussian-like envelope and its Fourier-transformed electric field, respectively. Note that an electric field component, appearing in the leaky region in (c), disappears, then resulting in an increased Q-factor of 45,000 and Q/V of 6.4×10^{17} cm^{-3}. Experimental confirmation of such a high-Q value is implemented using a 2-D PC composed of an input/output line-defect waveguide located near the cavity mentioned above [19]. Furthermore, these high-Q nano-cavities with tuned 2-D PC air rods are described for more detail in [20–22].

Nano-cavity with Double Hetero-structure Cavity

In the previous Fig. 2.13d, further tuning positions of the second and third nearest-neighbor air rods achieve further improvement of the Q-value. Nevertheless, the air-rod-tuning approach has a limitation to a large extent for realizing ultimately high-Q nano-cavities. Another new concept of photonic double hetero-structures for ultimate high-Q nano-cavity is introduced [23]: Triangular-lattice PC areas (I and II) with different lattice constants a_1 and a_2 are joined to construct a double hetero-structure. In the area I, a_1 is 410 nm, while a_2 is increased to 420 nm in the area II, thus forming a structure like a face-centered rectangular-lattice. In the PCs I and II, confinement of photon in the waveguide direction is conducted differently due to the mode-gap effect originated from difference of the lattice constant. As a result, penetration of an evanescent electric field in the PC II into the PC I can be changed by selecting a_1 and a_2. Light fields both in the input waveguide and in the cavity are coupled by the evanescent component in each field distribution. The double hetero-structure PC cavity was prepared using a SOI material on a Si substrate. A measured resonant spectrum results in a tremendously narrow line width of 2.8 pm, indicating an increased Q-factor of 600,000. Surprising to say, further increase in the Q-factor is theoretically certified by adding another PC (II′) between PC I and PC II to form a multistep double hetero-structure, resulting in the maximum theoretical Q-factor exceeding 24,000,000, for instance. A dramatically high Q-factor PC nano-cavity is a powerful tool for a nano-photonic device capable of high performance based on the strong light–matter interaction.

2.7.3 Slow-Light Effect in Photonic Crystal Slab Waveguide

As shown in Fig. 1.5, a general light wave in practical use, amplitude-modulated wave, propagating in the optical waveguide accompanies two sorts of propagation velocities: a phase velocity v_{p} for the carrier wave velocity and a group velocity v_{g} for the envelope velocity, the latter defined by the slope on the $\omega(k)$ point in the dispersion curve. Furthermore, basic dispersion curves in the band diagram of the PC waveguide often accompany singular zones where the group velocity asymptotically approaches zero, as shown near the band edge in Figs. 2.4 and 2.5.

"Slow light" is a light wave specified to such low group velocity light and is generated in the modified air-hole arrays in the PC [24]. For instance, a triangular-lattice 2-D PC waveguide with air-holes in three differently chirped arrangements was proposed, as shown in Fig. 2.14 [24]. The chirped PC waveguide is composed of gradually changing structural parameters such as air-hole diameters, lattice constants, and background refractive index along the waveguide, resulting in the smooth shift of the photonic band diagram. Figure 2.14a shows three different upward bands shifting to a high-frequency range due to an index-chirping. In the upper panel, input light decreases the speed first, reaches the band-edge state, stays and is reflected soon, then finally propagates backward with some delay. Figure 2.14b shows a 2-D PC directional coupler with upward and downward-chirped waveguides A and B. Input light in the waveguide A, reaching the band-edge region and staying there, transfers energy into the waveguide B owing to an efficient directional coupling effect and propagates forward in the waveguide B, increasing the group velocity owing to the chirping effect. A localization-induced delay might be applicable to a delay device. Note that, if upward and downward band profiles of A and B are symmetric, a group velocity dispersion (GVD) near the band edge is completely canceled. Then, a group delay device is available. Actual designs of slow-light waveguides involve several proposals [24]: The waveguide width w' reduced to 0.85 w, where w is a waveguide width with single-line missing air-holes, presents an upward band diagram. The waveguide with a normalized air-hole diameter reduced to $2r'/a$ at the center of the waveguide presents upward and downward bands with a local gap. In Sect. 6. 2, a variety of applications using the slow-light effect as well as other slow-light waveguide structures are described.

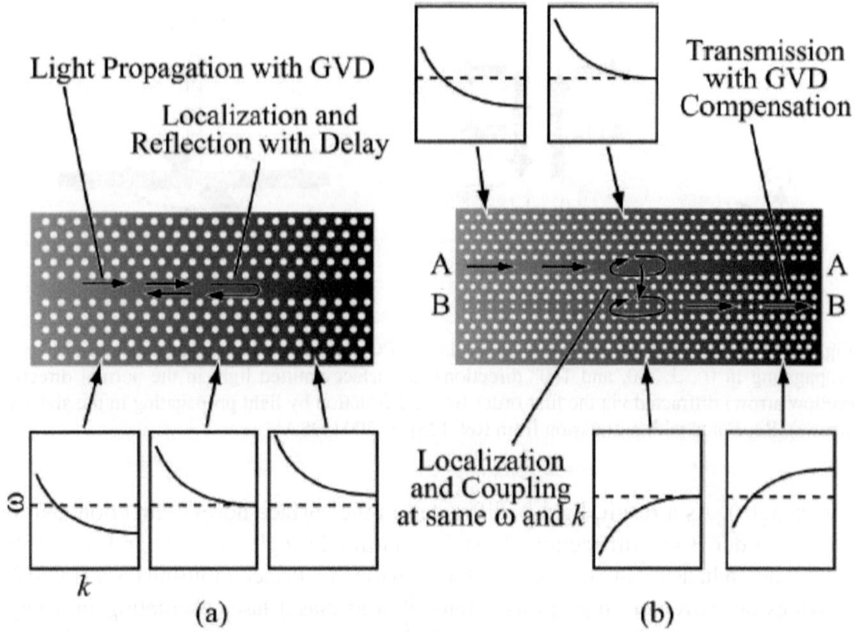

Fig. 2.14 Triangular-lattice 2-D PC waveguides with differently chirped air-hole arrangement. **a** Three different upward bands shifting to a high-frequency range. Simple chirped waveguides. **b** 2-D PC directional coupler with upward- and downward-chirped waveguides, A and B. Reprinted from Ref. [24] with the permission of the American Institute of Physics, © 2004 AIP

2.7.4 Band-Edge Emission in Two-Dimensional Photonic Crystal

As discussed above, a photonic crystal (PC) exhibits an extremely unique behavior guided by a specific photonic band (PB) diagram like an electronic band in the semiconductor. In the 2-D PC slab with a missing air-hole line defect, light wave propagates along a line-defect waveguide controlled by the defect-mode band, in particular, presenting a "slow light" with a very small or zero-group velocity near the band edge or on the particularly designed PB (Fig. 2.14). The line defect with a limited number of missing air-hole in the PC slab provides a "high-Q nano-cavity" where the wavelength-dependent light wave is strongly confined to show a sharp resonance. Meanwhile, a concept of the Γ-point, often appearing at the PB edge in the 3-D PC with no missing defects (Figs. 2.4 and 2.5), provides an extremely unique behavior in a uniform lattice PC slab: "coherent surface emission of light". Figure 2.15a shows a top view of a square-lattice PC slab [25]. Two arrows at the bottom left indicate two proper propagation directions Γ-X and Γ-M, while black bold arrows indicate propagating light waves, causing a resonant-wave mode caused by in-plane first-order Bragg diffraction (the lattice period corresponds to the light

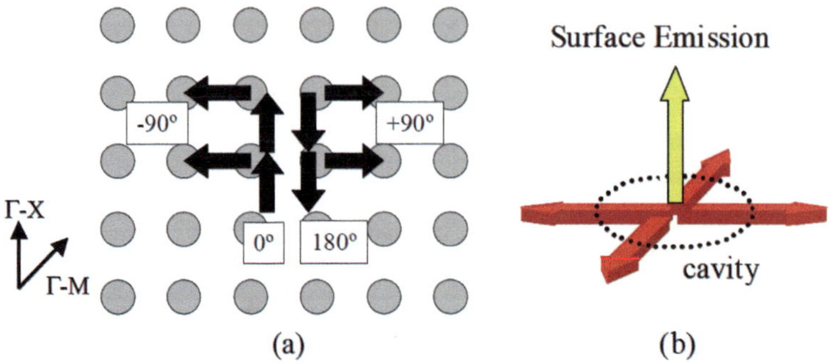

Fig. 2.15 **a** Schematic top view of a square-lattice PC slab. Black bold arrows mean light waves propagating in 0°, ± °90, and 180° directions. **b** Surface-emitted light in the normal direction (yellow arrow) diffracted via the first-order Bragg diffraction by light propagating in the slab (red arrows). Reprinted with permission from Ref. [25], © 2004 OSA

wavelength). As a result, light is diffracted in the surface-normal direction also by the first-order Bragg diffraction, "surface-emitting lasing", as shown in Fig. 2.15b.

As shown in detail in Sect. 6.5, the topic on the PC surface-emitting laser (PCSEL) provides an extremely high-power (tens of watt-class) laser oscillating in a large area with excellent beam quality (small divergence). The PCSEL is much more superior than ever reported 2-D PC lasers, defect-mode lasers using a PBG, artificially introduced defects [26] and multi-directional DFB lasers [27], etc., as well as any conventional compact laser including vertical-cavity surface-emitting lasers (VCSELs).

2.8 Summary

Typical fundamental properties of the PC were surveyed: Basic classification of the PC corresponds to index contrast-based 1-D to 3-D lattice structures. Light states in the PC originated from the lattice-dependent dispersion relations in the band diagram and full bandgap. Point/line defects are introduced into the uniform PC lattice, thus emerging defect-mode bands in the bandgap of the PC. These extra-bands cause nano-scale cavities and waveguides in the PC by localizing energy of light there, resulting in a variety of PC devices. 2-D PC waveguides are formed in the slab structure. Besides in-plane confinement due to the PC effect, light is localized in the normal direction by way of a high refractive-index contrast between core (e.g., dielectric) and clad (e.g., air) regions. Concepts of light cone and leaky mode specific to the 2-D PC waveguide are discussed also for guiding careful attention to designing the 2-D PC device. As preliminary works that characterize features of the PC waveguides, theoretical simulation of a 2-D PC bend waveguide and experimental demonstration of a low-loss 2-D PC waveguide were introduced. Finally, peculiar phenomena of the PCs that straightforwardly guide to unique PC applications were introduced as an introductory part of Chap. 6 that shows various PC applications: A super-prism effect based on the extraordinary refraction in the 3-D PC structure is one of the most characteristic phenomena that never occur in the ordinary dielectric media other than the PC. An extremely high-Q nano-cavity in the 2-D PC defects is applicable to nano-scale optical memory elements, while proposal of slow light caused by the low-/zero-group velocity suggests tunable delays and dispersion devices. Coherent surface emission of light at the Γ-point in the 2-D PC lattice leads to a principle of an innovative PC surface-emitting laser with a single-mode, high-power and an extremely narrow beam divergence. Applications of these phenomena are described in Chap. 6 as innovative integrated nano-scale PC devices that are definitely promising in the advanced digital signal processing networks and next-generation smart societies.

References

1. Inoue K, Ohtaka K (eds) Photonic crystals—physics, fabrications and applications. Springer Series in Optical Sciences. ISSN 0342-4111
2. Yablonovitch E (1987) Inhibited spontaneous emission in solid-state physics and electronics. Phys Rev Lett 58:2059–2062
3. John S (1987) Strong localization of photons in certain disordered dielectric super-lattices. Phys Rev Lett 58:24862489
4. Ohtaka K (1979) Energy band of photons and low-energy photon diffraction. Physic Rev B19:5057
5. Joannopoulos JD, Johnson SD, Winn JN, Meade RD (1995) Photonic crystals. In: Molding the flow of light, 1st edn. Princeton University Press, Princeton. Also Joannopoulos JD, Johnson SG, Winn JN, Meade RD (2008) Photonic crystals. In: Molding the flow of light, 2nd edn. Princeton University Press, Princeton
6. Sakota K, Shiroma H (1997) Numerical method for localized defect modes in photonic lattices. Phys Rev B 56:4830–4835
7. Chern RL, Chung Chang C, Chang CC, Hwang RR (2003) Large full band gaps for photonic crystals in two dimensions computed by an inverse method with multigrid acceleration. Phys Rev E 68:026704
8. Benisty H, Labilloy D, Weisbusch C, Smith CMJ, Krauss TF, Benaud A, Cassagne D, Jouanin C (2000) Radiation losses of waveguide-based two-dimensional photonic crystals: positive role of the substrate. Appl Phys Lett 76:532
9. Mekis A, Chen JC, Kurand I, Fan S, Villeneuve PR, Joannopoulos JD (1996) High transmission through sharp bends in photonic crystal waveguides. Phys Rev Lett 77:3787–3790
10. Mekis A, Fan S, Joannopoulos JD (1998) Bound states in photonic crystal waveguides and waveguide bends. Phys Rev B 58:4809
11. Fan S, Johnson SG, Joannopoulos JD, Manolatou C, Haus HA (2001) Waveguide branches in photonic crystals. J Opt Soc Am B 18:162–165
12. Inoue K, Sugimoto Y, Ikeda N, Tanaka Y, Asakawa K, Maruyama T, Miyashita K, Ishida K, Watanabe Y (2004) Ultra-small GaAs photonic-crystal slab-waveguide based near-infrared components: fabrication, guided-mode identification, and estimation of low-loss and broad-band-width in straight-waveguide, 60°-bends and Y-Splitters. Japan J Appl Phys 43:6112–6124
13. Kosaka H, Kawashima T, Tomita A, Notomi M, Tamamura T, Sato T, Kawakami S (1999) Photonic crystals for micro light-wave circuits using wavelength-dependent angular beam steering. Appl Phys Lett 74:1370–1372
14. Kosaka H, Kawashima T, Tomita A, Notomi M, Tamamura T, Sato T, Kawakami (1998) Super-prism phenomena in photonic crystal. Phys Rev B 58:R10096–R10099

15. Notomi M (2000) Theory of light propagation in strongly modulated photonic crystals: refraction like behavior in the vicinity of the photonic band gap. Phys Rev B 62:10696–10705
16. Ochiai T, Sanchez-Dehesa J (2001) Super-prism effect in opal-based photonic crystals. Phys Rev B 64:245113
17. Baba T, Nakamura M (2002) Photonic crystal light deflection devices using the super-prism effect. IEEE J Quantum Electron 38:909–914
18. Bravo-Abad J, Ochiai T, Sanchez-Dehesa J (2003) Anomalous refractive properties of a two-dimensional photonic band-gap prism. Phys Rev B 67:115116
19. Akahane Y, Asano T, Song B-S, Noda S (2003) High-Q photonic nano-cavity in a two-dimensional photonic crystal. Nature 425:944–947
20. Srinivasan K, Barclay PE, Painter O, Chen J, Cho AY, Gmachl C (2003) Experimental demonstration of a high quality factor photonic crystal micro-cavity. Appl Phys Lett 83:1915–1917
21. Yanik MF, Fan S, Soljacic M (2003) High-contrast all-optical bi-stable switching in photonic crystal micro-cavities. Appl Phys Lett 83:2739–2741
22. Soljacic M, Joannopoulos JD (2004) Enhancement of nonlinear effects using photonic crystals. Nat Mater 3:211–219
23. Song B, Noda S, Asano T, Akahane Y (2005) Ultra-high-Q photonic double hetero-structure nano-cavity. Nat Mater 4:207–210
24. Mori D, Baba T (2004) Dispersion-controlled optical group delay device by chirped photonic crystal waveguides. Appl Phys Lett 85:1101–1103
25. Ohnishi D, Okano T, Imada M, Noda S (2004) Room temperature continuous wave operation of a surface-emitting two-dimensional photonic crystal diode laser. Opt Express 12:1562–1568
26. Painter O, Lee RK, Scherer A, Yariv A, O'Brien JD, Dapkus PD, Kim I (1999) Two-dimensional photonic band-gap defect mode laser. Science 284:1819–1821
27. Meier M, Mekis A, Dodabalapur A, Timko A, Slusher RE, Joannopoulos JD, Nalamasu O (1999) Laser action from two-dimensional distributed feedback in photonic crystals. Appl Phys Lett 74:7–9

Chapter 3
Survey of Surface Plasmon Waveguides

Abstract Nano-photonics in the previous chapters was based on the optical behaviors within the dielectric materials, while metal optics-based surface plasmon (SP) in this chapter is involved in another nano-photonic family—a composite electromagnetic (EM) wave that interacts with coherent free electron motions in the vicinity of the metal surface. It allows a small amount of energy loss within the narrow vicinity of the metal surface. Nevertheless, modern interest in the SP is rapidly growing because the EM field in the SP is strongly confined within several tens of nm at the metal/dielectric interface. The result allows the SP to be free of diffraction limit with minimum energy loss, thus leading to a contemporary favorite together with Si photonics and PCs for the next-generation nano-photonic devices. This chapter surveys fundamentals of the SP: *first*, Drude model-based metal optics leading to a plasmon-polariton (PP) and its dielectric function and dispersion relation, *second*, basic physics of the SP including a dispersion relation and several physical phenomena originated from the PP, *third*, an excitation method of the SP, and *fourth*, a localized surface plasmon (LSP) confined in the metallic nano-particle and nano-gap. Applications of the SPs are introduced in Chap. 7.

3.1 Introduction

Surface plasmon (SP) is a composite electromagnetic (EM) wave interacting with coherent free electron motions (plasmons) on the metal surface. First, we introduce concepts of plasma in general, a plasmon as a particular mode of plasma, and a plasmon-polariton (PP) as a particular mode of the plasmon, as follows [1, 2]:

Plasma is a collection of randomly distributed ionized gas (positive ions and free electrons), as shown in Fig. 3.1a. For instance, plasma exists in ionosphere in earth's upper atmosphere, discharge in a fluorescent tube, plasma chambers in semiconductor device fabrication equipment, etc. Incidentally, a plasmon is a collective wave of electron gas: coherent oscillation of an electron gas density in a bulk metal. There, electric field $\boldsymbol{E}$ in the plasmon is parallel to the wave vector $\boldsymbol{k}$, namely, a longitudinal wave like a sound wave, as shown in Fig. 3.1b, while electric field in the EM wave

K. Asakawa et al., *Nano-photonics for Advanced Networks*,
https://doi.org/10.1007/978-3-031-80637-7_3

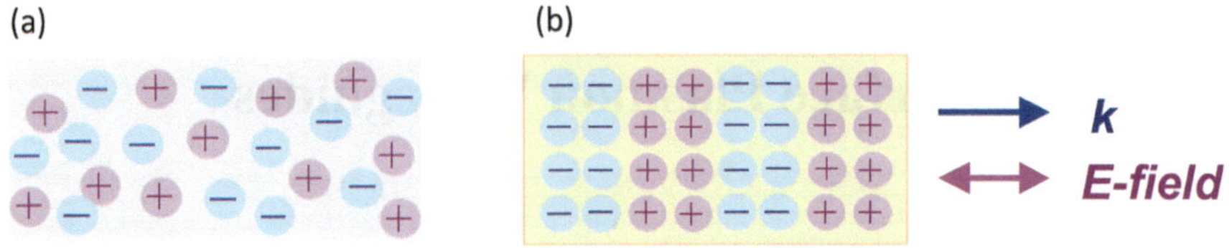

Fig. 3.1 Schematics of electron (negative charge) and ion (positive charge) distributions in **a** plasma and **b** bulk plasmon

is a transverse wave (vertical to the wave vector). This means that the plasmon is unable to couple to the EM wave in general. The plasmon exists mainly in metals, where electrons are weakly bound and movable randomly. The plasmon is also called a bulk plasmon, distinguishing from a PP, as shown next.

The state where the plasmon is coupled to the photon is called the PP. In general, a polariton is defined as a coupled state between a photon and elementary excitations such as plasmon, exciton, phonon. As a result, they are called a plasmon-polariton, an exciton–polariton, a phonon-polariton, etc. As shown later in Fig. 3.3b, the plasmon is presented as a particular state of the PP: a bulk (or volume) plasmon with a plasma frequency (ω_p). At ω_p, light cannot couple to the plasmon due to the polarization mismatch, as mentioned above. However, above the ω_p, (modified) light propagates through the metal as a state of the PP under the specific dispersion relation originating from the dielectric behavior of the metal. The SP is an aspect of the PP restricted to the metal surface only. Consequently, for a preliminary study of the SP, we investigate the physical properties of the PP using the Drude model-based metal optics, as follows.

3.2 Metal Optics of Plasmon-Polariton

3.2.1 Drude Model

A dielectric function of the PP presents optical properties of the metal. So, we introduce classical Maxwell's equations in a free electron gas theory based on the Drude model. In the early 1900s, Drude established a theory on electrical as well as thermal phenomena, connecting a kinetic theory on gas to the metal as electron gas. The Drude theory is based on the following: In the metal brought together by atoms of metallic elements, valence electrons separate from the metallic element and move in the metal, while metallic ions stay motionless as positive particles. Single isolated atom forms the following three species [1]:

- Nucleus of charge *eZa*: where *Za* is an atomic number, *e* a magnitude of an electronic charge, thus -*eZa* a total charge surrounding the nucleus.
- Weakly bound valence electrons with –*eZ*.
- Tightly bound electrons with remaining –*e* (*Za* –*Z*) (core electrons).

Meanwhile, metal is formed resulting from concentration of isolated atoms mentioned above: Core electrons are bound to a nucleus, while valence electrons separate from parent atoms and form conduction electrons. The Drude model uses a kinetic theory on a neutral rare gas to express the dense metallic electrons under the following conditions:

- Collisions of a given electron with other electrons and ions are neglected.
- Based on a kinetic theory, the electron instantly changes the velocity in collision.
- The electron behaves at a collision rate $1/\tau$, where τ is a relaxation time.
- Electrons collide with surroundings to reach thermal equilibrium.

3.2.2 Dielectric Function of Metal

Next, a dielectric function $\varepsilon_r(\omega)$ for presenting optical properties of the metal is derived. For an equation of motion on a free electron in an electric field, $\boldsymbol{E}$, the Lorentz model with no spring factor (Drude model) is considered as follows:

$$m\frac{d^2\boldsymbol{r}}{dt^2} + m\gamma\frac{d\boldsymbol{r}}{dt} = -e\boldsymbol{E}, \tag{3.1}$$

where m shows an effective electron mass, while γ ($= 1/\tau$) shows a collision rate as described above. For a driving field based on a harmonic time-dependence $\boldsymbol{E}(t) = \boldsymbol{E}_0 e^{-i\omega t}$, an oscillation of the electron is given by $\boldsymbol{r}(t) = \boldsymbol{r}_0 e^{-i\omega t}$. Then, from Eq. (3.1), $\boldsymbol{r}(t)$ is given from $\boldsymbol{E}(t)$ using a phase shift relating to the driving field and response as:

$$\boldsymbol{r}(t) = \frac{e}{m(\omega^2 + i\gamma\omega)}\boldsymbol{E}(t). \tag{3.2}$$

A dipole moment per unit volume showing polarization is:

$$\boldsymbol{P}(t) = -ne\boldsymbol{r}(t) = -\frac{ne^2}{m(\omega^2 + i\gamma\omega)}\boldsymbol{E}(t), \tag{3.3}$$

where n is an electron density. Since an electron flux density $\boldsymbol{D}$ is given by $\boldsymbol{D} = \varepsilon_r \varepsilon_0 \boldsymbol{E} = \varepsilon_0 \boldsymbol{E} + \boldsymbol{P}$, a dielectric function $\varepsilon_r(\omega)$ of free electron gas with a frequency ω is:

$$\varepsilon_{\mathrm{r}}(\omega) = 1 - \frac{\omega_{\mathrm{p}}^2}{\omega^2 + i\gamma\omega}, \ \omega_{\mathrm{p}} = \sqrt{\frac{ne^2}{\varepsilon_0 m}}, \tag{3.4}$$

where ω_{p} is a plasma frequency. In a high-frequency region: $\omega\tau \gg 1$, $\gamma \to 0$, so that

$$\varepsilon_{\mathrm{r}}(\omega) = 1 - \frac{\omega_{\mathrm{p}}^2}{\omega^2}. \tag{3.5}$$

Figure 3.2 shows a schematic of the dielectric function vs frequency. When $\varepsilon_r(\omega) > 0$, incident light propagates through the metal without damping, while light is totally reflected when $\varepsilon_r(\omega) < 0$. The reflected light accompanies an evanescent wave with a penetration depth δ (called a skin depth) in the metal.

We now consider the $\varepsilon_r(\omega)$ in both low- and high-frequency regions. A complex dielectric function is expressed as $\varepsilon_r = \varepsilon_1 + i\varepsilon_2$, where ε_1 and ε_2 are real and imaginary parts, respectively. A complex refractive index is also expressed as $n_r = n_1 + in_2$, where n_1 and n_2 are real and imaginary parts, respectively. n_r is also defined as $n_r = \sqrt{\varepsilon_r}$, from which $\varepsilon_1 = n_1{}^2 - n_2{}^2$, $\varepsilon_2 = 2\, n_1\, n_2$. Here, $n_2 = \varepsilon_2/2n_1$ is given as an extinction coefficient, whereby an optical absorption of light α is determined as $\alpha(\omega) = 2\, n_2\, \omega/c$.

In a low-frequency range, $\omega\tau \ll 1$, hence $\varepsilon_2 \gg \varepsilon_1$ and $n_1 \approx n_2$. Then, from Eq. (3.4), $\varepsilon_2(\omega) = \omega_p{}^2\tau/\omega$. Since metals are absorbing in this range, the above α is expressed as:

$$\alpha(\omega) = \left(\frac{2\omega_p^2\omega\tau}{c^2}\right)^{1/2}. \tag{3.6}$$

Furthermore, the field decays in the metal as $e^{-z/\delta}$, where δ is a skin depth: $\delta = 2/\alpha = c/n_2\omega$. In the visible frequency range, a free electron model mentioned above does not fit on account of inter-band transitions [1]. Finally, the skin depth δ comes close to zero when the conductivity increases infinitely, while δ nearly equals tens of nm when λ is in a visible range. Refer to [1] for detail.

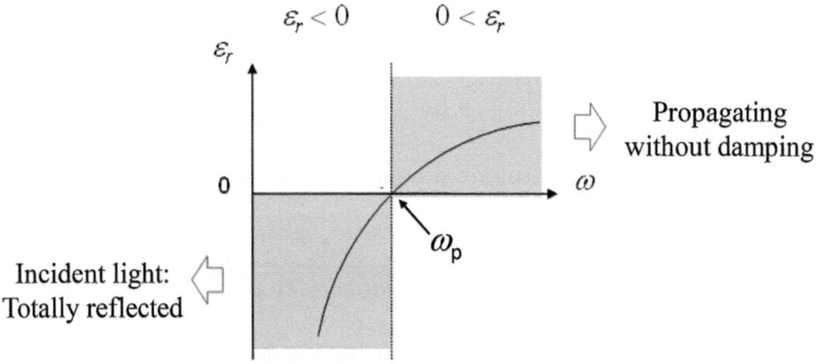

Fig. 3.2 Dielectric function of a free electron gas and behavior of $\varepsilon_r(\omega)$ in air and in the metal (ε_r: as real number here). When $\varepsilon_r < 0$, incident light is totally reflected, while incident light propagates through the metal without damping when $\varepsilon_r > 0$

3.2.3 Dispersion Relation of Plasmon

Here, we derive a dispersion relation of the PP as a result of the coupling between plasmons and photons. Now we consider a light wave equation as:

$$\frac{\varepsilon_{\mathrm{r}}}{c^2}\frac{d^2\boldsymbol{E}(r,t)}{dt^2} = \nabla^2\boldsymbol{E}(r,t). \tag{3.7}$$

A harmonic time-dependent electric field of light is also considered as:

$$\boldsymbol{E}(r,t) = \boldsymbol{E}_0(r,t)e^{i(kr-\omega t)}. \tag{3.8}$$

From Eqs. (3.5), (3.7), and (3.8), the dispersion relation on the PP is rewritten explicitly as:

$$\omega^2 = \omega_{\mathrm{p}}^2 + c^2k^2, \, or\, \omega = \sqrt{\omega_{\mathrm{p}}^2 + c^2k^2}. \tag{3.9}$$

Figure 3.3a shows a dielectric function of the electron gas, while Fig. 3.3b shows a dispersion relation of the PP. The dispersion relation implies that a PP frequency is always higher than the plasmon frequency ω_{p}. This means that no transverse EM wave with frequencies lower than the plasmon frequency can travel through the medium. Indeed, at the plasmon frequency ω_{p}, the wave vector as a transverse EM wave is zero and the PP becomes a bulk plasmon, i.e., a longitudinal wave, as schematically shown in Fig. 3.1b. Here $\hbar\omega_{\mathrm{p}} \approx 10$ eV in most metals, while $\hbar\omega_{\mathrm{p}} <$ 0.5 eV in the semiconductors depending on the dopant concentration.

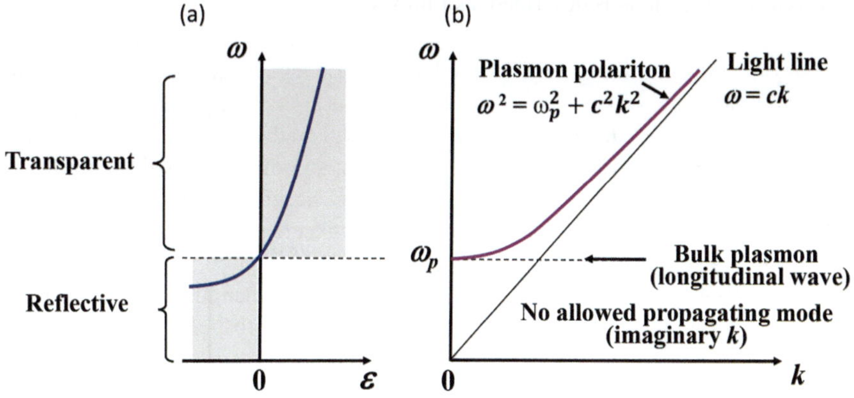

Fig. 3.3 **a** Dielectric function in the metal. **b** Dispersion relation of the PP

3.3 Surface Plasmon

3.3.1 Dispersion Relation of Surface Plasmon

A surface plasmon (SP) is a sort of electro-magnetically excited state, confined at a dielectric-metal interface. The SP field propagates along the interface while penetrating into both metal and dielectric media exponentially from the interface. Since it is coupled to the light wave on the metal surface, it can be called a surface plasmon-polariton (SPP) by a definition of the polariton mentioned in Introduction (3.1). In this book, however, we call it simply a surface plasmon, following many other examples. Figure 3.4a shows a schematic cross-sectional picture of the SP fields and coherent oscillatory distribution of electrons and positive ions below the metal–dielectric interface.

Maxwell's Equations and Boundary Condition

Electromagnetic (EM) characteristics of the SP in the vicinity of the dielectric-metal interface are derived from Maxwell's equations. Figure 3.4a shows a schematic of electric (***E***)-field and coherent electron gas charge distributions in dielectric and metal regions, respectively. An upper region ($z > 0$) is comprised by a dielectric material with a dielectric constant of ε_d, while a lower half-space ($z < 0$) is comprised by a metal with a complex dielectric function of ε_m. Now a transverse magnetic (TM) wave propagating in the x-direction is considered. Figure 3.4a shows that an ***E***-field in the dielectric medium involves components both normal and parallel to the surface, E_z and E_x, while an ***E***-field in the metal involves a parallel component E_x only. Detailed derivation of the SP dispersion relation from Maxwell's equations and related boundary conditions is described in Sect. 3.7 (Appendix). An essential part in Maxwell's equations is described, as follows:

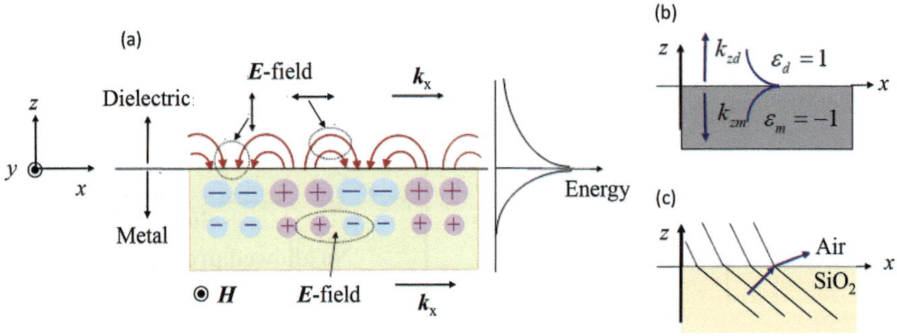

Fig. 3.4 **a** Schematic cross-sectional views of ***E***-field and coherent electron gas charge distributions of the SP at the metal–dielectric interface. **b** Decaying k_{zi} profiles in the metal and dielectric media. **c** Wave front of the refractive light across the dielectric–dielectric interface for continuity of E_x and H_y

- Maxwell's equation (Ampere's circuital law):

$$\nabla \times H_i = \varepsilon_i \frac{\partial E_i}{\partial t} (i = \mathrm{m, d}). \tag{3.10}$$

- ***E***-field wave equation:

$$\frac{\varepsilon_\mathrm{r}}{c^2} \frac{d^2 \boldsymbol{E}(r, t)}{dt^2} = \nabla^2 \boldsymbol{E}(r, t). \tag{3.11}$$

- Solution for the electric field:

$$\boldsymbol{E}(r, t) = \boldsymbol{E}_0 \mathrm{e}^{i(k-r-\omega t)}. \tag{3.12}$$

In addition to these equations, boundary conditions regarding continuities on tangential E_x and H_y, normal D_z and tangential k_x should be reserved at the metal–dielectric boundary. Figure 3.4b shows decaying profiles of k_z components at both sides of the metal–dielectric interface, while Fig. 3.4c shows a wave front of the refractive light across the dielectric–dielectric interface in general, which verifies the continuity of the tangential k_xs at both dielectric media.

Dispersion Relation

From these equations and boundary conditions, k_x and k_zi are calculated as:

$$k_x = \frac{\omega}{c} \sqrt{\frac{\varepsilon_\mathrm{m} \varepsilon_\mathrm{d}}{\varepsilon_\mathrm{m} + \varepsilon_\mathrm{d}}} = \frac{\omega}{c} \sqrt{\frac{\left(\omega^2 - \omega_\mathrm{p}^2\right) \varepsilon_\mathrm{d}}{(1 + \varepsilon_\mathrm{d})\omega^2 - \omega_\mathrm{p}^2}}, \tag{3.13}$$

$$k_\mathrm{zi} = \frac{\omega}{c} \sqrt{\frac{\varepsilon_i^2}{\varepsilon_\mathrm{m} + \varepsilon_\mathrm{d}}}, \tag{3.14}$$

where $i = m$ (metal) or d (dielectric). See Sect. 3.7 (Appendix) for their derivations. Equation (3.13) represents a dispersion relation ($\omega - \boldsymbol{k}$) of the SP, while Eq. (3.14) gives decaying k_{zi} profiles in the metal ($i = m$) and dielectric ($i = d$) media, as shown in Fig. 3.4b. In Eq. (3.13), as $k_\mathrm{x} \to \infty$, the frequency ω gets close to $\omega_\mathrm{sp} = \omega_\mathrm{p} / \sqrt{1 + \varepsilon_\mathrm{d}}$. In the air ($\varepsilon_\mathrm{d} = 1$), the frequency ω_sp equals $\omega_\mathrm{p} / \sqrt{2}$, which is called the SP frequency.

Based on these results, the dispersion relation ($\omega - \boldsymbol{k}_\mathrm{x}$) of the SP is depicted, as shown in Fig. 3.5, where the dispersion curve of the PP is also plotted for comparison. As a wave vector $\boldsymbol{k}$ increases from zero, a frequency ω of the PP varies from the plasma frequency ω_p, while the frequency ω of the SP approaches from zero to ω_sp asymptotically. As indicated above, note that the PP is a radiative mode (both k_x and k_z are real), while the SP is a bound mode (k_x is real but k_z is imaginary) confined

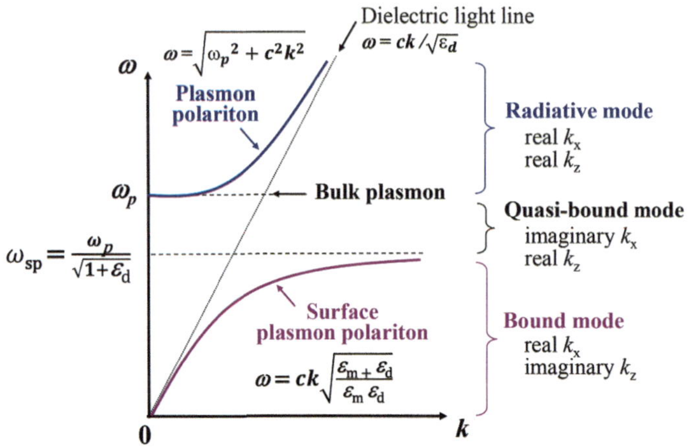

Fig. 3.5 Dispersion relation of the SP (in pink) as compared with the PP (in blue). Although both are categorized into the polariton, the former is a bound mode where light is strongly confined to the metal–dielectric interface, while the latter is a radiative mode where light propagates through the metal

to the metal–dielectric interface. Between ω_p and ω_{sp}, a plasmon is a quasi-bound mode with imaginary k_x and real k_z, and no propagating plasmon exists anymore.

3.3.2 Physical Properties of Surface Plasmon

The SP is a bound mode tightly confined to the metal–dielectric interface, so that it exhibits unique propagation properties and field profiles which seriously influence its applications. Consequently, physical properties of the SP are investigated in the following in more detail.

Dispersion Function Versus Dielectric Function

Equations (3.13) and (3.14) on k_x and k_{zi} can be presented in complex expressions as:

$$k_x = \sqrt{\frac{\varepsilon_m \varepsilon_d}{\varepsilon_m + \varepsilon_d}} k_0 = k_x' + ik_x'', \tag{3.15}$$

$$k_{zi} = \frac{\omega}{c}\sqrt{\frac{\varepsilon_i^2}{\varepsilon_m + \varepsilon_d}} = k_{zi}' + ik_{zi}''. \tag{3.16}$$

Since the SP is a TM wave in the dielectric region, the profile of k_{zi} decays perpendicular to the interface, so that k_{zi} must be imaginary (evanescent) and $k_x{}'$ must be real. These results also lead to relations between dielectric constants ε_m and ε_d as:

$$\varepsilon_{\mathrm{m}} + \varepsilon_{\mathrm{d}} < 0(\text{ or } \varepsilon_{\mathrm{m}} < -\varepsilon_{\mathrm{d}}) \tag{3.17}$$

$$\varepsilon_{\mathrm{m}} < 0 \tag{3.18}$$

Consider the low ω region. Since $\varepsilon_{\mathrm{m}\rightarrow} - \infty$,

$$k_{\mathrm{x}} = \frac{\omega}{c}\sqrt{\frac{\varepsilon_{\mathrm{m}}\varepsilon_{\mathrm{d}}}{\varepsilon_{\mathrm{m}} + \varepsilon_{\mathrm{d}}}} \rightarrow\approx \frac{\omega}{c}\sqrt{\varepsilon_{\mathrm{d}}}. \tag{3.19}$$

This means that a dispersion curve does not cross a dielectric light line but is tangent to it only at $\omega = 0$. It also means that the SP is not excited by light in a direct coupling but in a particular method. See Sect. 3.4 for excitation of the SP.

The dielectric function of the metal derived in Fig. 3.2 and the dispersion curve of the SP derived above in Fig. 3.6 are compared, as shown in Fig. 3.6a and b, respectively. It is worthwhile to compare them from the viewpoint of their proper features and applications, dividing them into the low, middle, and high k_{x} regions: A low k_{x} region corresponds to an infra-red frequency region where an electric field is weakly confined and most of the energy is guided in the clad medium, so that the propagation loss is comparatively low. Accordingly, clad-sensitive SP waveguide applications are promising. A middle k_{x} region corresponds to a visible-light frequency region and field enhancement is moderate. This frequency region exhibits an extraordinary transmission of light as a unique phenomenon of the SP [3]. The extraordinary transmission of light is an extremely unique phenomenon that has proposed novel plasmonic behaviors/devices on the periodic metal configuration (such as sub-wavelength holes array) and contributed to producing novel sensors and display applications [3]. Finally, a high k_{x} region corresponding to an ultra-violet (UV) region reveals strong field confinement and a very low group velocity. It also fits nano-lithography and SP-enhanced light-emitting diodes.

For SP waveguide devices applied to photonic networks as this book aims, propagation loss of the SP should be as low as possible, so that the low (to middle) k_{x} region is preferable for low propagation losses. This problem is discussed in the next topics "propagation length" and "field penetration depth".

Propagation Length

On the metal surface, the SP field decays exponentially with a propagation length L_{SP} due to ohmic losses while propagating along a smooth surface, where the L_{SP} is defined as the propagation length needed before the intensity of the SP field decreases to 1/e. It is given by the imaginary part of the SP wave vector. The complex wave vector k_{x} is rewritten from Eq. (3.15) as:

$$k_x = \sqrt{\frac{\varepsilon_{\mathrm{m}}\varepsilon_{\mathrm{d}}}{\varepsilon_{\mathrm{m}} + \varepsilon_{\mathrm{d}}}} k_0 = k_x' + ik_x''.$$

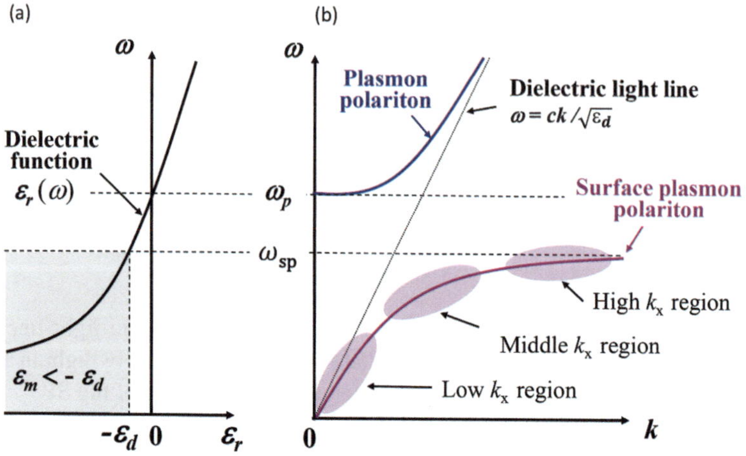

Fig. 3.6 Relationship between **a** the dielectric function of the metal and **b** the dispersion curve of the SP. The SP is considered to represent proper features and applications depending on the low, middle, and high k_x regions. See the text for detail

When an $\boldsymbol{E}$ (x) is expressed as:

$$\boldsymbol{E}(x) = \boldsymbol{E}_0 \exp(ik_x \cdot x)\cdot = \boldsymbol{E}_0 \exp\left(ik_x' \cdot x\right) \exp\left(-k_x'' \cdot x\right), \tag{3.20}$$

the propagation length L_{SP} is given by:

$$L_{sp} = \frac{1}{2k_x''} \tag{3.21}$$

For example, the L_{sp} of silver is 22 μm at λ = 514.5 nm, while that of gold is 500 μm at λ = 1060 nm. Low-loss SP waveguides involving hybrid metal–dielectric structures are discussed in Sect. 7.2.

Field Penetration Depth

As schematically shown in Fig. 3.4b, an electric field penetration depth δ_i is given by $\delta_i = 1/|k_{zi}|$, where k_{zi} is a complex wave vector ($i = m$ or d) given by Eq. (3.16) as:

$$k_{zi} = \frac{\omega}{c}\sqrt{\frac{\varepsilon_i^2}{\varepsilon_m + \varepsilon_d}} = k_{zi}' + ik_{zi}''.$$

So, δ_i s are given as:

$$\delta_d = \frac{1}{|k_{zd}|} = \frac{\lambda}{2\pi}\sqrt{\varepsilon_m + 1}, \tag{3.22}$$

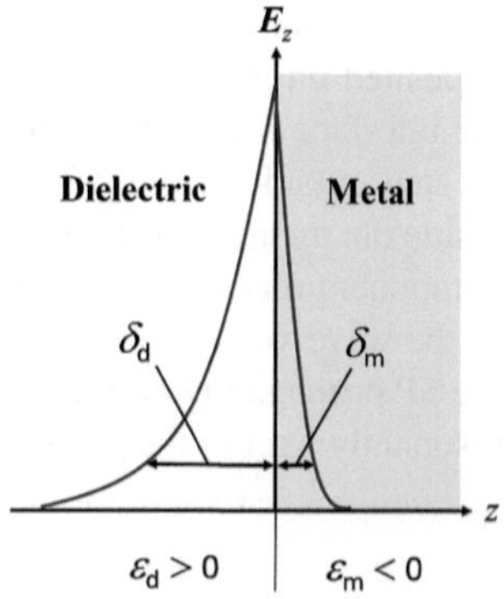

Fig. 3.7 Evanescent field distribution on the metal–dielectric interface

$$\delta_m = \frac{1}{|k_{zm}|} = \frac{\lambda}{2\pi}\frac{\sqrt{\varepsilon_m + 1}}{\varepsilon_m}, \quad (3.23)$$

where $\varepsilon_d = 1$ (in air) is assumed here. From Eq. (3.17) and Fig. 3.6a, $|\varepsilon_m| > |\varepsilon_d| \geq 1$, so that $\delta_d > \delta_m$. That is, a penetration depth of the ***E***-field in the dielectric medium is always larger than the penetration depth in the metal. For example, $\delta_d = 390$ nm and $\delta_m = 24$ nm for silver ($\lambda = 600$ nm), while $\delta_d = 280$ nm and $\delta_m = 31$ nm for gold ($\lambda = 600$ nm). In addition$_i$, at a large k_x ($\varepsilon_m \rightarrow -\varepsilon_d$), $\delta \approx 1/k_x$, that is, field concentration is strong near the surface in both media. However, at a low k_x ($|\varepsilon_m| \gg 1$), $E_z = \pm i E_x$, so that E_z relative to E_x is as:

$$E_z/E_x = -i\sqrt{|\varepsilon_m|} \text{ in air; larger } E_z \text{ component,}$$
$$E_z/E_x = i/\sqrt{|\varepsilon_m|} \text{ in metal; smaller } E_z \text{ component.}$$

Consequently, a low k_x provides a good condition for a low-loss SP waveguide. Figure 3.7 shows a schematic of evanescent field distributions on the metal–dielectric interface.

3.4 Excitation of Surface Plasmon

3.4.1 Configurations of SP Excitation

As seen from the SP dispersion relations, the SP wave vector is larger than the light wave vector in free space for a given light wavelength. Thus, light in free space cannot excite the SP directly. Instead, light and SP wave vectors can be matched using particular geometries represented by Kretschmann and Otto configurations [4], as shown in Fig. 3.8. Figure 3.8a shows a Kretschmann-type SP excitation configuration: Incident light impinges on a dielectric prism at an incidence angle θ_{sp}, larger than the total internal reflection (TIR) in the dielectrics [5]. Here, a thin metal film is deposited on the prism. Then, the light incident on the prism is totally reflected at the interface between the metal and prism. Reflected light is detected at a detector located outside of the prism. We consider here the penetrated component of light inside the metal. When the thickness of the metal film is thin enough, the penetrated component reaches the metal–air interface. At a certain incidence angle θ, matching of the wave vectors occurs between the light propagating in-plane in the prism and the SP propagating on the air–metal surface. In this way, light incident on the prism resonantly excites the SP with the wave vector of k_{sp} through the metal film, as:

$$k_{sp} = \frac{\omega}{c}\sqrt{\varepsilon_{prism}}\sin\theta. \tag{3.24}$$

Under these resonant conditions, θ -dependent reflectivity of the incident light shows a sharp dip. The dip bottom becomes minimum as light is coupled to SPs with near 100% efficiency. A detailed principle is explained in Fig. 3.9.

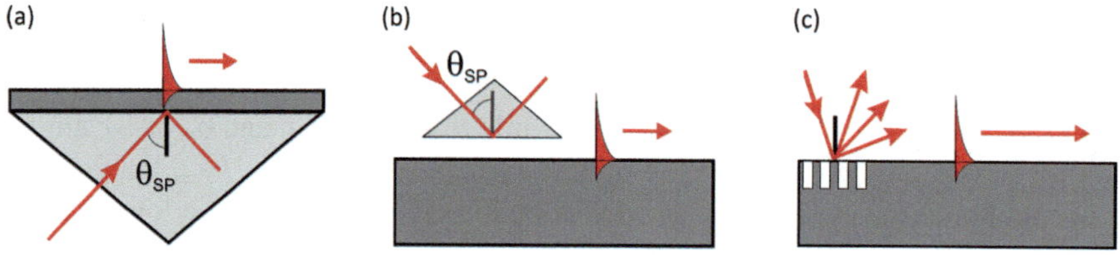

Fig. 3.8 SP excitation methods. **a** Kretschmann configuration. **b** Otto configuration. **c** Configuration using diffractive grating. Reprinted with permission from Ref. [4]. © 2003 IOP Publishing Ltd

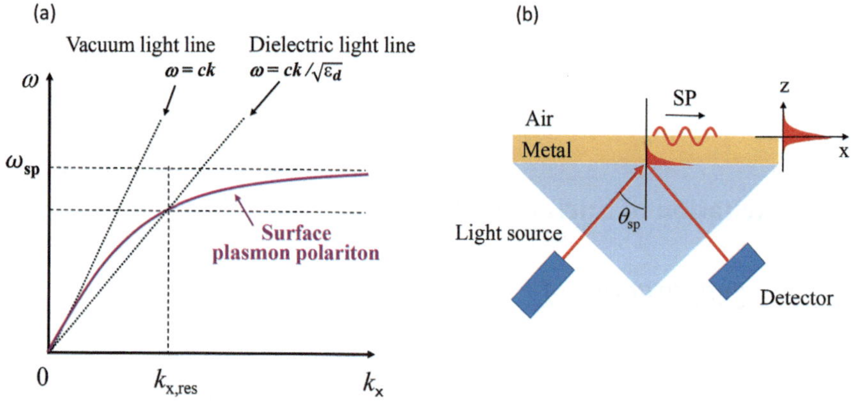

Fig. 3.9 Principle of SP excitation by Kretschmann configuration. **a** Dispersion relation with the vacuum light line and the dielectric light line. $k_{x,res}$ is a wave vector at which external light can excite the resonance with SP. **b** Experimental arrangement in Kretschmann configuration to realize the condition sketched in (**a**)

When the thickness of the metal film is larger than the penetration depth of the incident light, the SP excitation is weakened and the Kretschmann configuration becomes useless. Instead, the Otto configuration, as shown in Fig. 3.8b, is available for it. Here, the prism is located near the metal surface (air-gap of sub-micron). Incident light on the prism, reflected by TIR at the inner bottom surface of the prism, induces evanescent wave in the air-gap and excites the SP on the metal surface by tunneling of the incident light through the air-gap [6]. A resonant principle is similar to the case in the Kretschmann configuration, Eq. (3.24).

Another method for excitation of SP is to use a diffraction grating. As shown in Fig. 3.8c, a diffraction grating is fabricated on the metal film. Incident light causes several diffraction components, as shown by several arrows. One of the components whose wave vector equals the SP wave vector couples to the SP based on the wave vector conservation [7]. See [4] for detailed coupling conditions.

3.4.2 *Kretschmann Configuration*

The Kretschmann configuration is experimentally convenient for exciting the SP from free-space light [8]. The principle of the Kretschmann configuration is explained in more detail by using a dispersion relation with the vacuum light line and the dielectric light line, as shown in Fig. 3.9a. For a given energy $\hbar\omega$, the wave vector k_x of the SP is always larger than the wave vector in vacuum light: There is no cross-point between the vacuum light line and SP dispersion curve. Consequently, the SP on the metal surface cannot be excited by vacuum light of any frequency. However, the SP can be excited by light if the wave vector of the exciting light is increased over its free-space value. As illustrated in Fig. 3.9b, this condition is given by the configuration with the metal film deposited on the dielectric prism. In this way, the SP on the metal surface (air-metal interface) and light incident into the prism are resonantly coupled through the evanescent component accompanied by the TIR. Here, the refractive index of the prism is $n = \sqrt{\varepsilon_d} > 1$. Then, the slope of the light line is decreased by a factor of n since $\omega = ck/n$, as shown by the dotted dielectric light line in Fig. 3.9a: The light line and SP dispersion curve come to cross each other, meaning the resonant coupling between the SP and light, namely successful excitation of the SP by free-space light. Figure 3.9b shows a schematic of an experimental arrangement in the Kretschmann configuration. Light incident into the prism is reflected at the metal–prism interface by the TIR, as mentioned above, and reflected light intensity is detected as a function of incident angle θ_{sp}. By the resonant coupling at particular θ_{sp}, input light energy is transferred to the excited SP energy, exhibiting the dip in the measured reflectivity curve as a result of the resonant excitation of the SP.

3.5 Localized Surface Plasmon

The surface plasmon (SP) involves two modes depending on the dynamic behavior of the SP on the metal–dielectric interface: One is a propagating SP (PSP) that propagates along a flat metal–dielectric interface, as discussed above. Another is a localized SP (LSP) that is concentrated in a metallic nano-particle surface. This section discusses on the electromagnetic behaviors of the LSP based on the resonance phenomenon and theoretical optics supporting such a resonance in the metal nano-particle.

3.5.1 *Localized Surface Plasmon Resonance*

A SP resonance is caused by an interaction between free electrons in a tiny metal particle and light incident into the metal particle. Electrons localized in a metallic nano-particle are capable of in-phase oscillation with the incident light, namely a localized SP resonance (LSPR). These resonances are often associated with noble metal nanostructures and cause strong near-field EM enhancements. They also induce sharp spectral absorption and scattering peaks, thus resulting in optical extinction. These images are schematically illustrated, for instance, in Ref. [9, 29]. Here, a simplified schematic image of the LSPR is illustrated in Fig. 3.10. An electric field in LSPR is strongly enhanced on the surface of the metal nano-particle and rapidly attenuated with departure from the surface. Maximum optical extinction in the LSPR occurs at the LSPR frequency, often at a visible wavelength for a noble metal nano-particle. An optical behavior of the LSPR based on the scattering in the metal nano-particle is discussed later in more detail.

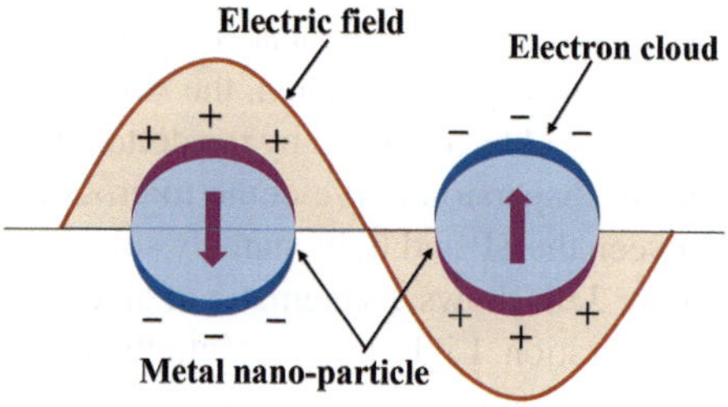

Fig. 3.10 Schematic of LSP in a metal nano-particle with electron clouds excited by incident light

3.5.2 *Metal Nano-particle Optics*

An optical behavior of the LSPR primary depends on the size, shape, and composition of the particle together with its surrounding situation. Here, we discuss optics on the spectral property of a nano-scale spherical noble metal particle (typically gold). When the metal nano-particle is illuminated by light, electrons on the particle surface

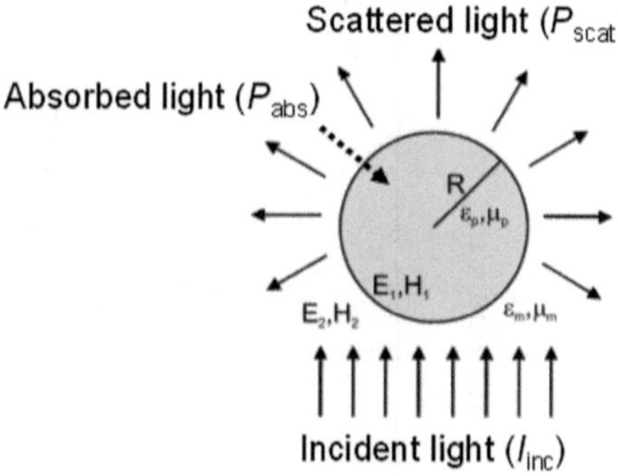

Fig. 3.11 Schematic of a metal particle illuminated by incident light, which generates electric and magnetic fields inside the particle and induces absorption and scattering of light inside and outside the particle, respectively

oscillate and generate radiation: scattering. When the particle transfers part of energy in incident light to another energy (heat, for instance), light is absorbed. Sum of the scattering and absorption equals extinction of light, i.e., extinguished energy of light.

Mie Scattering

Absorption and scattering of light by a small particle are solved by Maxwell's equations. They are derived according to the schematic picture in Fig. 3.11. An optical behavior of spherical particles was exactly solved first by G. Mie in the early 1900s [10]. An important concept that characterizes the Mie theory is a "cross-section": a geometrical parameter that indicates the scattered, absorbed, or extinguished powers (P_{sca}, P_{abs}, P_{ext}, respectively) normalized by the incident light power (I_{inc}) as:

$$\sigma_{sca} = \frac{P_{sca}}{I_{inc}}, \quad \sigma_{abs} = \frac{P_{abs}}{I_{inc}}, \quad \sigma_{ext} = \frac{P_{ext}}{I_{inc}}. \tag{3.25}$$

Equation (3.25) gives cross-sections against absorption, scattering, and extinction: σ_{sca}, σ_{abs}, and σ_{ext}, respectively, for the spherical particle with a dielectric constant ε_p. Since an extinguished power is sum of scattered and absorbed powers, the absorption cross-section is expressed as:

$$\sigma_{abs} = \sigma_{ext} - \sigma_{sca}. \tag{3.26}$$

See [10] for the mathematical derivation of these parameters.

LSPR-Based Absorption Spectra

According to the Mie theory, electronic and optical properties of metal nano-particles as well as absorption, scattering, and extinction cross-sections are derived by their sizes and shapes [11, 12]. The principle of the theory is based on the dynamic behavior of the SP: the collective motion of free electrons caused by incident light [13, 14]. In the spherical particle, the total extinction cross-section is brought by summarizing oscillation of all electric and magnetic multipoles. The concept of the Mie theory is simple and involves wide applicability for explaining experimental results. Figure 3.12 shows an example of absorption spectra for gold nano-particles with four

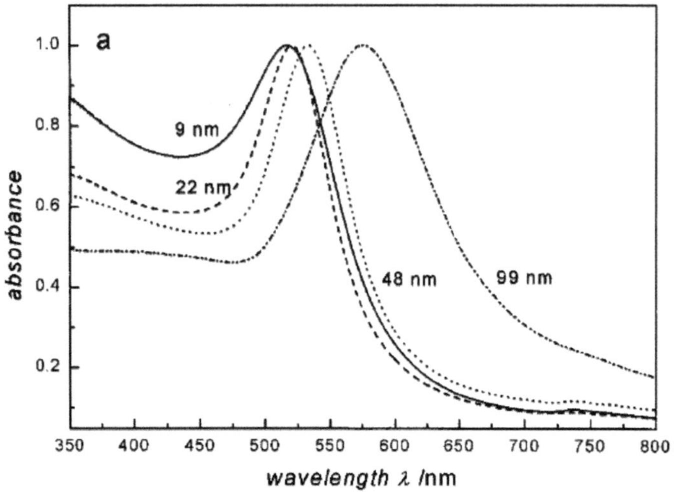

Fig. 3.12 Absorption spectra for gold nano-particles with four different sizes. Reprinted with permission from Ref. [15]. © 1999 American Chemical Society

different sizes [15]. Absorption peaks are present in the visible wavelength range and indicate red shifts with increasing particle diameters. In the figure, wavelength peaks of $\lambda_{\mathrm{max}} = 517$, 521, 533, and 575 nm correspond to 9, 22, 48, and 99 nm particles, respectively. In addition, a bandwidth at a minimum particle size of 9 nm decreases to the minimum first at 22 nm size and then increases again toward the 99 nm size.

Group of Proper Plasmon Frequencies

According to the Mie theory described above, absorption, scattering, and extinction cross-sections, σ_{abs}, σ_{sca}, and σ_{ext} defined in Eqs. (3.25) and (3.26) can be derived as:

$$\sigma_{\mathrm{sca}} = 8\pi k^4 R^6 \left| \frac{\varepsilon_{\mathrm{m}} - \varepsilon_{\mathrm{d}}}{\varepsilon_{\mathrm{m}} + 2\varepsilon_{\mathrm{d}}} \right|, \tag{3.27}$$

$$\sigma_{\mathrm{abs,}} \approx \sigma_{\mathrm{ext}} = 4\pi k R^3 \mathrm{Im}\left(\frac{\varepsilon_{\mathrm{m}} - \varepsilon_{\mathrm{d}}}{\varepsilon_{\mathrm{m}} + 2\varepsilon_{\mathrm{d}}} \right), \tag{3.28}$$

where R is a radius of the sphere nano-particle. See [10, 16] for mathematical derivation of these equations. Let us consider the particular case. These cross-sections are maximized when the denominators in Eqs. (3.27) and (3.28) are minimized, when $\varepsilon_{\mathrm{m}} = -2\varepsilon_{\mathrm{d}}$. Now, consider the concept of the LSPR on the metal sphere in the air where $\varepsilon_{\mathrm{d}} = 1$ and $\varepsilon_{\mathrm{m}} = -2$. Under these conditions, the frequency ω_{LSPR} in a dielectric function $\varepsilon_r(\omega)$ of the metal in Eq. (3.5) ($\varepsilon_r(\omega) = \varepsilon_{\mathrm{m}}(\omega)$) can be derived as:

$$\omega_{\mathrm{LSPR}} = \frac{\omega_{\mathrm{p}}}{\sqrt{1 + 2\varepsilon_{\mathrm{d}}}} = \frac{\omega_{\mathrm{p}}}{\sqrt{3}}. \tag{3.29}$$

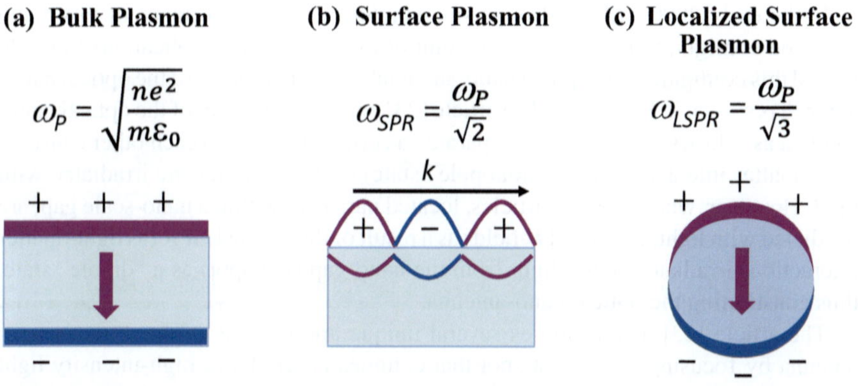

Fig. 3.13 Group of plasmonic frequencies. **a** Bulk plasmon. **b** Surface plasmon. **c** Localized surface plasmon

This frequency is called a localized SP resonance (LSPR) frequency specific to the metal nano-sphere. Group of plasmonic frequencies ever surveyed including the present ω_{LSPR} is summarized in Fig. 3.13, where figures (a), (b) and (c) correspond to bulk plasmon frequency ω_{p}, SP resonance (SPR) frequency ω_{SPR}, and localized SP resonance (LSPR) frequency ω_{LSPR}, respectively [16].

Extinction and absorption spectra of nano-particles were experimentally investigated using several-shape samples having sharp features fabricated by chemical syntheses and lithography methods [16, 17]. For instance, a sample with a sharp tip was found to present a red shift in LSPR, the result of which means increase of the refractive-index sensitivity and suggests that a sharp tip is preferable for molecular detection by means of highly enhanced $\boldsymbol{E}$-field on the sharp tip. For more detailed experiments, refer to articles [18–22], for samples with silver nano-cubes, gold nano-stars/-rods/-bipyramids, etc.

3.5.3 *Metal Nano-antenna Optics*

Optical Nano-antenna

Thus far, principles and physical properties of the LSP and LSPR were explained as phenomena specific to the metal nano-particle, as shown in Fig. 3.10. On its surface, time-varying electric fields induced by light waves cause a collective oscillation specific to the SP; in particular, the LSPR is caused at the LSPR frequency ω_{LSPR} determined by the radius of a sphere nano-particle (monopole).

When double nano-particles (dipole) come close to each other, the gap region at the dipole (called *hot spot*) greatly enhances the electric field due to electron clouds confined at the edge of the particle and positive ions at the other edge.

In the early 2000s, M. L. Brongersma emphasized the importance of this metal nano-rod configuration from the viewpoint of many potential applications [23]. He named this configuration "optical nano-antenna". A typical shape of the optical nano-antenna is shown in Fig. 3.14a. The article [23] explains an origin of the optical nano-antenna as follows: Two particles separated a certain distance to each other cause no light–matter interaction (as a "monopole" state) even when they are irradiated with light. However, when double particles, located narrowly within a nano-scale gap, are irradiated with light, enhanced ***E***-fields as a result of the strong LSPR (as light–matter interaction) irradiate intense light from the nano-gap (hot spot) as a "dipole" state, thus constituting the optical nano-antenna.

The article [23] also proposes several unique applications of the optical nano-antenna by focusing on the hot spot that confines or irradiates high-intensity light as a result of LSPR mentioned above, as follows: The hot spot buried with semi-conductors realizes low-noise photodetectors [24]. The hot spot buried with other

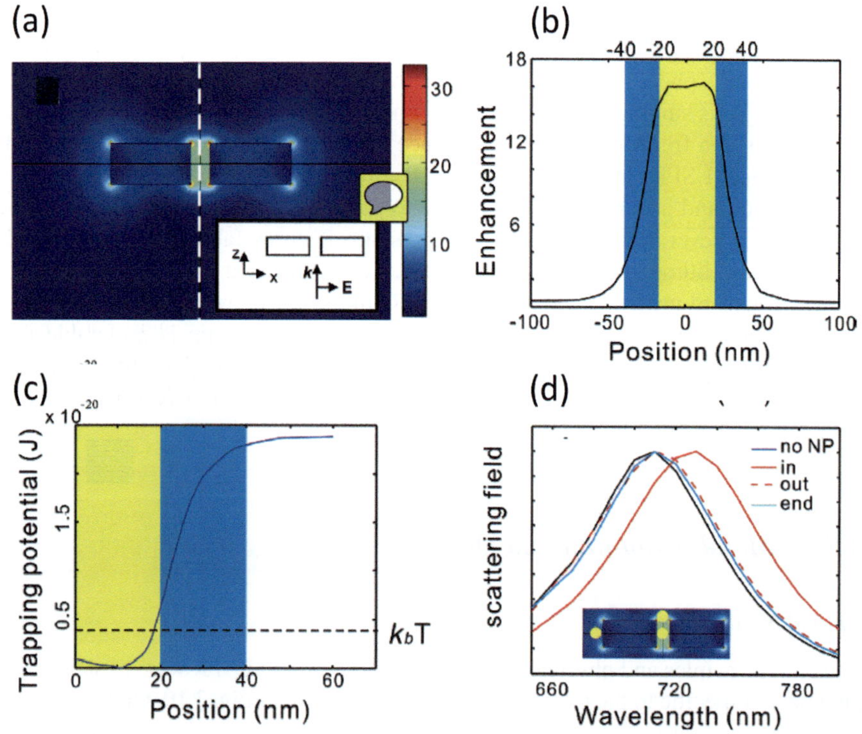

Fig. 3.14 **a** Calculation of ***E***-field intensity distributed in the vicinity of an antenna with an 80 nm arm length and 25 nm gap. The inset shows a shape of the antenna. **b** Distribution of the enhanced ***E***-field vs position along the dashed line in (**a**). **c** Distribution of trapping potential vs position originated from the localized ***E***-field. **d** Spectra of Rayleigh scattering at different positions indicated by dots in yellow in the inset. Reprinted with permission from Ref. [28]. © 2010 American Chemical Society

metals creates barcodes serving as optical tags for various purposes [25]. The hot spot with high optical intensities is capable of nonlinear frequency converters as well as low-capacitance and high-speed optical switching devices. Other than these photonic network devices, the hot spot in the dipole antenna (optical nano-antenna) contributes to ultra-high sensitivity biosensors as miniaturized optical spectroscopy tools for single-molecule studies. These ideas will bring about many proposals in large nano-photonic application areas such as advanced photonic network devices, novel optical nonlinear signal processing configurations, miniaturized chemical/medical tools, etc. In this book, indeed, Sects. 7.3 and 7.4 overview recent various applications of optical nano-antennas to valuable high-speed and broad-band plasmonic devices such as mode converter, modulator, detector, etc. Refer to them for detail.

Family of Optical Nano-antennas

As indicated above, enhancement of the ***E***-field at the hot spot in the dipole antenna strongly depends on several parameters such as metal species, shape of dipole particles, gap width, and surrounding condition. The article [26] reviews actual several optical nano-antennas involving monopoles, two-nano-particle-based dimer nano-antennas, bowtie, and dipole antennas. Among them, they emphasize that dipole nano-antennas have features of simplicity in fabrication, fine tunability in operation, and highly confined ***E***-field enhancement in the gap. The article [27] compares monopole and dipole nano-antennas from viewpoints of confinement factors squared δ^2 as a function of the exciting radiation wavelength. As an example, the maximum confinement factor δ^2 for a dipole nano-antenna with a gap size of 10 nm has proved to be 60 times larger than that of the nano-monopole. This value becomes more noticeable with decreasing gap size. Yet, challenges of an optical nano-antenna for nano-photonic applications include controllability of direction in the radiation of the field. One of the solutions for this problem is a proposal on the analogy of Yagi-Uda radio-frequency antennas. See [26] for more detail on Yagi-Uda nano-antennas.

Field Enhancement in Nano-gap

An example of calculation on the ***E***-field distribution of an optical nano-antenna is shown in Fig. 3.14, where calculated ***E***-fields at several points in dipole nano-antenna, in particular, highly enhanced fields in a nano-gap are shown [28]. In the calculation, the antenna is irradiated with a light wave polarized parallel to the antenna. In Fig. 3.14a, ***E***-field amplitudes are distributed in the gap and around the antenna. Length, width, and gap of the antenna are 80 nm, 40 nm, and 25 nm, respectively. ***E***-field distribution along the dashed line through the gap is shown in Fig. 3.14b, where the yellow zone corresponds to the field within the width of the arm. Figure 3.14c shows a trapping potential derived from the ***E***-field distribution. A potential well indicates an abrupt wall (in blue) corresponding to the edge of the nano-gap. Figure 3.14d shows spectra of Rayleigh scattering at different positions indicated by dots in yellow in the inset. A spectrum in black corresponds to that in the antenna with no particle (NP), while spectra in blue, dotted red, and red correspond to those in the antennas at an edge, a side edge in the gap, and the center of the gap, respectively, as indicated in the inset. Note that the ***E***-field at an edge of the antenna (blue) is enhanced similarly to the field at a side edge in the gap of the antenna (dotted red). These simulated results greatly contribute to the decision of applicability in device performance in proposing a novel nano-antenna device.

In Chap. 7, several metal–insulator–metal (MIM) slot devices are overviewed as applications of novel plasmonic nano-antennas devices proposed here. Typical examples include a mode converter (Fig. 7.8), a modulator (Fig. 7.10), and a detector (Fig. 7.11). Refer to them for detail.

3.6 Summary

A SP is a composite transverse EM wave coupled to a longitudinal motion of free electron gas on the metal surface. Due to enhanced EM fields within several tens of nm at the metal/dielectric interface, the SP exhibits diffraction-limit free propagation with negligible energy loss. Consequently, it plays a contemporary favorite in nano-scale integrated waveguide devices. This chapter started with general behaviors of the PP as a preliminary step to introduction of the SP: The Drude model-based metal optics was used to derive a dispersion relation of the PP. Then, a dispersion relationship of the SP was derived from Maxwell's equations under boundary conditions at the metal/dielectric interface as the most specific origin of the SP. Other specific properties of the SP such as propagation length and field penetration depth were also derived as well as several excitation technologies of the SP. Such SP is a propagating SP (PSP), that is, the SP propagating along the waveguide applicable to integrated devices for the advanced nano-photonic network. Besides the PSP, a localized SP (LSP), that is, the SP resonating in the wavelength-scale metal nano-particle or nano-gap (nano-antenna) was also discussed from the Mie scattering-based metal nano-particle optics. Highly enhanced optical intensity of the LSP localized in the MIM slot was verified by theoretical simulation, the result of which suggests significantly attractive nano-scale SP devices such as nano-antenna, plasmonic mode converter, modulator, and detector. Applications of the PSP and LSP are overviewed in Chap. 7.

3.7 Appendix

This section derives a dispersion relation of the SP based on Maxwell's equations and some boundary conditions specific to the SP, with short-cut description for reaching the target.

The Maxwell's equations in the medium i, where i = m (metal) or d (dielectric), are given by:

$$\nabla \times \boldsymbol{H} = \frac{\partial \boldsymbol{D}}{\partial t} = \varepsilon_i \frac{\partial \boldsymbol{E}}{\partial t} (i = m, d). \tag{3.30}$$

E: Electric field vector ***H***: Magnetic field vector.
D: Electric flux density, where $\boldsymbol{D} = \varepsilon_\mathrm{i}\, \boldsymbol{E}$ (ε_i: dielectric constant):
B: Magnetic flux density, where $\boldsymbol{B} = \mu_\mathrm{i}\, \boldsymbol{H} = \boldsymbol{H}$ (assume that $\mu_\mathrm{i} = 1$).

Here, only the Ampere's circuital law is given as a representative equation. Consider that ***H*** involves H_y only while ***E*** involves E_x and E_z both in metal and dielectric regions. Then boundary conditions are as:

$$E_\mathrm{x}, H_\mathrm{y}(\text{ tangential })\text{continuity:} E_\mathrm{xm} = E_\mathrm{xd}, H_\mathrm{ym} = H_\mathrm{yd}, \tag{3.31}$$

$$D_\mathrm{z}(\text{ normal })\text{continuity : } \varepsilon_\mathrm{m} E_\mathrm{zm} = \varepsilon_\mathrm{d} E_\mathrm{zd}, \tag{3.32}$$

$$k_x(\text{ tangential })\text{continuity:} k_\mathrm{xm} = k_\mathrm{xd}. \tag{3.33}$$

Equation (3.30) is rewritten as:

$$\nabla \times H_i = \varepsilon_i \frac{\partial E_i}{\partial t} (i = m, d). \tag{3.34}$$

Assume that harmonic time-dependent ***E*** (r, t) is shown as:

$$\boldsymbol{E}(r,\ t) = \boldsymbol{E}(r,\ t)_0 e^{i(kr - \omega t)}. \tag{3.35}$$

Equation (3.34) is extended as:

$$k_\mathrm{zi} H_\mathrm{yi} = \omega \varepsilon_i E_\mathrm{xi}, \text{ i.e., } k_\mathrm{zm} H_\mathrm{ym} = \omega \varepsilon_\mathrm{m} E_\mathrm{xm} \text{ and } k_\mathrm{zd} H_\mathrm{yd} = \omega \varepsilon_d E_\mathrm{xd}. \tag{3.36}$$

Combination of Eqs. (3.31) and (3.36) is as:

$$\frac{k_{zm}}{\varepsilon_m} = \frac{k_{zd}}{\varepsilon_d}. \tag{3.37}$$

From Eq. (3.30), the wave equation is as:

$$\frac{\varepsilon_\mathrm{r}}{c^2} \frac{d^2 E(r,t)}{dt^2} = \nabla^2 E(r,t). \tag{3.38}$$

Under the current boundary conditions, combination of Eqs. (3.35), (3.37), and (3.38) is as:

$$\varepsilon_i k_0^2 = k_x^2 + k_\mathrm{zi}^2 = k^2 (k_0 = 2\pi/\lambda = \omega/c). \tag{3.39}$$

In the respective metal and dielectric, Eq. (3.39) is as:

$$\varepsilon_\mathrm{m} k_0^2 = k_x^2 + k_\mathrm{zm}^2, \tag{3.40}$$

$$\varepsilon_{\rm d}k_0^2 = k_x^2 + k_{\rm zd}^2. \tag{3.41}$$

Subtraction of Eq. (3.40) – Eq. (3.41) is as:

$$k_{\rm zm}^2 - k_{\rm zd}^2 = k_0^2(\varepsilon_{\rm m} - \varepsilon_{\rm d}). \tag{3.42}$$

Addition of Eq. (3.40) + Eq. (3.41) is as:

$$k_x^2 = \left\{k_0^2(\varepsilon_m + \varepsilon_{\rm d}) - \left(k_{\rm zm}^2 + k_{\rm zd}^2\right)\right\}/2. \tag{3.43}$$

Elimination of ${k_{\rm zd}}^2$ from Eqs. (3.37) and (3.42) is as:

$$k_{\rm zm}^2 = \frac{\varepsilon_m^2}{\varepsilon_m + \varepsilon_{\rm d}}k_0^2. \tag{3.44}$$

Elimination of ${k_{\rm zm}}^2$ from Eqs. (3.37) and (3.42) is as:

$$k_{\rm zd}^2 = \frac{\varepsilon_{\rm d}^2}{\varepsilon_{\rm m} + \varepsilon_{\rm d}}k_0^2. \tag{3.45}$$

Insertion of Eqs. (3.44) and (3.45) into (3.43) is as:

$$k_x^2 = \frac{\varepsilon_{\rm m}\varepsilon_{\rm d}}{\varepsilon_{\rm m} + \varepsilon_{\rm d}}k_0^2, \tag{3.46}$$

$$k_x = \sqrt{\frac{\varepsilon_{\rm m}\varepsilon_{\rm d}}{\varepsilon_{\rm m} + \varepsilon_{\rm d}}}k_0 = \frac{\omega}{c}\sqrt{\frac{\varepsilon_{\rm m}\varepsilon_{\rm d}}{\varepsilon_{\rm m} + \varepsilon_{\rm d}}} = \frac{\omega}{c}\sqrt{\left(\frac{\left(\omega^2 - \omega_{\rm p}^2\right)\varepsilon_{\rm d}}{(1 + \varepsilon_{\rm d})\omega^2 - \omega_{\rm p}^2}\right)}. \tag{3.47}$$

In Eq. (3.47), insertion $k_{\rm x} \to \infty$ is as:

$$\omega = \omega_{\rm sp} = \frac{\omega_{\rm p}}{\sqrt{1 + \varepsilon_{\rm d}}}. \tag{3.48}$$

In air ($\varepsilon_{\rm d} = 1$):

$$\omega_{\rm sp} = \frac{\omega_{\rm p}}{\sqrt{2}}. \tag{3.49}$$

Equation (3.47) can be expressed by complex $k_{\rm x}$ as:

$$k_x = \sqrt{\frac{\varepsilon_{\rm m}\varepsilon_{\rm d}}{\varepsilon_{\rm m} + \varepsilon_{\rm d}}}k_0 = k_x' + ik_x''. \tag{3.50}$$

Equations (3.44) and (3.45) can be expressed by complex $k_{\rm zi}$ as:

$$k_{zi} = \frac{\omega}{c}\sqrt{\frac{\varepsilon_i^2}{\varepsilon_m + \varepsilon_d}} = k'_{zi} + ik''_{zi}. \tag{3.51}$$

k_{zi} must be imaginary (evanescent):

$$\varepsilon_m + \varepsilon_d < 0 \text{ or } \varepsilon_m < -\varepsilon_d. \tag{3.52}$$

$k_x{}'$ must be real:

$$\varepsilon_m < 0. \tag{3.53}$$

Equation (3.47) shows the targeted dispersion relation of the SP, while Eqs. (3.52) and (3.53) show properties of the dielectric constant of the metal surface when the electron cloud oscillation is excited by the SP.

References

1. Margapoti E, Plasmonics: Fundamentals and Applications (Lecture notes, SS2012). Summer semester 2012 – TU München
2. Ha DT, Thuy DT,. Th. Hoa V, Van TTT, Viet NA (2017) On the theory of three types of polaritons (phonon, exciton and plasmon polariton). IOP Conf Ser: J Phys Conf Ser 865:012007, 1–8
3. Ebbesen TW, Lezec HJ, Ghaemi HF, Thio T, Wolff PA (1998) Extraordinary optical transmission through sub-wavelength hole arrays. Nature 391:667–669
4. Zayats AV, Smolyaninov II (2003) Near-field photonics: surface plasmon polaritons and localized surface plasmons. J Opt A: Pure Appl Opt 5:S1–S35
5. Kretschmann E, Raether H (1968) Radiative decay of non radiative surface plasmons excited by light. Z Naturforsch A 23:2135–2136
6. Otto A (1968) Excitation of non-radiative surface plasma waves in silver by the method of frustrated total reflection. Zeitschrift für Physik A: Hadrons Nuclei 216:398–410
7. Raether H (1988) Surface plasmons on smooth and rough surfaces and on gratings. Part of the Springer Tracts in Modern Physics book series, vol111,Springer, Berlin
8. Kretschmann E (1971) Die Bestimmung optischer Konstanten von Metallen durch Anregung von Oberflächenplasmaschwingungen. Z Phys 241:313–324
9. Bozzola A, Perotto S, De Angelis F (2017) Hybrid plasmonic–photonic whispering gallery mode resonators for sensing: a critical review. Analyst 142:883–898
10. Mie G (1908) Beitrage zur Optik truber Medien. Ann. d. Physik 25:377–445
11. Henglein A (1989) Small-particle research: physicochemical properties of extremely small colloidal metal and semiconductor particles. Chem Rev 89:1861–1873
12. Alivisatos AP (1996) Perspectives on the physical chemistry of semiconductor nanocrystals. J Phys Chem 100:13226–13239
13. Papavassiliou GC (1979) Optical properties of small inorganic and organic metal particles. Prog Solid State Chem 12:185–271
14. Bohren CF, Huffman DR (1983) Absorption and scattering of light by small particles. Wiley, New York
15. Link S, El-Sayed MA (1999) Size and temperature dependence of the plasmon absorption of colloidal gold nanoparticles. J Phys Chem B 103:4212–4217
16. Mayer KM, Hafner JH (2011) Localized surface plasmon resonance sensors. Chem Rev 111:3828–3857

17. Huang X, El-Sayed MA (2019) Gold nanoparticles: Optical properties and implementations in cancer diagnosis and photo-thermal therapy. J Adv Res 1:13–28
18. Sherry LJ, Chang S-H, Schatz GC, van Duyne RP (2005) Localized surface Plasmon resonance spectroscopy of single silver nano-cubes. Nano Lett 5:2034–2038
19. Nehl CL, Liao HW, Hafner JH (2006) Optical properties of star-shaped gold nanoparticles. Nano Lett 6:683–688
20. Liu M, Guyot-Sionnest P (2005) Mechanism of silver (I)-assisted growth of gold nanorods and bipyramids. J Phys Chem B 109:22192–22200
21. Nusz GJ, Marinakos SM, Curry AC, Dahlin A, Hook F, Wax A, Chilkoti A (2008) Label-free plasmonic detection of biomolecular binding by a single gold nanorod. Anal Chem 80:984–989
22. Hao F, Nehl CL, Hafner JH, Nordlander P (2007) Plasmon resonances of a gold nanostar. Nano Lett 7:729732
23. Brongersma ML (2008) Engineering optical nano-antennas. Nat Photonics 2:270–272
24. Yu Z, Veronis G, Fan S, Brongersma ML (2006) Design of mid-infrared photodetectors enhanced by surface plasmons on grating structures. Appl Phys Lett 89:151116
25. Nicewarner-Pena SR, Freeman RG, Reiss BD, He L, Pena DJ, Walton ID, Cromer R, Keating CD, Natan MJ (2001) Sub-micrometer metallic barcodes. Science 294:137–141
26. Krasnok AE, Maksymov IS, Denisyuk AI, Belov PA, Miroshnichenko AE, Simovski CR, Kivshar YS (2013) Optical nanoantennas. Phys Uspekhi 56:539–564
27. Biagioni P, Huang J-S, Hecht B (2012) Nano-antennas for visible and infrared radiation. Rep Prog Phys 75:024402
28. Zhang W, Huang L, Santschi C, Martin OJF (2010) Trapping and sensing 10 nm metal nanoparticles using plasmonic dipole antennas. Nano Lett 10:1006–1011
29. Juan ML, Righini MR, Quidant R (2011) Plasmon nano-optical tweezers. Nat Photonics 5:349–356

Chapter 4
Applications of Silicon Photonic Waveguides (I) Network Transceivers and Switches

Abstract Si photonics has significantly progressed recently owing to small size, low cost, and high power efficiency, all brought about by SOI CMOS process compatibility. Applications of Si photonics began in the data center and core network in the photonic network society. They are now expanding markets to new areas in sensing mobility and artificial neural network in the Internet of Things (IoT) society. This chapter begins with progress of Si photonics platform and then introduces latest applications to optical transceivers in the data centers and node switches in the core networks. Early activities of Si photonics platform are focused on heterogeneously integrated nano-lasers and hybrid nano-lasers between III-V compound and silicon materials using chip/wafer bonding and hetero-epitaxy. New topics on the sub-wavelength pitch Si nano-structures are introduced also. Topics on the optical transceiver are focused on three types; a monolithically integrated Si photonic transceiver using front-end SOI CMOS processes, a fingertip-size compact optical I/O core transceiver mounted on FPGAs, and a Si photonic transceiver for the 5G mobile network. Topics on the optical switch include a multi-cast Si photonic transponder aggregator (TPA) switch for ROADM with polarization-independent high practicality, and a multi-port (point-to-point) Si PILOSS optical switch for data-com markets. Challenges for further improvements in bandwidth, optical coupling, polarization dependence, etc., are discussed together with a possible higher (~ ns) switching speed using an EO (electro-optic) modulator. Novel applications to light detection and ranging (LiDAR) and optical neural network (ONN) are introduced in Chap. 5.

4.1 Introduction

In recent decades, explosively progressing traffic in the optical communication system requests networks with ultra-fast, broadband, low cost, and low power consumption. Historically, developments of TV broadcasting (1936), Internet (1969), and World Wide Web (WWW) (1993) have long urged the transport in the communication network to become modernized rapidly. Traffic explosions caused by these

K. Asakawa et al., *Nano-photonics for Advanced Networks*,
https://doi.org/10.1007/978-3-031-80637-7_4

demands of the times have accelerated various technological innovations in communication systems. Integrated photonic and electronic devices/circuits for high-speed data processing can provide one of the solutions and is expected urgently. Figure 4.1 shows innovative systems/technologies developed in the optical telecom network in four decades, 1980–2020 [1]. Time domain multiplexing (TDM) technologies in the 1980s–early 1990s mainly involved optical fiber amplifiers, thus reaching 10 Gbit/s in total transmission capacity. In the 1990s to 2010s, an innovative wavelength-division multiplexing (WDM) system drastically raised the total transmission capacity up to 10 Tbit/s, considerably supported by arrayed waveguide grating (AWG) and reconfigurable optical add/drop multiplexers (ROADMs) technologies, both based on silica-based planar light wave circuits (PLCs). Furthermore, in the 2010s, another attractive technology—digital-coherent technology—brought a 100-Gbit/s-based WDM system, thus leading to the era of 100 Tbit/s in total communication. Recently, NTT group, Japan is developing a future communication platform named "IOWN" (innovative optical and wireless network) based on the all-photonic network (APN) for targeting low power consumption, high quality and capacity, and less delay [2]. In the roadmap viewing Step 1 to Step 3 for development of a photonic-electronic convergence technology as a key to the APN, Si photonics chip is situated in Step 1. Application images of nano-photonic waveguide devices including Si photonics, photonic crystals, and surface plasmons to be discussed in this book are viewed in the integrated circuit chips in Step 2 and Step 3. Refer to [2] for detail.

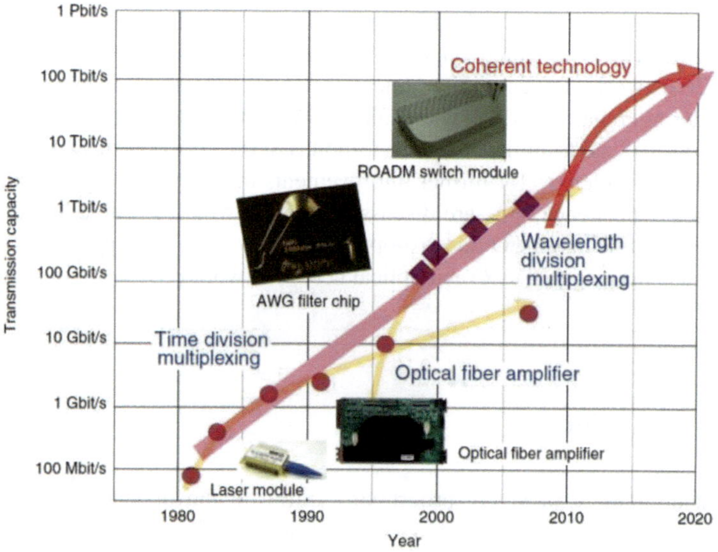

Fig. 4.1 Progresses in the optical communication network in four decades, 1980–2020. Reprinted with permission from Ref. [1], © 2011 NTT Technical Review

Meanwhile, III-V compound semiconductor (hereinafter referred to simply as III-V) lasers and photo-detectors have supported rapid growth of optical communication in Fig. 4.1. As for integration of optoelectronic semiconductor devices, III-V optoelectronic integrated circuits (OEICs) were once developed in the 1980s. However, their matured integration started with appearance of Si photonics in the 1980s–1990s. Si photonic nano-wires using SOI (silicon-on-insulator) wafer exhibit intense confinement of optical fields on account of high index contrast between Si and SiO_2. Combination of SOI-Si wafer is compatible with the mature CMOS process with low-cost mass production. It also enables scaling photonics down to nano-scales. Luxtera developed unexplored integrated photonics by CMOS-compatible fabrication technologies with a 90–130-nm node, as shown in Sect. 4.3.1. Refer to Fig. 4.11 for detail.

Now Si photonics is a nano-scale killer photonic device in advanced telecom and data-com networks. This chapter puts emphases on recent applications of Si photonic transceivers and switches for communication networks and data center markets. Rapid development of Si photonics in recent years has been overviewed by several review articles [1–6, 55].

4.2 Progress of Silicon Photonic Platform

4.2.1 CMOS-Compatible Silicon Photonic Platform

Advanced Si Photonic Nano-wire Waveguide

Si/SiO_2 waveguides on a SOI substrate form an excellent photonic integrated platform on account of a large contrast in refractive index of Si ($n = 3.45$) core and SiO_2 ($n = 1.45$) cladding regions, resulting in intense confinement of optical fields in nano-scale waveguides, typically 50 ~ 400 nm in heights and widths [4]. Around the 2000s, appearance of photonic crystals (see Chaps. 2 and 6) considerably inspired pioneering of Si photonic platform for practical applications in the photonic network. In such nano-scale waveguides, implementation of smooth nano-fabrication is essential for suppression of optical propagation loss.

In the early 2000s, smooth Si photonic single-mode waveguides with small sidewall surface roughness were realized on a SOI substrate by a standard CMOS fabrication process using a 200-mm wafer [7]. Figure 4.2a–c shows different-orientation SEM photographs of a cross-sectional strip waveguide. Widths and heights are 445 nm and 220 nm, respectively, while minimal propagation loss for the transverse electric (TE) mode is 3.6 dB/cm at 1.5 μm wavelength. Moreover, 90°-bending losses for radii of 1–2 μm are 0.086–0.013 dB. Figure 4.2d and e shows spectra of bend loss for respective TE and transverse magnetic (TM) modes. Each mode is measured with bending radii of 1, 2, and 5 μm. In the TM mode, bend loss is degraded even at 1200-nm wavelength for bending radii of 1–2 μm, while, in the TE mode, the loss spectra shows almost flat in a full-wavelength region, in particular, for radii of 2

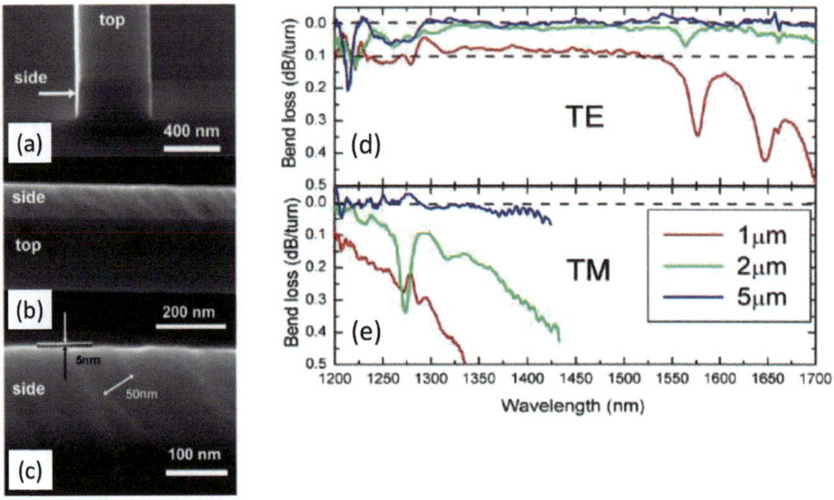

Fig. 4.2 **a–c** Cross-sectional SEM photographs of single-mode strip waveguides viewed from different angles. **d** and **e** Spectra in bend loss with different bend radii for TE and TM polarizations, respectively. Reprinted with permission from Ref. [7], © 2004 OSA

and 5 μm. The results developed here are considered to establish a solid foundation of highly integrated Si photonics based on the SOI CMOS process.

Advanced Si Photonic Fiber-to-Chip Coupler

Realization of a smooth Si photonic nano-wire waveguide using a standard SOI CMOS fabrication process, mentioned previously, was followed by fabrication of several types of passive Si photonic waveguides such as couplers and resonators [5, 7]. Above all, a fiber-to-chip light coupler is inevitable for introduction of light beam into an optical integrated circuit, yet often accompanies difficulty in a high coupling efficiency. The reason is due to a large discrepancy in mode fields and effective refractive indices at the boundary between a tens-of-μm scale optical fiber and nano-scale Si photonic waveguide. As introduced by standard couplers in Sect. 1.4.6, an inverse taper coupler for edge-coupling and a grating coupler for surface-coupling have been realized so far. Nevertheless, reduction of coupling loss is still a challenging issue. Recently, an advanced inverse taper coupler between a tapered optical fiber to a 450 nm × 250 nm Si strip waveguide was implemented using a compact cantilever coupler [8]. In the 1477 nm ~ 1580 nm wavelength range, coupling losses for TE and TM polarizations are 1 dB at most. Thanks to this type of coupler, a broadband and highly efficient fiber/chip coupling has become possible to access directly to any chip surface of photonic integrated circuits without dicing or cleaving the chip. Refer to [8] for more detail.

Sub-λ Grating Si Photonic Platform

Fiber-to-Si waveguide coupling was largely improved using another new concept; a sub-wavelength grating (SWG) with a sub-wavelength-pitch periodic waveguide indicating its peculiar anisotropic and dispersive behaviors [9]. Figure 4.3a shows a schematic of a cross-sectional SWG coupler. The SWG has a periodicity with less than a period of 1st-order Bragg diffraction, typically showing 0.9-dB coupling loss for a SWG with a length of 50 μm, 0.07 dB excess loss due to misalignment of ± 1 μm to the fiber in a transverse direction, and 0.24 dB excess loss due to angular misalignment of ± 2°. According to the effective-medium theory [10], a structure composed of inhomogeneous media in sub-wavelength scale is thought to be equivalent to a homogeneous medium. In addition, by an effective index of the waveguide is modified gradually, diffraction and reflection loss are minimized owing to the SWG effect. Figure 4.3b shows calculated field distributions of H_y and E_x components in a SWG coupler with a length of 10 μm. Note that a wave front of input light is changed rapidly to a focusing shape owing to the sub-wavelength grating action. Calculated coupling efficiencies are 65.2% at $\lambda = 1550$ nm, for example.

Another advanced SWG is a vertical fiber-to-Si waveguide light coupler on a Si layer with a thickness of 250-nm deposited on a buried oxide (BOX) layer with a

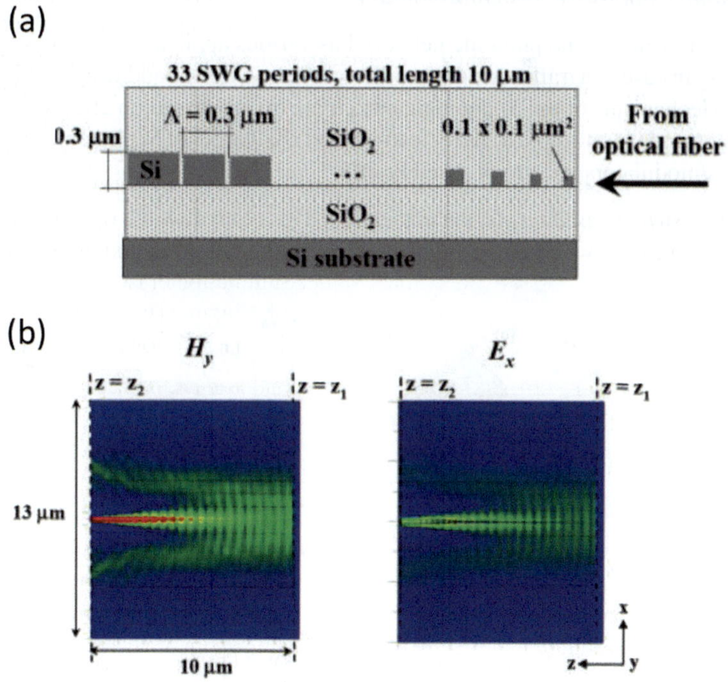

Fig. 4.3 **a** Schematic of cross-section in the SWG waveguide with a taper in the thickness direction. **b** Calculation of H_y and E_x field components in the SWG coupler in SOI platform. Reprinted with permission from Ref. [9], © 2006 OSA

thickness of 1 μm [11]. Figure 4.4a shows a schematic of the SWG. The grating is a sub-wavelength nano-structure with 2-D periodic rectangular (80 nm × 343 nm) air-hole arrays on a Si layer ($n_{si} = 3.476$). In a direction to coupled waveguides, the pitch and length of the air-hole are indicated with Λ_G and L_{sub}, while the pitch and width in the lateral direction are indicated with Λ_{sub} and W_{sub}, respectively. According to the aforementioned effective-medium theory [10], a sub-wavelength region is thought to be equivalent to a homogenous medium with effective refractive indices of n_{si} and n_{sub} for two regions without and with an air-hole like a conventional grating coupler (GC), as shown in Fig. 4.4b. The uniform material is then replaced by a sub-wavelength nano-structure with Λ_{sub}, W_{sub}, and L_{sub}, as shown in Fig. 4.4c. Values of the Λ_G and n_{sub} are sought to optimize the coupling efficiency between free-space light and guided wave in the SWG. Figure 4.4d shows calculated coupling efficiencies and back-reflection of the SWG coupler versus wavelength, involving coupling efficiencies to air (red) and to fiber (blue). For instance, when a width, length, and period of the SWG are 10 μm, 17.1 μm, and 25 μm (50% duty cycle), respectively, then a maximum coupling efficiency and 3-dB bandwidth for a TE mode are 59% at 1551.6 nm and 60 nm, respectively. Note that, in comparison with conventional gratings, the performance is comparable, while the structure is available with much more simplified fabrication processes.

Si Photonic Nano-wire Switch/Modulator

An optical switch in the photonic network has various applications such as by-pass switching in case of a traffic accident in the network as well as add/drop multiplexing nodes. Mechanisms of the Si photonic switch are mainly based on effects of plasma dispersion and thermo-optics. Here, features and challenges of Si photonic nano-wire switches/modulators are summarized briefly.

- *Optical switch based on plasma dispersion*: This type of switch operates with low power and fast switching speed of < 10 ns. Although propagation loss is added

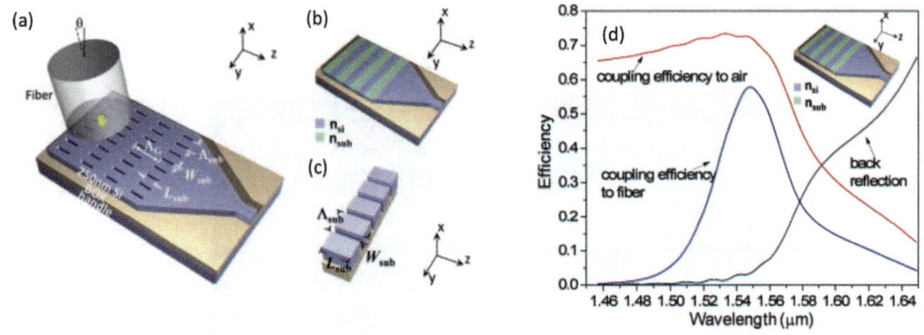

Fig. 4.4 **a** Schematic of a SWG structure. **b** Equivalent SWG like conventional GC based on the effective-medium theory. **c** Schematic of the periodic sub-wavelength rectangular air-hole array with a pitch Λ_{sub}, width W_{sub}, and length L_{sub}. **d** Calculated coupling efficiencies to air and to fiber and efficiency in back-reflection based on the equivalent homogeneous structure (inset). Reprinted from Ref. [11] with the permission of the American Institute of Physics, © 2012 AIP

due to introduced dopants, consumption power of as low as sub-mW is available with a Mach–Zehnder interferometer (MZI) configuration. Various articles on this type have been reviewed in [12, 13].

- *Optical switch based on a thermo-optic (TO) effect*: Owing to a large TO coefficient of Si at 1.55 μm wavelength (~ 1.8 × 10^{-4}/K), very compact switches based on a MZI and a micro-ring resonator (MRR) with low loss and low power have been developed, as shown schematically in Fig. 4.5a and b, respectively. A heater for inducing the TO effect in the MZI is placed on either parallel arm, while it is often placed on the whole MRR waveguide to secure a large phase shift. Compared to MRRs, MZIs provide a larger bandwidth. Nevertheless, extinction ratios (ERs) of conventional 2 × 2 MZIs are limited to 20–25 dB. This is caused by small phase errors in the two arms, thus leading to a nonzero output in an OFF state. Since the late 1990s, unique TO switches such as slow-light-based photonic crystal (PC) switch and ultra-short length (~ 5 μm) directional coupler have been demonstrated.
- *Optical modulator based on plasma dispersion*: Due to absence of a linear EO effect (electro-optic Pockels effect), a high-speed Si modulator is inherently based on other mechanisms represented by a plasma dispersion effect. It provides a fast response based on the refractive-index perturbation, resulting in the high-speed operation. Intel first demonstrated a modulator with a speed of Gbps in 2004–2005 using a MOS capacitor in a Si rib waveguide.
- *Optical modulator based on p-i-n and p-n junctions:* As shown in Fig. 4.5c and d, *p-i-n* and *p-n* junctions in reverse bias are widely used for optical modulators. For a MZI structure, typically 60 Gbps operation was demonstrated with an optimized doping condition in a phase shifter with a length of 750 μm. In Sect. 6.2.3, a

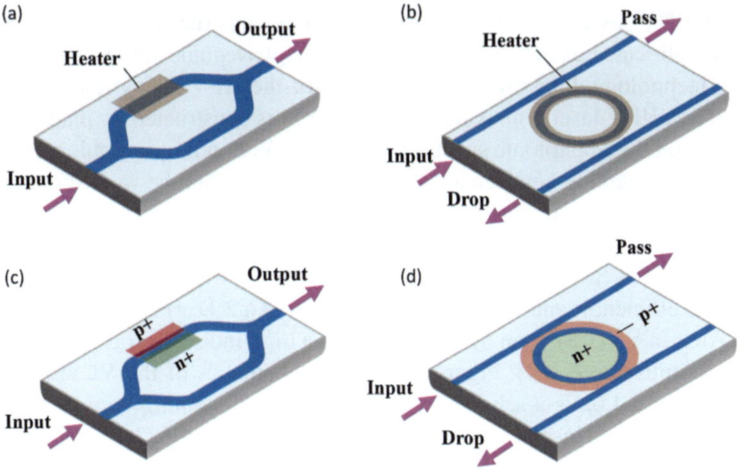

Fig. 4.5 Schematics of various optical switches and modulators. **a** TO MZ switch. **b** TO MRR switch. **c** *p-i-n* MZ modulator. **d** *p-i-n* MRR modulator. The *p-i-n* configuration is shown for illustration purposes only

compact MZI modulator with a p-n phase shifter has been described as an application of the PC slow-light device to the compact modulator. For the MRR structure (Fig. 4.5d), on the other hand, a *p-i-n* or *p–n* junction is often formed along the MRR waveguide.

4.2.2 Advanced Laser-on-Silicon Platform

A Si single crystal for photonic applications has following excellent features: crystal qualities with high purities and low defects, SOI CMOS process compatibility, and high thermal conductivity suitable for integration of active devices. These features provide significant benefits to integrate active Si photonic devices such as modulators, switches, and photodetectors. Nevertheless, Si is difficult in lasing with high efficiency because of its indirect bandgap material, where effective radiation recombination needed for efficient lasing is inherently absent: Gallium arsenide (GaAs), a representative of a direct bandgap III-V semiconductor, is capable of efficient radiative recombination due to matching of a $\boldsymbol{k}$ vector at the minimum in conduction band and maximum in valence band. So far, three kinds of alternative on-chip semiconductor lasers for Si photonics have been developed [5]:

(i) III-V laser with chip bonding or wafer bonding on Si.
(ii) III-V laser with direct epitaxial growth on Si or SOI.
(iii) Combination of these approaches.

Three types of III-V-on-Si laser are briefly introduced, as follows.

III-V Compound Laser with Chip Bonding or Wafer Bonding On-Si

A hybrid III-V laser with chip bonding on Si is constructed by placing a III-V laser chip fabricated externally on the Si photonic waveguide substrate using a chip bonding technology, thus minimizing an area of the III-V lasing material. On the other hand, a III-V laser with wafer bonding on Si is constructed by placing a III-V laser wafer on the Si photonic substrate using a wafer bonding technology, resulting in an efficient coupling between an optical mode pumped in the wafer-bonded III-V laser and confined mode guided in the underlying Si waveguide. As compared to an edge emitting laser with a standard longitudinal cavity, a vertical-cavity surface-emitting laser (VCSEL) is fit for Si photonic platform owing to small size, low cost, high power efficiency, and high integration ability of a 2-D array. Here, an advanced hybrid AlGaInAs VCSEL on Si substrate with a high index contrast grating (HCG) reflector is introduced [14]. Figure 4.6 shows schematics of the VCSEL with the Si HCG. Figure 4.6a is a schematic HCG picture for coupling between a vertical input plane wave and in-plain propagation output wave. The HCG presents both large reflectivity at the bottom of the VCSEL and high coupling into the Si channel waveguide. Figure 4.6b and c shows calculated vertical- and in-plane-coupling spectra, and power-ratio spectra between the vertical- and in-plane-coupling, respectively. At a 1550 nm wavelength, the reflectivity of > 99.4% and in-plain coupling of 47% are

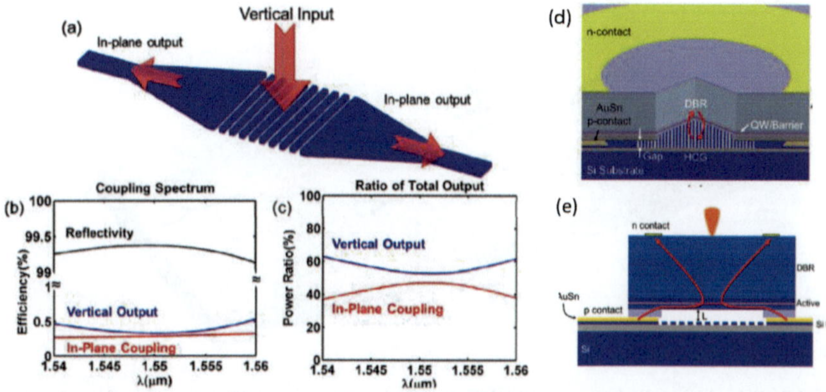

Fig. 4.6 **a** Schematic of a HCG structure. **b** Calculated spectra of coupling efficiencies in vertical output, in-plane coupling, and reflectivity. **c** Calculated total output power ratios in vertical output and in-plane couplings. **d** Schematic tilted view of the cross-sectional VCSEL structure with circulating red arrows indicating an optical cavity. **e** Schematic of cross-sectional HCG VCSEL with curved red lines indicating current confinement in the proton-implant-defined cavity. Reprinted with permission from Ref. [14], © 2015 OSA

obtained. Figure 4.6d and e shows schematics of tilted and cross-sectional views of the VCSEL, respectively. Circulating red arrows in (d) indicate an optical cavity, while curved red lines between contacts in (e) indicate current confinement in the proton-implant-defined cavity. Note that the Fabry–Perot cavity is formed between a top mirror with a distributed Bragg reflector (DBR) in a III-V material and a bottom mirror with a Si HCG. The VCSEL with the HCG operates with power of > 1 mW in continuous wave (CW) and single mode up to 65 °C. Furthermore, 5 Gb/s direct modulations with a 3-dB bandwidth of > 2.5 GHz were demonstrated under an error-free transmission condition with a single-mode fiber over a 2.5 km. Refer to [14] for basic properties and design of the HCG.

III-V Compound Laser with Direct Epitaxial Growth on Si or SOI

In general, III-Vs are difficult in direct epitaxial growth on Si substrate due to a large lattice mismatch between III-V and Si materials and their different thermal expansions. Yet, growth of III-Vs on Ge-on-Si (Ge/Si) is an alternative method because lattice mismatch is reduced considerably. For instance, an on-Si laser with InAs/GaAs quantum dots (QDs) on a Ge/Si substrate showed CW operation at room temperature (RT) [15]. Another on-Si laser with III-V QDs monolithically grown on a Ge substrate showed J_{th} of 163 A/cm^2 at a wavelength of 1280 nm using cavities with a width of 20 μm and length of 3 ~ 3.5 mm [16].

Heterogeneously Integrated Narrow-Linewidth Laser

In the last decade, a heterogeneously integrated innovative narrow-linewidth laser on the Si substrate was developed [17]. Figure 4.7 shows a schematic of an on-Si laser monolithically integrated with phase- and wavelength tuners, fabricated using a

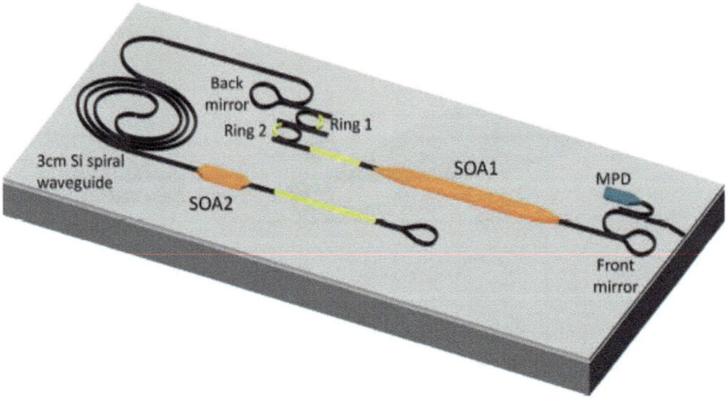

Fig. 4.7 Schematic of a heterogeneously integrated narrow-linewidth laser with phase tuners in straight line and wavelength tuners in ring (both yellow), additionally equipped with an external cavity. SOA1 (orange) is a laser active section, while SOA2 (orange) is an on–off switch or feedback control section. Reprinted with permission from Ref. [17], © 2015 IEEE

heterogeneous photonic integration process of indium phosphide (InP) active regions and Si photonic passive waveguide regions. The whole laser chip is composed of two parts: One is a laser active section (SOA1 in orange) connected to a thermal phase-tuner section (yellow), wavelength-tuning ring resonators (Ring1 and Ring2), and front/back loop mirrors, totally forming a 2-mm long cavity. The other is an external cavity section (4-cm length in total) involving an on–off switch and feedback-level control gain section (SOA2 in orange) connected to another thermal phase-tuner section (yellow) and a spiral waveguide cavity (3 cm). At the laser output section, monitor photodiode: blue (MPD) measures a laser output power to adjust laser parameters. As a result, a wavelength of the laser is tunable in the range of 1237.7–1292.4 nm. Note that a demonstrated linewidth in full tuning ranges 50–100 kHz: a record linewidth in monolithic integration with tunable lasers. An important achievement of the narrow linewidth is brought by the long external Si waveguide (4 cm) with low propagation loss; the large benefit of the Si photonic platform.

Necessity of an integrated narrow-linewidth (e.g., < 300 kHz) laser is rapidly increasing as an optical signal modulation format in the network system is shifting to a more complex method such as 16-, 64-, and 256-QAM (quadrature amplitude modulation), while wavelength-tunable lasers are inevitable in the dense wavelength-division multiplex (DWDM)-based system. Application of Si photonics is now growing steadily.

4.2.3 *Forecast of Silicon Photonics*

Current Foundries and Market Forecast

For manufacturing integrated Si photonic devices, limited companies (supply chains) having special technological fields undertake their specified processes classified into design, manufacturing with CMOS process, etc. [18]. For instance, a company named "fabless" only designs semiconductor chips and other hardware, but does not fabricate such chips and outsources them. On the other hand, a "foundry" (or just "fab") does not design integrated devices but manufactures them only. In the late 2000s through early 2010s, Luxtera, Intel, Cisco, etc. were typical Si photonics manufacturing foundries. As introduced in Sect. 4.3.2, GlobalFoundry (GF) is a representative foundry for manufacturing Si photonic transceivers using monolithic CMOS processes particularly named "zero-change". According to an article [19], GF is the second-largest foundry in the world, whose revenue in 2017 is estimated as \$5.86 billion. As demand of Si photonic platform is ever-growing, mission of the foundry will also become more important. The fluid reports in part by these future-estimating market research companies are reflecting these situations. Si photonic platform engineers cannot take their eyes off from them.

Si Photonic Transceiver Market

Thomson et al. reported a widely reviewed roadmap on Si photonics [20]: Through the initial developments of SOI waveguides in the 1980s and practical applications to

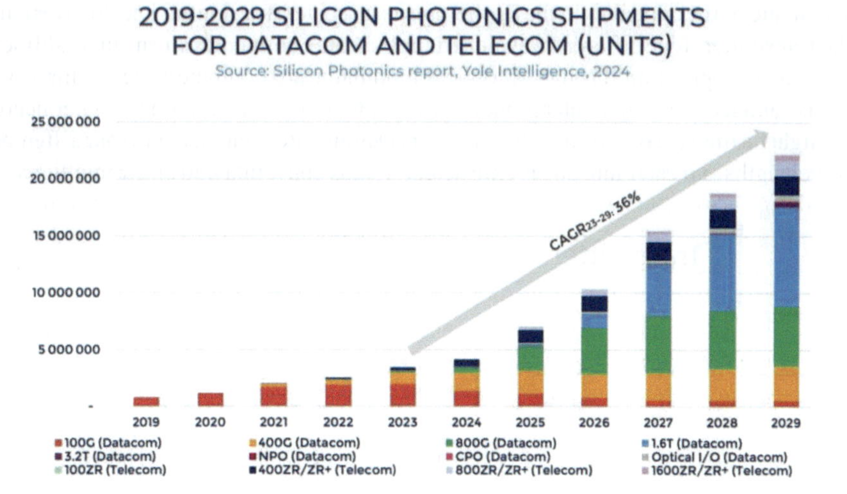

Fig. 4.8 Si photonics forecast for data center in 2019–2029. Reprinted with permission from Ref. [21], © Yole Group 2024

the WDM telecom products in the 1990s, telecom and data-com systems necessitate Si photonics more as the next-generation style of optical devices. Figure 4.8 shows a Si photonics forecast (2019–2029) for the data center [21]. It reflects that mega companies for the data center are deploying Si photonic interconnects in a large scale by replacing the hardware beginning around ~ 2019. The forecast is performed per typical photonic transceiver modules such as 100G PSM-4 (parallel single-mode 4-lane), 100G CWDM-4, etc. Reference [21] reports that metrics for introduction of Si photonic interconnects in the data center are low cost and high power efficiency as well as capability of CMOS-compatible fabrication and reliability.

4.3 Silicon Photonic Interconnect for Data Center and Short Optical Links

4.3.1 Front Runners in Silicon Photonic Transceiver Research

In the data center based on an electronic hardware, an electrical interconnection using copper wires has a significant problem of electrical resistance in the wire, thus degrading signal intensities seriously, and is forced to be limited in maximum lengths of the copper wire. For this problem, Si photonics can provide a solution by replacing an electrical interconnection with an optical link: a Si photonic interconnect transceiver capable of higher data rates over long distances. Figure 4.9 shows a schematic of the optical link: the Si photonic transceiver (transmitter and receiver). In the transmitter, four different wavelength light beams emitted from an external laser array and propagating through an optical modulator are combined to a multiplexer (Mux) and led to a single interconnect waveguide. In the receiver, on the other hand, the light beams received by a de-multiplexer (Demux) are separated into four different wavelengths and enter into an on-chip detector, thus converted into electronic signals.

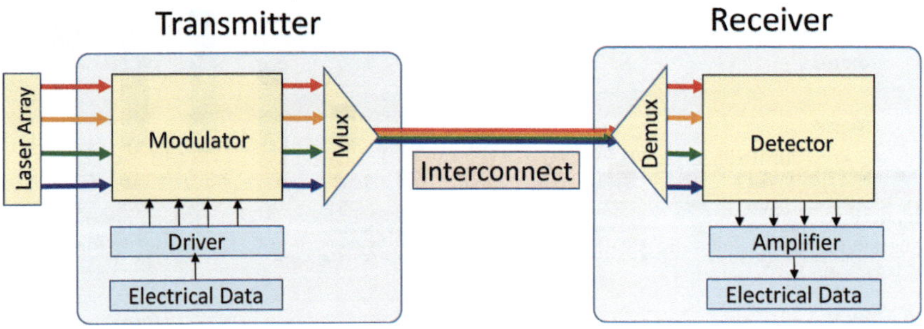

Fig. 4.9 Schematic of a Si photonic transceiver (transmitter and receiver)

In the late 2000s through early 2010s, primary three front runners, Luxtera, Intel, and IBM, have contributed to development of Si photonic transceivers. Their considerable activities are introduced in the following.

Luxtera

Luxtera—now a subsidiary of Cisco Systems—is a pioneering company that developed Si photonics platform using CMOS process manufacturing plants. Luxtera was founded in 2001 by a group of the California Institute of Technology, USA. In 2006, the company first integrated several sorts of photonic devices together with electronic elements using a SOI CMOS process in a 130-nm node [4, 22]. Figure 4.10 shows a SEM photograph of a demonstrated input/output optical grating coupler and Si waveguides, surrounded by various LSI elements on a Si substrate. Si waveguides are found to be combined to a micro-ring filter for mux/de-mux procedure. A grating coupler between an optical fiber and Si waveguide exhibits a coupling efficiency of 66% at 1550 nm. The demonstration can be considered as world-wide beginning of "Si photonics". After this, Si photonic interconnects with high data rates and transmission over longer distances in the telecom and data-com center was realized at CMOS-compatibility-based low manufacturing costs. Luxtera also developed a 10 Gb/s transceiver with a WDM transceiver with four channels, fabricated by SOI CMOS process with a 90-nm node [22]. Four-channel InP lasers were mounted on Si substrate by flip-chip bonding, while other components such as filters, amplifiers, and photodetectors were monolithically integrated on the same substrate. Afterward, development of these on-chip Si photonic transceivers was globally expanded by many groups.

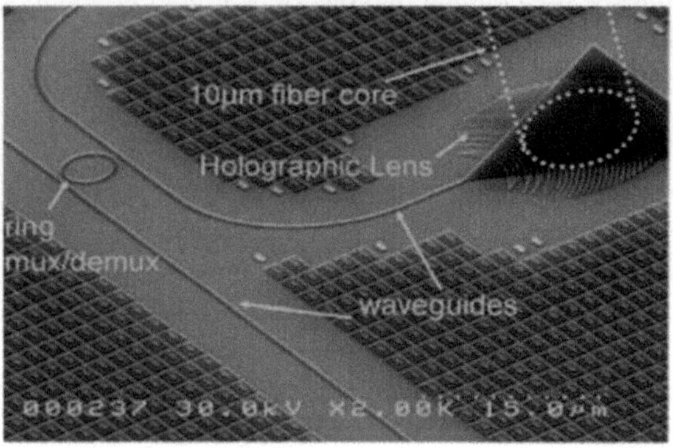

Fig. 4.10 SEM photograph of an optical grating coupler connected to Si waveguides monolithically integrated on the Si substrate, demonstrated first by Luxtera. Reprinted with permission from Ref. [22], © 2006 IEEE

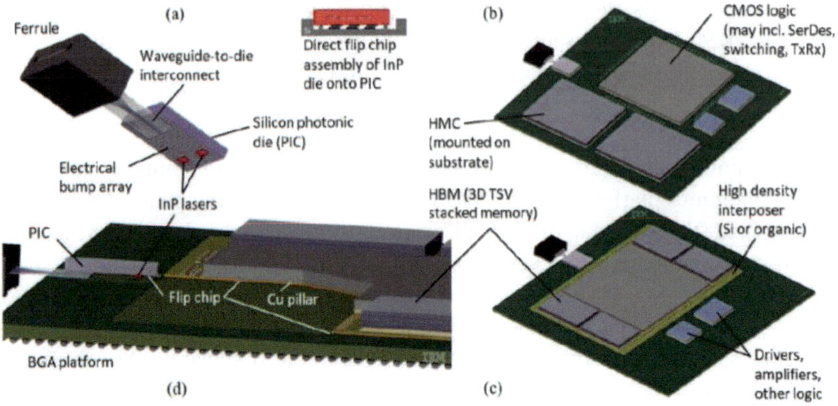

Fig. 4.11 Schematic of Si photonic MCM. **a** Si photonic die with InP lasers mounted by flip-chip bonding and waveguide-to-die interconnect coupled from a ferrule. **b** BGA with CMOS logic, HMC, and PIC. **c** High-density interposer with CMOS logic, HBM, and PIC. **d** Cut-off schematic of BGA platform with Cu pillar, flip-chip, InP lasers, and PIC. Reprinted with permission from Ref. [25], Reprint Courtesy of IBM Corporation ©

Intel

In 2007, Intel announced a Si photonic transceiver coupled to an optical fiber for a 50 Gbps optical link [23, 24]: The transmitter is composed of four InP lasers flip-chip bonded on Si and optical modulators. Four optical beams, emitted from lasers with four different wavelengths, encoded at 12.5 Gb/s by modulators, and combined through a multiplexer, are coupled into one optical fiber at 50 Gb/s. In the receiver chip, incident light beams coupled into a de-multiplexer in the chip are separated into four beams with different wavelengths and impinging on each photodetector. The chip with a data rate of 100 Gb/s or higher is under development.

IBM

IBM has long developed packaging techniques using Si photonic platform to complete a processor module with photonics integrated into electronics. In the previous data center, optical transmitters and receivers were located at the input/output ports of each server. Thus, an optical component was a certain distance from the processor chip. In 2017, IBM announced a new strategy on packaging with Si photonics: Si photonic MCM (multi-chip-module) [25]. Figure 4.11 illustrates a schematic of a Si photonic integrated circuit (PIC) with the Si photonic MCM. Figure 4.11a shows the PIC with InP lasers (red) mounted by flip-chip bonding and a waveguide-to-die interconnect coupled from a ferrule. An upper panel indicates a schematic of the InP laser die. On the die surface, photonic waveguides are prepared for large coupling between optics and electronics by an electrical flip-chip bump array to support electrical signals. Figure 4.11b shows the PIC (left edge), a CMOS logic (central), two HMCs (hybrid memory cubes), etc., all mounted directly onto a BGA (ball grid array) substrate. Figure 4.11c and d show alternative designs: In

Fig. 4.11c, CMOS logic dies and high bandwidth memories (HBMs)—3-D TSV (through silicon via) stacked memory—are mounted on an interposer to connect logic, memory, and PIC closely, while Fig. 4.11d shows a cut-off schematic of a copper pillar, flip-chip, and PIC mounted on a BGA platform.

In 2015, IBM first demonstrated a four-channel Si photonic transceiver chip fabricated by a 90-nm CMOS process through GlobalFoundries [25]. The chip includes input/output ports for external lasers, optical modulators, Ge photodetectors, and wavelength-division multiplexers, combined with electronic circuits. Four-wavelength signals with each data rate of 25 Gb/s are multiplexed to 100 Gb/s at fiber output ports. The result showed clear eye diagrams: a high-quality response.

4.3.2 Monolithically Integrated Silicon Photonic Transceiver

In the Si photonic platform, loss-minimum optical waveguides require sufficient isolation between optical waveguides and surrounding media (Si substrate). In the typical Si photonic chip fabricated by a SOI CMOS process, the optical isolation is given by a buried oxide (BOX) layer with a thickness of 1 μm. Nevertheless, the BOX thickness in the 45-nm SOI process is less than 200 nm. For fully depleted SOI transistors, in particular, it is reduced to 5–10 nm due to electrostatic conditions. Here, the following two examples are introduced from some attempts for solving these issues:

(i) *Back-end SOI CMOS photonics*: Photonics are drawn on top of the back-end in the CMOS.
(ii) *Front-end SOI CMOS photonics:* part of the substrate near the front-end processed Si photonic platform is removed to minimize penetration of light field from the waveguide.

Back-End SOI CMOS Photonics

Si photonics platform deposited on the back-end in the CMOS is formed with combination of poly-Si regions annealed with an excimer laser at low temperature for active devices and low-loss Si nitride (SiN) regions for photonic waveguides, thus, photonics being deposited on the back-end region of the underlying CMOS [26]. In this structure, a poly-Si modulator is formed similar to a single-crystalline Si modulator with careful design of parameters such as doping level, etc. For instance, an optical modulator with a data rate of 2.5 Gb/s and a modulation depth of 10-dB are available with this scheme. Therefore, back-end CMOS photonics can use an area for photonics effectively, free from the constraint footprint independently from CMOS.

Front-End SOI CMOS Photonics

Thereafter, a world-wide first microprocessor with a record-high energy efficiency and speed photonic transceiver was realized [27]. This is thought to be one of the most

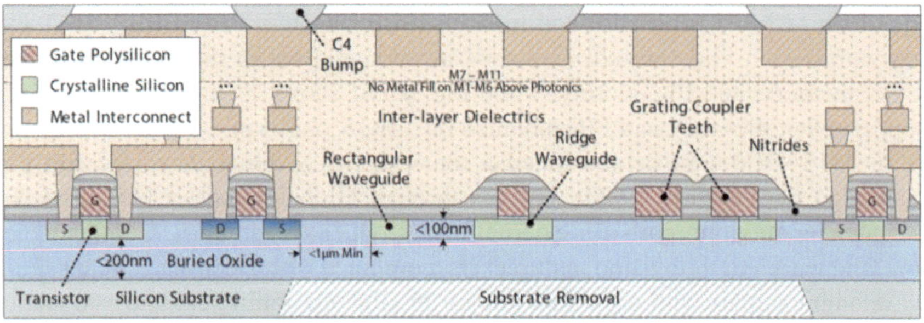

Fig. 4.12 Schematic of cross-sectional Si photonic CMOS circuit showing front-end with both photonic and electronic devices fabricated by 45-nm SOI CMOS process. Reprinted with permission from Ref. [28], © 2016 IEEE

advanced monolithic "zero-change" Si photonic platform because it was manufactured by a commercial CMOS SOI process technology on the 45-nm node without any extra-change of process, i.e., with "unchanged" process. It also features the highest level of integration with electronic and photonic circuits. Figure 4.12 shows a cross-sectional schematic picture of the Si photonic SOI CMOS circuit including both photonic and electronic components at the front-end. Note that all-photonic devices are fabricated according to the design of electronic devices fabricated in the crystalline Si (c-Si) layer with a thickness of sub-100-nm without modifying the native process [28]. For preserving sufficient optical isolation from the surrounding region, several metal layers above the photonic elements are removed. Furthermore, a Si substrate under the photonics structures is removed also since the BOX layer was thin (200 nm). Active devices such as MRMs and PDs are formed with an original source and drain and available SiGe in this process. Rectangular and ridge waveguides as well as transistors are fabricated in the crystalline Si regions, while grating coupler teeth are fabricated in the gate polysilicon regions.

Figure 4.13 presents a top-view photograph of the 45-nm Si photonic die fabricated by "zero-change" SOI CMOS process. Figure 4.13a shows a system-on-chip (SoC) with transceiver and receiver sets, RISC-V processor, cache memory based on SRAM, and optical I/Os for DWDM system. Figure 4.13b shows optical transceivers with a transmitter bank (upper photo) and receiver bank (lower photo). A magnified transmitter photo includes MRM, drop port photodetector, waveguide, and vertical grating coupler (VGC) as well as electronic elements, while a magnified receiver photo includes a photodetector and receiver input VGC as well as electronic elements. Figure 4.13c shows key photonic elements of the optical link; VGC (left), SiGe photodetector (middle), and MRM (right).

Structures and performances of the primary passive and active photonic elements implemented in the aforementioned SOI CMOS processed Si photonic platform are explained, as follows [27]:

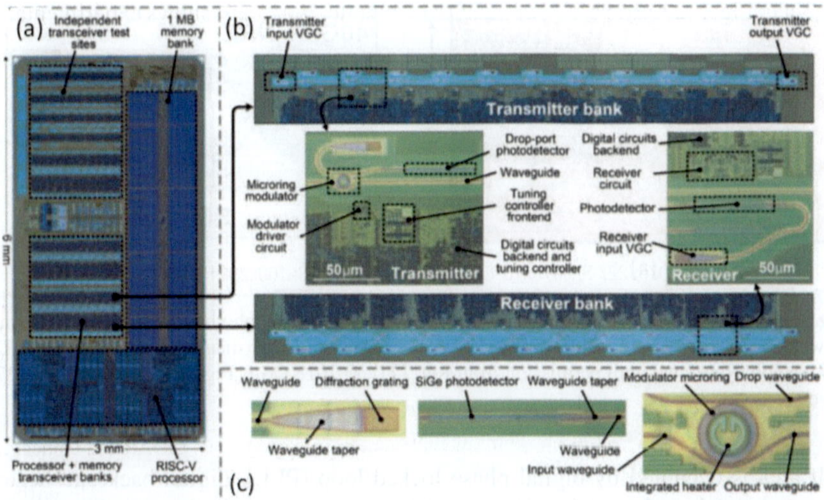

Fig. 4.13 Top-view photograph of the 45-nm Si photonic die. **a** SoC with transmitter and receiver sets, RISC-V processor, cache memory, and optical I/Os. b Transmitter and receiver banks with MRM, VGC, waveguides, etc. **c** Main photonic devices of an optical link. Reprinted with permission from Ref. [27], © 2018 OSA. Development timeline in Fig. 3 in Ref. [27] has been omitted here

- *Waveguides*: Waveguides as passive optical elements are fabricated in the c-Si layer with a thickness of sub-100 nm. Measured loss is 3.7–4.6 dB at wavelengths in the range 1280–1550 nm.
- *Input/output VGC*: VGCs between an optical fiber and on-chip waveguide are formed using c-Si and poly-Si layers for double-layered unidirectional VGCs. Coupling loss is 1.5 dB.
- *Optical modulator*: MRMs are formed by arranging interleaved *p* and *n* junctions along the ring cavity. They feature 3-D structured spoked ring modulators. Active MRMs with a radius of 5 μm achieve *Q*-factors of 18,000 and transmission of PAM-4 with a signal rate of 40 Gb/s.
- *Photodetector (PD)*: PDs are fabricated in the embedded epitaxial SiGe via advanced CMOS processes of 45 and 32 nm nodes. PN junctions are formed in the ring resonator's cavity like optical modulators. The resonated epitaxial SiGe PD in the 45-nm CMOS node presents a responsivity of ~ 0.5A/W in the 1180–1270 nm wavelength range.

Figure 4.14a shows a photograph of a NRZ transmitter die with a data rate of 40 Gb/s composed of MRM fabricated by the SOI CMOS process in the 45-nm node. Owing to compact footprints, the transmitter based on the MRM is improved in energy efficiency by two orders of magnitude compared to the conventional MZM-based transmitter. It is capable of DWDM for a LSI system with a high density in bandwidth in a range of Tb/s/mm^2. Note that modulator and driver areas of 0.03 mm^2 spend only 40 fJ/b, in spite that energy consumption in total is 0.7 pJ/b: More than

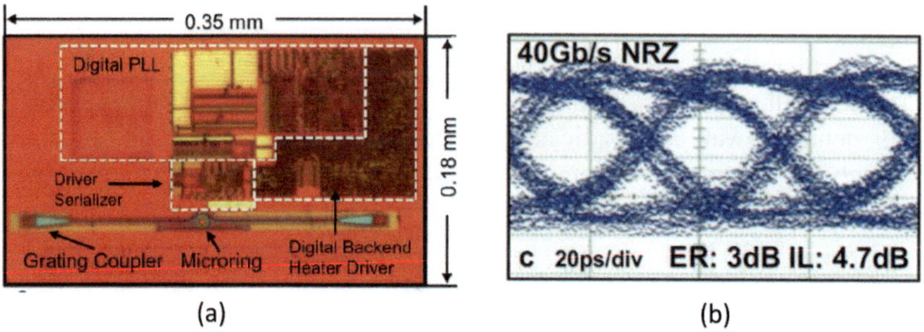

Fig. 4.14 **a** Photograph of 40 Gb/s NRZ transmitter die including photonic VGC and micro-ring as well as electronic digital PLL, heater driver, etc. **b** Measured eye diagram in the 40 Gb/s NRZ modulation. Reprinted with permission from Ref. [27], © 2018 OSA. Figure 7b and d in Ref. [27] have been omitted here

half area is occupied by digital phase locked loop (PLL), digital back-end heater driver, and other electronics. Figure 4.14b shows measured eye diagrams in the NRZ modulation with a data rate of 40 Gb/s, extinction ratio (ER) of 3 dB, and insertion loss (IL) of 4.7 dB. The transmitter in the PAM-4 modulation at 40 Gb/s was demonstrated also [29]. The aforementioned SOI CMOS processes using 45–32 nm nodes have certified great powers both in the current Si photonic technology and in the electronic/photonic device performance for future optical interconnects [27, 29].

4.3.3 Ultra-compact Silicon Photonic Transceiver

Fingertip-Size Si Photonic Transceiver

Field programmable gate arrays (FPGAs) are integrated configurable logic circuits connected via programmable interconnects. Owing to a high-bandwidth density, high energy efficiency, and low latency, in addition to such reprogrammability based on the desired application after manufacturing, FPGAs are capable of supplementing an inefficient processing speed of CPUs, GPUs, etc., thus attracting much attention in the recent IoT/edge computing system as well as in the data center.

Recently, the world-smallest Si photonic transceiver—optical I/O core—with a fingertip-size (5×5 mm^2) was developed [30]. It is mounted to the FPGA, thus exhibiting low power consumption and the maximum transmission capacity of 300 Gb/s with 25 Gb/s $\times$ 12 channels. Figure 4.15a shows a photograph of the whole chip with optical/electrical I/O ports and integrated circuit mounting areas [31]. Figure 4.15b and c shows schematics of a cross-sectional transmitter and receiver, respectively. The transmitter involves a Si photonic MZI modulator, grating coupler, and optical waveguides, while the receiver involves a Ge photo detector. In the transmitter, a laser diode (LD) with the type of Fabry–Perot is mounted, while optical

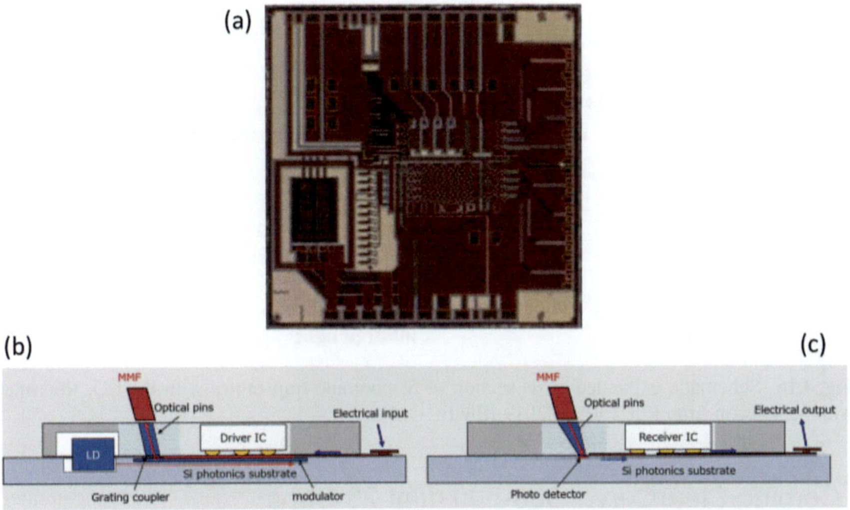

Fig. 4.15 Module of an optical I/O core. **a** Photograph of the whole module with optical and electrical I/Os and integrated circuits. **b** and **c** Schematics of a cross-sectional transmitter and receiver, respectively. Reprinted with permission from [31], © AIOCORE

pins coupled to the external multi-mode fiber (MMF) are assembled both in the transmitter and receiver. Red and blue arrows show optical and electrical signal flows, respectively. The whole Si photonic die is fabricated by the SOI CMOS process with a 300-mm wafer.

The LD and optical pins are integrated after completing the core chip. Figure 4.16 shows a schematic of the expanded cross-sectional Si photonic transmitter with the assembled LD [30]. Both alignment marks on the LD and Si substrate for an infrared camera are designed in advance, where horizontal and vertical misalignments between them are kept within ± 0.5 μm and smaller than ± 0.1 μm, respectively. An InAs/GaAs quantum-dot LD is adopted because it exhibits lasing operation at higher temperatures than an InGaAsP quantum-well LD.

As shown in Fig. 4.15b and c, optical pins made of resin-polymer connect a GC and Ge-photodiode to external MMFs instead of optical lenses. Note that the optical pin as a 3-D fine waveguide has an ability of strong optical beam confinement due to a high contrast between core/cladding refractive indices. Accordingly, optical pins minimize fluctuation in wavelength and radiation angle of the output laser beam, resulting in stable radiation of the output beam: Calculated tolerance of miss-alignment between the optical pin and GI50 MMF within 1-dB excess loss is larger than 10 μm at – 45 to 85 °C.

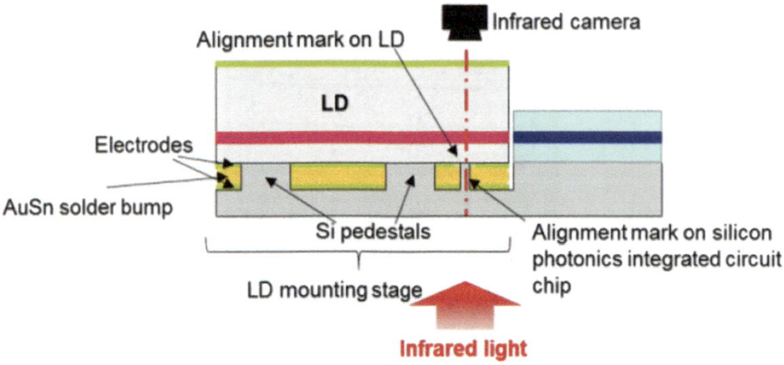

Fig. 4.16 Schematic expanded cross-section of Si photonic transmitter with the LD. Reprinted with permission from Ref. [30], © 2019 IEICE

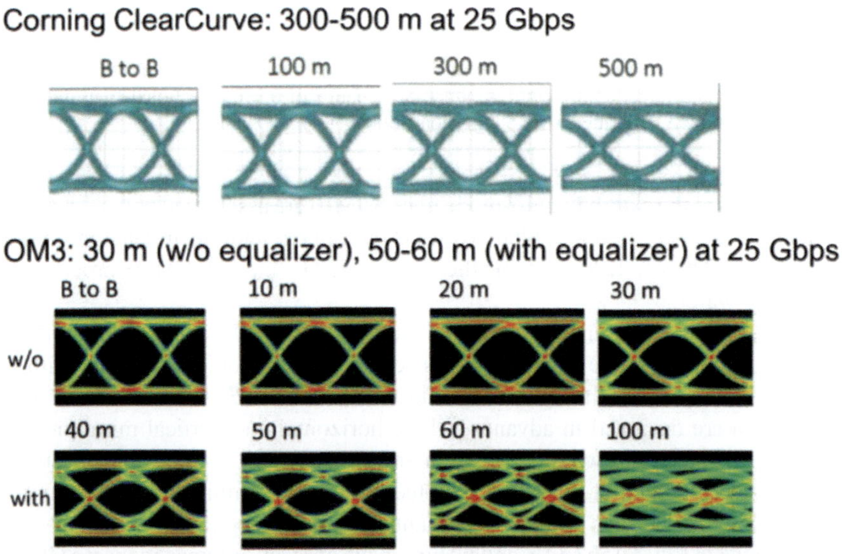

Fig. 4.17 Clear eye patterns in transmission measurement of the optical I/O core connected to two kinds of MMFs. Reprinted with permission from Ref. [30], © 2019 IEICE

Transmission characteristics of the optical I/O core were measured before and after mounting around the FPGA [30]. Before mounting, eye patterns were observed clearly at 25–85 °C and bit error rates (BERs) were less than 10^{-12} at 25 Gb/s at 20 and 85 °C. Figure 4.17 shows observed results for two types of MMFs: a corning LX MMF (upper panel) and conventional OM3 fiber (lower panel), both showing clear eye patterns. The former shows excellent transmission at a distance of 500 m at 25 Gb/s data rate, while the latter shows successful results also at 30-m and 60-m distance without and with an equalizer, respectively, both at the same 25 Gb/s data rate. After mounting optical I/O cores (24 transceiver and receiver channels) around the FPGA, good eye patterns and BERs of less than 10^{-12} with almost no degradation were successfully obtained, where FPGA boards were connected to a 300-m MMF. In summary, an aforementioned ultra-compact optical I/O core based on the Si photonic platform exhibits transmission up to 300 m at 25 Gb/s data rate under mounting around FPGAs. These results enable successful interconnects with high-bandwidth density and low consumption power among FPGAs in the data center and IoT/edge computing system.

Si Photonic Transceiver for 5G Mobile Network

Internet of Things (IoT)—network of physical objects—are now growing rapidly. At an endpoint of the IoT, 5G (the fifth generation) mobiles connect an enormous number of sensors and terminals via exchanging data over the Internet, as shown in Fig. 4.18a [32], where devices over there involve common household objects to high-performance industrial tools. In the 5G mobile terminal, a transmission rate increases to 10 Gb/s at maximum, while a number of connections reach about 10^6 terminals/km^2 around the front haul network having a large number of small cell antennas. To meet with the transmission capacity of the 5G mobile, a system of time and wavelength-division multiplexing (TWDM) and passive optical network (PON) is introduced, as shown in Fig. 4.18b, where up (subscriber to base station) and down (base station to subscriber) links are multiplexed with four wavelengths and at 10 Gbps per wavelength, resulting in the total transmission rate of 40 Gbps.

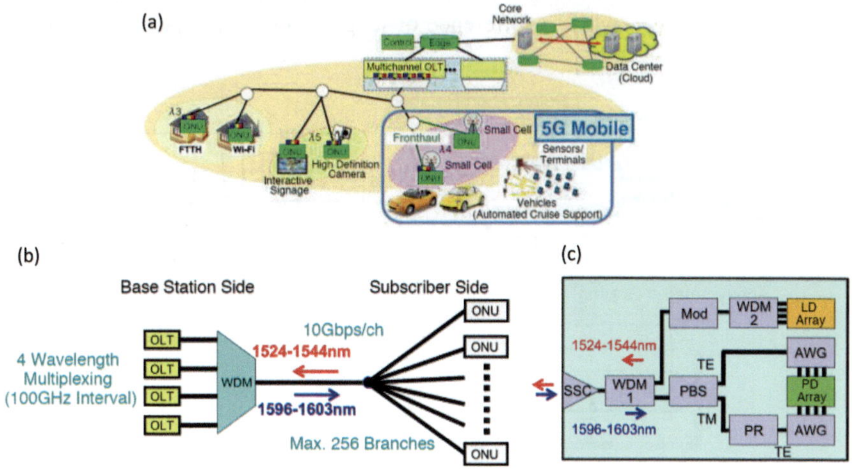

Fig. 4.18 **a** Schematic showing 5G mobile network for supporting IoT with various sensors and terminals over Internet. **b** Configuration of TWDM (e.g., four-wavelength and 100-GHz interval) and PON (e.g., max 256 branches). **c** Layout of an ultra-compact optical transceiver with polarization diversity and WDM functions. Reprinted with permission from Ref. [32], © 2017 OKI Corporate

For realizing the TWDM-PON optical network unit (ONU), an ultra-compact optical transceiver chip with a bidirectional Si photonics was developed recently [32]. Figure 4.18c shows a layout of the polarization-independent ultra-compact optical transceiver with a polarization diversity configuration: Light incident on a spot size converter (SSC) and WDM filter (WDM1) leads to a polarization beam splitter (PBS), where it is split into TE and TM modes. The TE mode directly leads to an arrayed waveguide grating (AWG) for wavelength separation, while the TM mode is sent to a polarization rotator (PR) for conversion to the TE mode and reaches the AWG. Both light beams are then received by the photodiode array.

Developed optical transceivers involve two key elements; a polarization-independent bandpass filter (WDM1) and an AWG with multi-mode interferometer (MMI)-rib waveguides (WDM2) [32]. In Fig. 4.19a, an upper panel shows a configuration of the WDM1 with a polarization-independent 3-dB coupler and a PR grating. A lower panel shows Si core waveguides with antisymmetric grooves on both sides in the PR grating for selective reflection of the specific wavelength light. Figure 4.19b shows reflection spectra between TE-TM and TM-TE mode conversions in the bandpass filter. The spectra show quite similar to each other, resulting in an excellent polarization-independent bandpass filter.

Figure 4.20a shows a microscope image of another key element; prototype AWG [32]. The structure involves slab waveguides with parallel arrangement connected to MMI-rib waveguides, where the MMI with a thinner slab terrace is formed at the interface between the slab waveguides and rib waveguides for minimizing an excess propagation loss [33]. An optimized terrace thickness is 130 nm for the slab thickness of 200 nm. A width and thickness of the arrayed waveguide are 1 μm and 200 nm, respectively, while a difference of lengths between adjacent waveguides

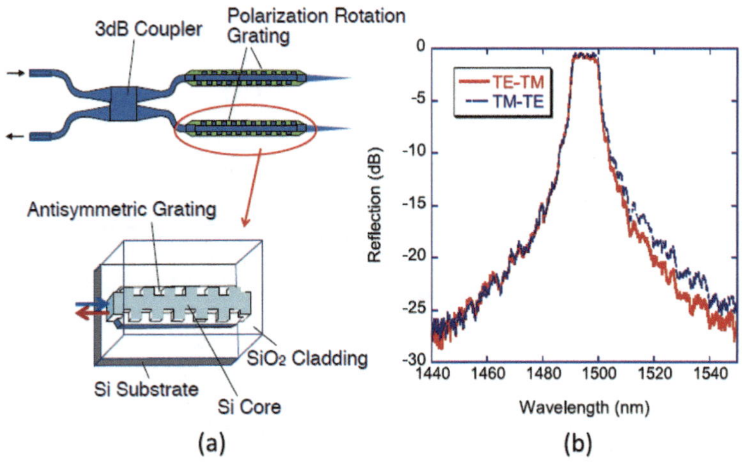

Fig. 4.19 **a** Schematic of polarization-independent bandpass filter with antisymmetric grooves in the PR grating. **b** Reflection spectra between the TE-TM and TM-TE mode conversions. Reprinted with permission from Ref. [32], © 2017 OKI Corporate

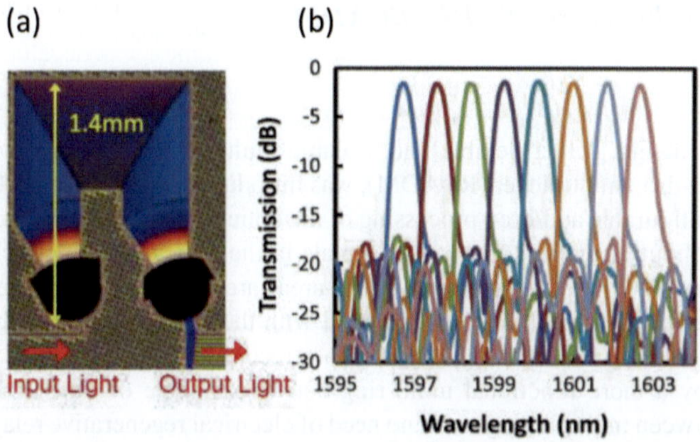

Fig. 4.20 **a** Microscope image of an AWG with slab waveguides connected to MMI-rib waveguides. **b** Transmission spectra of the fabricated AWG. Reprinted with permission from Ref. [32], © 2017 OKI Corporate

is 32.7 μm for an AWG with spacing of 100-GHz wavelength channel for the TE mode. Figure 4.20b shows measured wavelength responses. A minimum insertion loss, extinction ratio (ER), and output uniformity are 1.2 dB, 16 dB, and 0.5 dB, respectively.

These results have a potential to contribute to reducing cost and power consumption in the optical transceiver for the 5G mobile network.

4.4 Silicon Photonic Switch for Core Network and Data Center

In the photonic network, an optical switch, in particular, a multi-port switch is inevitable for switching optical paths at the node of the network. So far, several free-space optical switches such as micro-electromechanical system (MEMS) and liquid crystal on silicon (LCOS) have been commonly used. As network traffic is explosively growing recently, integrated optical switches, particularly Si photonic multi-port switches with excellent advantages such as high speed, broadband, small footprint, and low consumption power, have far advanced in the core network and data center like Si photonic transceivers mentioned above.

Physical switching mechanisms in the Si material represented by plasma dispersion effects of free carriers and TO effects are fortunately available for realization of Si photonic switches with configuration of the MZI and MRR. This section first reviews architectures of photonic switches; (i) flexible and scalable multiplexer systems enabling reconfigurable optical path processing, and (ii) 2-D topologies for switch fabrics. Then, a multi-cast optical switch for a core network node and various unicast (multi-port) optical switches for data centers are reviewed.

4.4.1 Architecture of CDC–ROADM

ROADM

As shown in Fig. 4.1, a flexible and scalable multiplexer system, reconfigurable optical add-drop multiplexer (ROADM), was introduced in the early 2000s for realizing reconfigurable add/drop processing of the optical path (specific to wavelength) signals without changing to electrical signals in the optical core network covering metro-areas [34]. Now, the core and metro-areas are basically constructed with the ring network, where each node is equipped with the ROADM system that enables division and insertion of optical signals. In a recent year, a single-ring network is replaced by a more functional multi-ring network capable of transferring optical signals between multiple rings with no need of electrical regenerative relay function.

Figure 4.21a shows a schematic of a two-ring (Ring A and Ring B) network controlled by 4-degree ROADM as an example of a multi-degree ROADM instead of a previous basic 2-degree ROADM, thus realizing mesh-based network topologies [35]. Figure 4.21b shows a detailed 4-degree ROADM at the node which connects

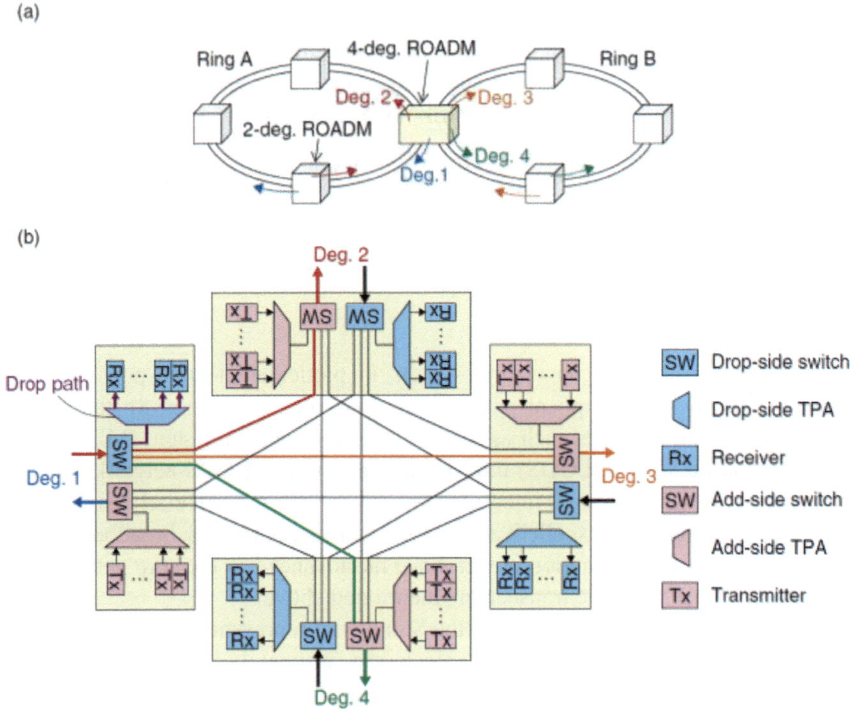

Fig. 4.21 **a** Schematic of a two-ring (A and B) network. **b** 4-degree (1–4) ROADM arrangement which connects two-ring networks. Reprinted with permission from Ref. [35], © 2014 NTT Technical Review

two-ring network. Input WDM signals coming through an input fiber at degree 1 are sent to an optical switch at the drop side (blue). For signals to be transferred to other ROADM nodes, an optical switch at the drop side chooses optical routes to degrees 2, 3, or 4 depending on each wavelength signal. For signals to be received at this node, an optical switch chooses the route (purple) to guide signals to another optical switch, i.e., the transponder aggregator (TPA). The TPA leads signals to the desired transponder that generates a different signal in response to receiving the signal. On the added side, on the other hand, signals launched from the transponder are collected at the TPA and guided to the add-side optical switch (pink), where they are launched to the neighboring node with other WDM signals.

In a recent year, more convenient functions, colorless, directionless, and contention-less (CDC) functions, were investigated for providing more flexible wavelength-routing/-assignment in the previous ROADM system, thus constructing a more advanced CDC–ROADM system in the core and metro area networks [36]. Each function and a whole CDC–ROADM system are shown in the following.

CDC (Colorless, Directionless, and Contention-Less) Functions

A colorless function is to provide a transponder with a wavelength-tunable function, which can connect input signals to any output port for a channel with each wavelength by not manual but software control. For this purpose, the conventional AWG is exchanged to a wavelength-selective switch (WSS) with a 1 × N switch structure. On the other hand, a directionless function is to assign a signal route to any input/output port for a channel with each wavelength by connecting the transponder to every degree of the ROADM node. For this purpose, two WSSs (1 × N switch structures) are introduced. Finally, a contention-less function is to allow multiple wavelength channels with the same wavelength at a single transponder, thus to enable assigning

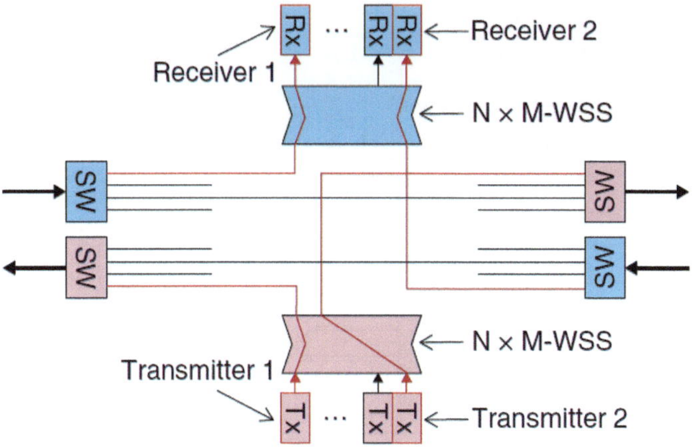

Fig. 4.22 Schematic configuration of contention-less TPA with M × N WSSs for CDC–ROADM. Reprinted with permission from Ref. [35], © 2014 NTT Technical Review

a transponder to any wavelength, as shown in Fig. 4.22. To this end, the contentionless function uses $M \times N$ WSSs, thus removing a wavelength restriction imposed in the colorless and directionless ROADM where two $1 \times N$ WSSs are cascaded. Meanwhile, realization of the $M \times N$ WSS involves several challenges by the current technologies, thus now under development by way of alternative technologies such as a multi-cast switch (MCS) [34]. Later (Sect. 4.4.2), recent development of the Si photonic MCS for CDC–ROADM is reviewed [39, 40].

Architecture of Photonic Switches

Architecture of photonic switches with integrated multi-port elements for advanced photonic networks involves unicast (point-to-point) and multi-cast switches in the data center and ROADM network core. They need to select appropriate switch topology. Topologies for unicast (multi-port) switches are overviewed in Refs. [37, 38], where various types of switch topologies (with $N \times N$ port counts) and their features have been surveyed. Here, typical four topologies are selected. Figures 4.23a–d show crossbar, path-independent-loss (PILOSS), Beneš, and switch-and-select (S&S) topologies, respectively. Metrics in their performances are: (i)

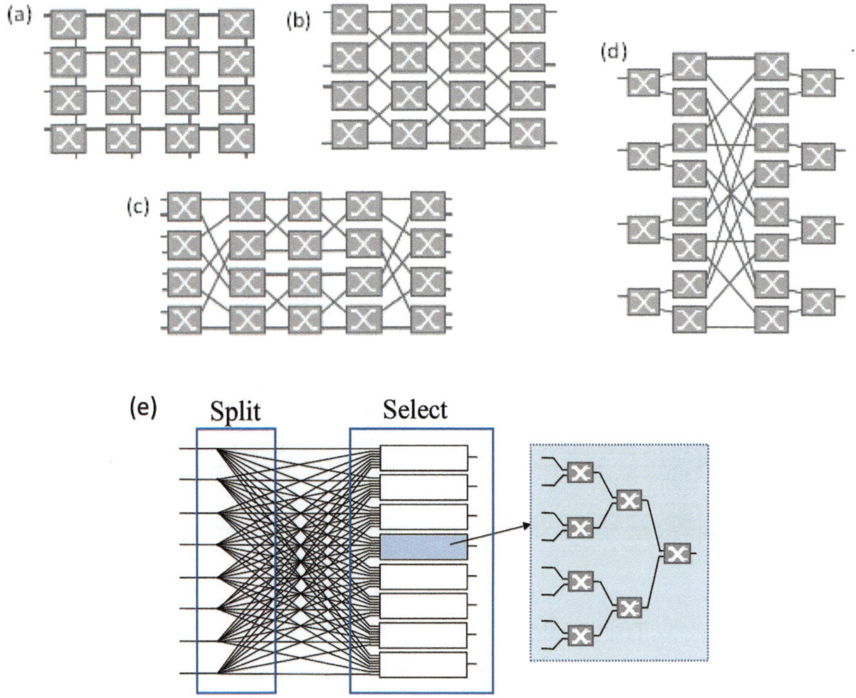

Fig. 4.23 Typical topologies for optical switches. **a–d** Crossbar, PILOSS, Beneš, and switch-and-select topologies, respectively. These are used for unicast (point-to-point) multi-port switches. **e** Split-and-select multi-cast switch topology [40]. **a–d** reprinted with permission from Ref. [37], © 2018 OSA

blocking characteristics, (ii) a number of cascading stages, (iii) a total number of switch cells, and (iv) crosstalk suppression. They have their own characteristics: Crossbar exhibits the most cascading stages, thus suffering from large path-dependent loss. PILOSS is desirable in case that uniform performances are requested for all paths, yet inevitable in the first-order crosstalk. Beneš exhibits the least cascading stages and switching cells, yet having a disadvantage of the first-order crosstalk due to two traversing signals. S&S can avoid the first-order crosstalk, yet including the most switching cells and complexity in central complicated network. These topologies are selected depending on the switching performance.

When multi-cast connection is required, on the other hand, these multi-port switches based on the point-to-point connection need additional complicated structures on each element. To meet this requirement, a novel multi-cast photonic switch with split-and-select connection is proposed, as shown in Fig. 4.23e [40]. An architecture of this new switch with strictly non-blocking and multi-cast capabilities for the CDC–ROADM system is reviewed in the next section.

4.4.2 Multi-cast Silicon Photonic Switch for CDC–ROADM

Recently, a Si photonic multi-cast switch for the CDC–ROADM application was developed [39, 40]. Figure 4.24 shows schematics of the CDC–ROADM system and configurations of the multi-cast photonic switches. Figure 4.24a shows a schematic of reconfiguring optical paths for two emergency cases in CDC–ROADM networks: Coming across an accidental intercept in one case, an original optical path (solid red arrow) between the nodes A and B is recovered by an alternative path (dashed blue arrow) on account of the CDC–ROADM function. In another case for the forced change of the path due to a rapid traffic accident, the new path (dashed blue arrow) between the nodes C and E is promptly provided on account of the CDC–ROADM function. In the optical switch system in the proposed CDC–ROADM, as shown in Fig. 4.24b, multi-degree transmission lines are connected to the TPAs through wavelength cross-connect parts. Figure 4.24c shows a unit of TPAs with compact 8×8 multi-cast switches. Transmission lines are divided through an external splitter into the TPA switch unit. Flexibility in the photonic network is improved by the compact TPA that can extend port counts and add switch modules and external splitters on the transponder side. Figure 4.24d shows a schematic configuration of the detailed 8×8 multi-cast switch (split-and-select) with eight splitters and selectors, each with 1×8 ports, connected in a full mesh way. As shown schematically in Fig. 4.24e, the proposed optical switch is capable of strictly non-blocking and multi-cast connection, thus minimizing excess optical switching loss and polarization-dependent loss (PDL) as well as a maximizing speed of switching time (e.g., ~ 1 ms). These performances were certified by implementing an integrated Si photonic switch with 152 TO (thermo-optic) Mach–Zehnder (MZ) elements, as shown later [40].

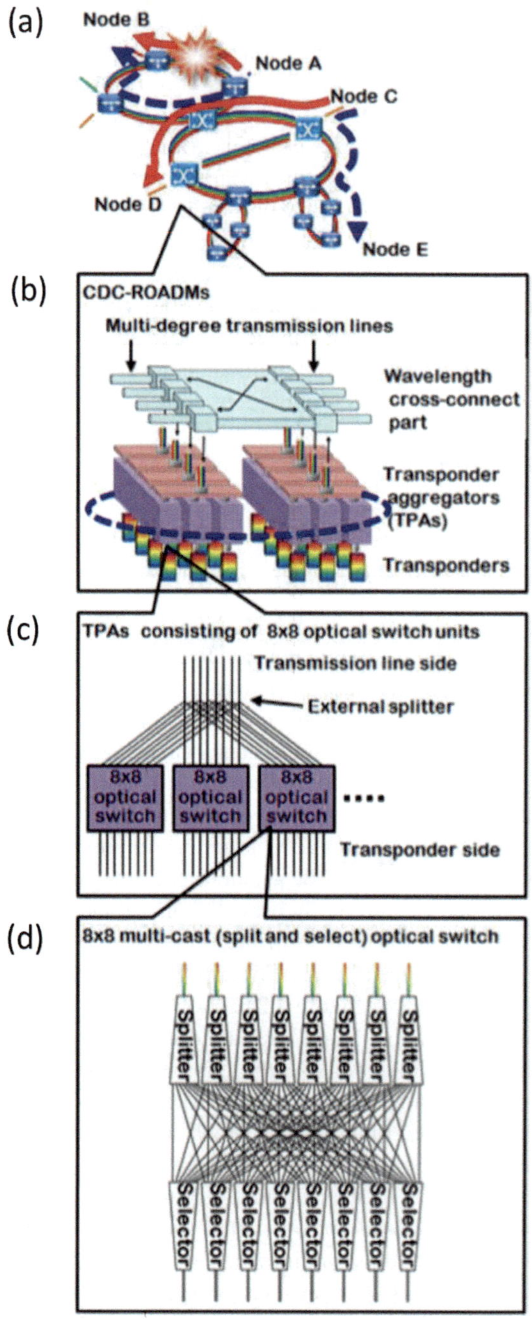

Fig. 4.24 **a** Optical paths to be reconfigured in emergency in CDC–ROADM system. **b** CDC–ROADM with multi-degree transmission lines connected to TPAs. **c** TPA with multiple optical switch modules. **d** 8×8 multi-cast optical switch with eight splitters and selectors, each with 1×8 ports. Reprinted with permission from Ref. [40], © 2016 IEEE

Polarization-Free Low-Loss and Tens-μsec Performance

Figure 4.25 shows a Si photonic switch element with MZI-based Si rib waveguides on a SOI wafer and transmission characteristics [40]. In general, a typical Si photonic waveguide is constructed with a nano-scale thickness (e.g., 220 nm) and width (e.g., 440 nm). In contrast, the multi-cast optical switch in this work focuses on the polarization-independent switch as well as low propagation loss. To this end, the thickness of the Si rib waveguide is designed as much thicker than 220 nm, i.e., 1.5 μm, as shown in Fig. 4.25a. In addition, for reduction of coupling loss between an external optical fiber and Si rib waveguide, facets of the rib waveguide are newly designed with spot size converters (SSCs), where SSCs are composed of two-layer Si tapered structures stacked on the rib waveguide. As a result, optical loss is less than 2 dB for both TE and TM polarizations. Moreover, the Si MZI switch with a TO-based phase shifter (100-μm length) is sandwiched by two MMI couplers, as shown in Fig. 4.25b. Figure 4.25c shows optical transmission spectra of the MZI switch for TE and TM modes at output ports 1 and 2. Switching contrast of about 25 dB, showing polarization insensitivity, is confirmed with a heater power of ~ 40 mW for a π-phase shift. The on–off switching time is 15 μs. These performances are sufficient for failure recovery applications, as mentioned in Fig. 4.25a. Figure 4.25d shows a configuration of an 8 × 1 selector in the 8 × 8 multi-cast optical switch based on the fundamental split-and-select topology, as shown in Fig. 4.24e. Note that, other than a selector part with seven 1 × 1 MZI switches (C1 to E1) as key elements in Fig. 4.24e, a gate part with twelve 1 × 1 MZI switches (A1 to B4) is added in the selector that results in 19 MZI switches in total. This configuration provides a more robust light-blocking system than the simple topology in Fig. 4.24e. That is, eight optical paths in the 8 × 1 selector include one optical path and other seven blocking paths. For instance, A1 and B1 are off when the light propagating them is to be blocked, while A2 and C1 are off when the light propagating them is to be blocked. Here, the on–off switch is operated with TO-based temperature control in the MZI switch element. This light-blocking system provides a switch with a high on–off contrast independent of surrounding temperature, as shown in Fig. 4.25e. Resultantly, the 8 × 8 multi-cast switch is constructed by eight sets of the 1 × 8 selector with total of 152 TO-based MZI switch elements.

Performance of Si Optical Switch Module

The Si multi-cast optical switch mentioned above has been integrated into an optical switch module with a chip size of 12 mm × 14 mm, as shown in Fig. 4.26a [40]. The module includes optical switch circuits, input and output SSCs, and electrical pads as well as TO-based MZI switch elements. An array of 16 single-mode fibers (SMFs) is connected to an end of a sub-mount on which the chip was wire-bonded. Transmission spectra for on- and off-states of an optical route are measured, as shown in Fig. 4.26b. Lines in two colors for two orthogonal polarizations are significantly similar to each other, indicating that the optical switching module is polarization-insensitive. Measured optical loss for total 64 paths at 1570 nm, as shown in Fig. 4.26c, results in an average optical loss of 15 dB with excess loss of 6 dB in module and intrinsic

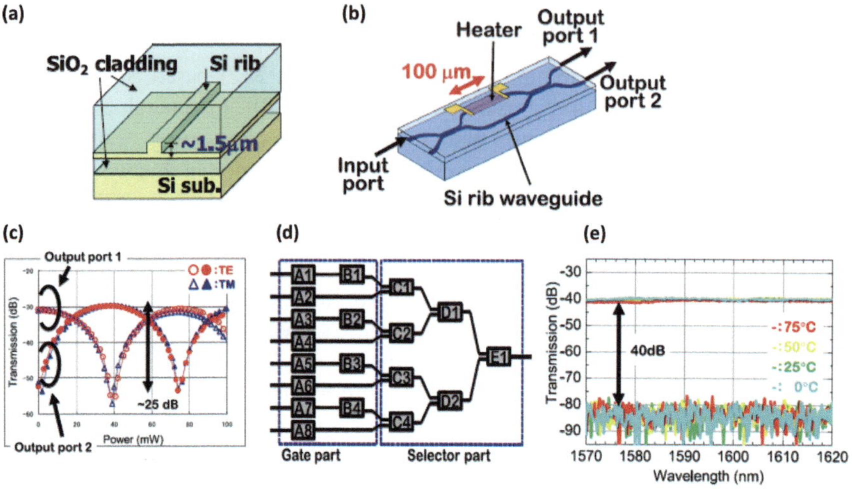

Fig. 4.25 **a** Schematic of polarization-independent Si rib waveguide. **b** Schematic of TO-based MZI switch. **c** Transmission versus power of the TO MZI switch. **d** Schematic of the 8 × 1 selector involving gate and selector parts with twelve and seven switch elements, respectively. **e** High on–off contrast by virtue of the temperature-insensitive 1 × 8 selector. Reprinted with permission from Ref. [40], © 2016 IEEE

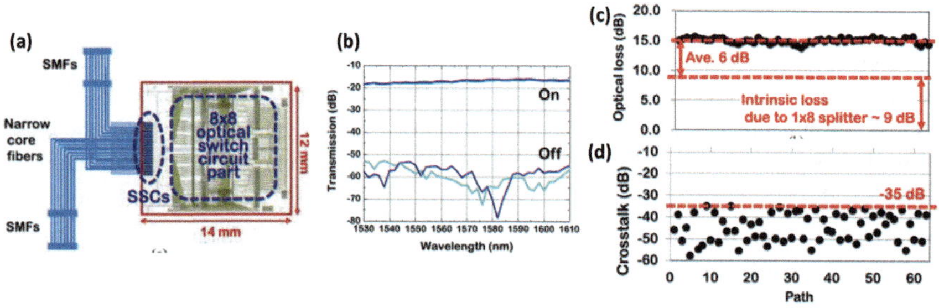

Fig. 4.26 **a** Schematic of a Si chip with an 8 × 8 optical switch circuit and optical fiber array. **b** Transmission spectra with more than 35 dB gap between on and off levels. **c** and **d** Optical loss and crosstalk for all 64 paths, respectively. Reprinted with permission from Ref. [40], © 2016 IEEE

loss of 9 dB due to the 1 × 8 splitters. Figure 4.26d shows a crosstalk of less than −35 dB.

Demonstration in TPA Sub-System

A TPA sub-system including the aforementioned Si chip modules with 8 × 8 optical switch circuits and optical fiber arrays was demonstrated, as shown in Fig. 4.27 [40]. Figure 4.27a shows a schematic of drop-side and add-side TPAs, each with twelve 8 × 8 optical switch modules. Each TPA on add and drop sides has 8 ports on the transmission line side and 48 ports on the transponder side, resulting in the

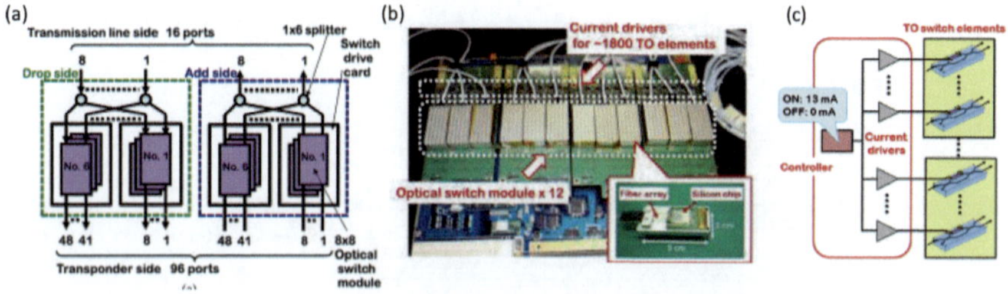

Fig. 4.27 **a** and **b** Schematic and photograph of the TPA with twelve 8 × 8 optical switch modules, respectively. In **b**, current drivers for 1800 TO elements are also indicated. **c** Control method for operating TO-based MZI elements. Reprinted with permission from Ref. [40], © 2016 IEEE

densely packed whole sub-system with 112 ports on the board. Three of twelve 8 × 8 optical switch modules are mounted each on an 8 cm × 18.5 cm switch drive card together with current drivers for ~ 1800 TO-based MZI switch elements. Figure 4.27b shows a photograph of the TPA sub-system. In the lower panel, the Si chip with a 2 cm × 5 cm package connected to a fiber array is shown to be embedded in the optical switch module. For operating many TO-based MZI switch elements, a simple control method easy to practically increase port count is used, as shown in Fig. 4.27c: A specific scheme to set the optical path resulted in an average optical loss of 25.1 dB for 48 optical paths, including a splitting loss of ~ 7.8 dB at the 1 × 6 optical splitter. The average polarization-dependent loss (PDL) was measured as 0.32 dB. In addition, a measured switching time in optical path change in the TPA sub-system mentioned above resulted in 0.15 ms.

The article [40] summarizes that measured values including the switching time of 0.15 ms using the TPA sub-system are promising for future faster ROADM networks.

4.4.3 Densely Integrated Silicon Photonic Switch

TPA Architecture

Another type of CDC–ROADM architecture was proposed in the FP7 IRIS project aiming at a new TPA with CMOS-integrated dense Si photonic devices as well as electronics for optical switching with low cost, low power consumption, high transmission, and mass-productivity [41]. Figure 4.28 shows an architecture of the TPA for dropping and adding WDM channel functions, each achieved simply by inverting the direction of the optical paths. First, dropping WDM channels are considered by four sets of 12 WDM channels with maximum 200 GHz spacing. Input channels coupled to single polarization grating couplers (SPGCs) are separated into odd and even channels by INTs (inter-leaver blocks). Thus, the wavelength spacing is increased and requirement for channel isolation is mitigated. Then, two types of MUXs (de-mux blocks) de-multiplex odd and even separated channels. Output signals are guided

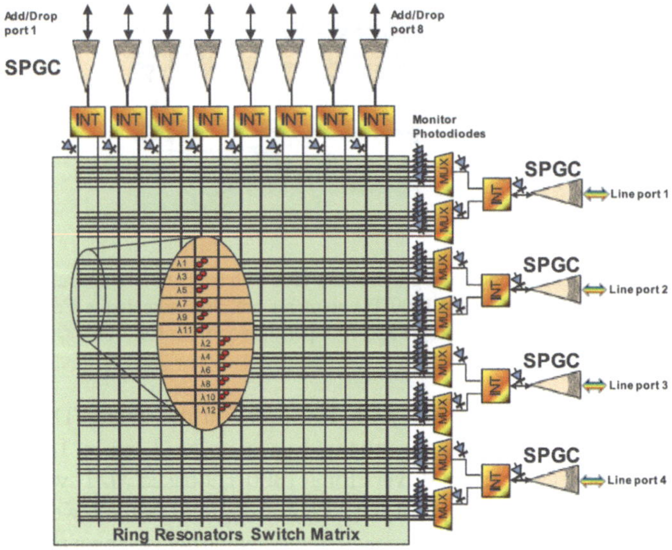

Fig. 4.28 Architecture of TPA for dropping and adding WDM channel functions proposed by the IRIS project. Reprinted with permission from Ref. [41], © 2016 IEEE

to a switching matrix with 6 × 8 rows and 2 × 8 columns as an optical crossbar including micro-ring resonators (MRRs). De-multiplexed signal goes on a row until a cross with a drop column. There, two MRRs transfer signals to corresponding drop columns by thermal tuning in resonance with the signal wavelengths, where two waveguides are connected to drop ports separated by even and odd wavelengths. The INT recombines odd and even wavelengths, guiding to the drop port, where the SPGC leads to an optical fiber. Next, adding WDM channels are considered. INTs at adding ports guide odd and even wavelengths to the right columns. Afterward, optical flows are achieved by inverting their directions to the output line ports against the flows in the add switch. Photodetectors for monitoring different blocks are set at input and output locations.

Design of Photonic Integrated Circuit and Devices

Design rules and simulated characteristics of the optical key devices in the aforementioned photonic switch were established, as follows [41]: A waveguide is arranged for a 220 nm × 480 nm cross-section, single mode at 1.55 μm and TE polarization. A grating coupler uses a SPGC with a depth of 70 nm and 33 trenches for a fiber angle of 8° at 1.55 μm wavelength. An inter-leaver consists of a MZI with a race track resonator in one branch. After channels are interleaved, arrayed waveguide gratings (AWGs) are designed with channel separation of 400 GHz for signals to be de-multiplexed to switch every channel independently. Single and double MRRs with different shapes/sizes were fabricated. Single micro-ring is a circular ring, while double micro-rings are composed of two race track resonators having straight couplers with lengths of

2.2 μm. The article [41] notes that design of the integrated reconfigurable Si photonics MRR switch has been achieved: The next stage is fabrication and demonstration of photonic and electronic integrated circuits at a system level. Further breakthroughs are also desired for scaling the integration level, managing the polarizations, and reducing crosstalk and losses.

4.4.4 Silicon Photonic Switch for Data Center

The amount of data in the contemporary data center is vastly increasing on account of tremendously increased various cloud services. Along with that, network switches connecting a variety of servers in the data center are forced to highly enhance speeds of data processing with sufficiently suppressed consumption power. To this end, highly integrated optical switches with non-blocking and high port count as well as low cost and low power consumption are strongly expected in high-performance data centers. Selection of Si materials enables Si photonic platform-based highly integrated optical switches by using a standard CMOS-compatible fabrication process. Main technical concerns are choices of thermo-optic (TO) and electro-optic (EO) switching effect and MZIs and MRRs in device structure. In this section, a recently demonstrated multi-port MZI Si photonic switch based on the TO effect is introduced first [42]. Then, characteristics in TO effect versus EO effect and MZI versus MRR based on recently demonstrated examples are discussed later.

Multi-port Si Photonic Switch

Recently, an ultra-compact Si photonic switch mounted on a land grid array (LGA) interposer with a strictly non-blocking architecture and record-high 32 × 32 port count at that moment was demonstrated, as shown in Fig. 4.29 [42]. Figure 4.29a shows a microscope photograph of the 32 × 32 switch chip with a PILOSS topology. The chip involves integrated 1024 MZ switches with 2 × 2 input/output ports, resulting in a chip size of 11 mm × 25 mm. Figure 4.29b shows an enlarged microscope photograph of photonic/electronic elements such as MZ switch, 3-dB DC (directional coupler), waveguide intersection, and TiN heater for TO phase shifter. Si waveguides with a width of 430 nm and height of 220 nm are formed in a 1.5-μm-thick SiO_2 cladding layer. The whole chip is manufactured by a 45-nm CMOS-compatible process on a 300-mm wafer. Figure 4.29c shows a schematic of selected six paths; 1–12, 1–20, 1–32, 32–1, 32–13, and 32–19, for measurement of transmission characteristics of the switch. Owing to adoption of the PILOSS topology, any path goes through 32 MZ switches and 31 intersections. For operating the whole path, 96 MZ switches function in total. Figure 4.29d shows measured on-chip loss for six optical paths at 1545 nm, resulting in 15.8 dB except for fiber-to-chip coupling loss. Access waveguide loss for the input and output was 2.1 dB, while Si waveguide loss in the switch part was 5.5 dB and other losses such as bending loss and intersection loss were 6.1 dB. In addition, the crosstalk and total power consumption were estimated as – 20 dB and 2.93 W, respectively. Optical switching with 43-Gb/s

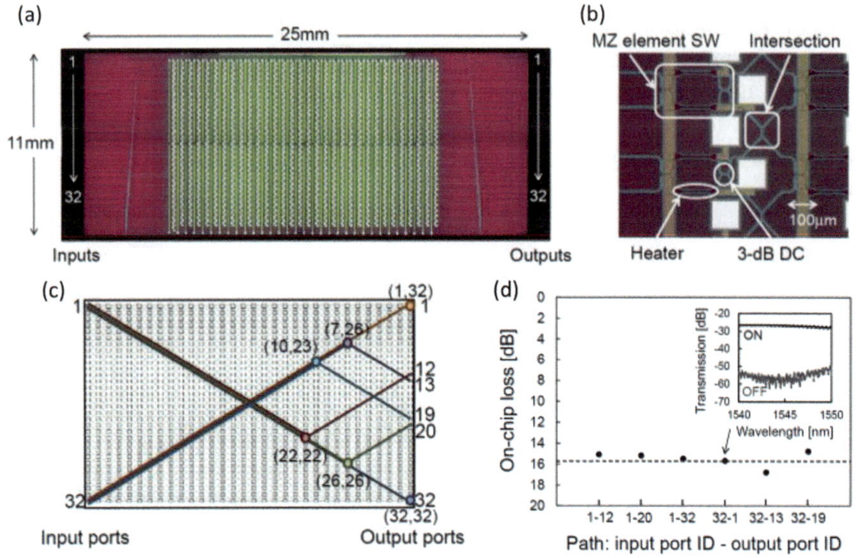

Fig. 4.29 **a** Microscope photograph of a 32 × 32 PILOSS switch chip. **b** Its magnified image showing key elements. **c** Path on the switch matrix including six optical paths as an example. **d** On-chip loss for the six sampled optical paths. Inset shows a transmission spectral of the path 32-1. Reprinted with permission from Ref. [42], © 2015 OSA

digital-coherent QPSK signals was also demonstrated successfully. Refer to the last part of this section for further advancing Si photonic PILOSS switch.

TO Versus EO and MZI Versus MRR

Typical semiconductor optical switches mostly depend on switching mechanisms of either a TO effect or EO effect (Pockels effect). A Si photonic switch usually uses the TO effect because of its large coefficient (1.8×10^{-4} K^{-1}) in spite that the switching speed is limited to μs-scale. If the higher switching speed on the ns-scale is needed, the EO effect is preferable in general. Unfortunately, Si photonic switch, having no linear EO effect, widely uses an EO effect based on plasma dispersion at a carrier-injection-based PIN junction. Nevertheless, when such an EO effect is used in one arm of the MZI configuration, a particular attention is needed to avoid power imbalance between two parallel arms in the MZI induced by an undesirable electro-absorption, thus resulting in degradation of the crosstalk. Indeed, a newly developed 16 × 16 Si MZI switch with an EO-based phase shift in one arm exhibited imbalances in loss and crosstalk between cross and bar states caused by free-carrier absorption [43]. One of the solutions to such imbalance is to adopt a new type of 32 × 32 Si photonic switch with a Beneš topology [44]. In the improved MZI switch, an additional optical phase bias of $\pi/2$ was intentionally given to the phase shift arm, thus successfully inducing the phase change in two arms and balancing the bar and cross-states. Other than introduction of such an optical phase bias, another developed method is a 2 × 2 nested MZI structure [45], where a variable optical attenuator was used for power balancing between two arms, thus performing sufficiently low crosstalk as low as – 35 and ~ 2 dB insertion loss. Yet, suppression of insertion loss as well as crosstalk is still challenging to manage for large-scale optical switches. In this regard, introduction of a built-in amplifier such as semiconductor optical amplifier (SOA) is a reasonable solution to add on-chip gain in a larger Si EO switch.

Multi-port MRR Photonic Switch

A micro-ring resonator (MRR), surveyed in Sect. 1.4.3 for basic characteristics and introduced in Sect. 4.4.3 for a Si photonic switch for the CDC–ROADM architecture [41], was first demonstrated for photonic switching in the data center by the group of Cornell University [46] and joint groups in Columbia and Cornell [47]. There the stimulated work such as hitless router was performed. For constructing wavelength-selective MRR, various methods including on-chip wavelength locking with high speed were demonstrated [48]. A typical crossbar-type switch matrix with the largest port count is a thermally tuned 8 × 7 Si MRR with series-coupled five rings [49]. Characteristics of a single MRR switch element were 100-GHz passband, 350-GHz FSR, 0.9-dB loss for off-state ring, and 2.0-dB loss for on-state ring. Yet, path-dependent insertion loss of 14–22 dB is a challenge of this kind of switch fabric. A recent topic includes MRR-based spatial multiplexers/de-multiplexers with a 1 × *N*/*N* × 1-assembled switch-and-select (S&S) module [50]. Although the 8 × 8 MRR switch showed excellent results, necessity of waveguide crossings at the complicated central area is desired to avoid as the port count increases.

Ever-Advancing Multi-port Si Photonic Switch

Recently, Si photonics-based optical path switches for telecom and data-com networks have been far advanced in sophisticated device structures, large chip integration scales, and polarization-insensitive signal transmission performances: A loss-less 4 × 4 Si photonic switch with flip-chip bonded 4-channel SOAs was demonstrated and succeeded in transmission of 800 Gb/s WDM signals through all the four paths of the switch [51]. Meanwhile, an improved 32 × 32 Si photonic PILOSS switch mentioned above has realized fiber-to-fiber insertion loss of 10.8 dB and on-chip power consumption of 1.9 W by employing an extremely high-Δ (refractive-index ratio) PLC (planar light wave circuit) fiber connector [52]. Such a multi-port Si photonic PILOSS switch was furthermore highly improved: A non-duplicate polarization diversity 32 × 32 Si photonic switch has been developed on a SiN/Si double-layer platform [53]. A Si photonic lower layer involved MZ matrix switches, while a SiN upper layer involved overpass access waveguides, thus, reducing the number of intersections between switching and access waveguides. A polarization-insensitive structure was composed of an ultra-compact polarization splitter-rotator (PSR) with an adiabatic taper and an asymmetric directional coupler, located at an input MZ switch matrix port [54]. The Si matrix switch chip was fabricated with a 300-mm SOI CMOS processing pilot line, comprising more than 7,600 components, then flip-chip bonded to a packaging ceramic interposer. Measured polarization-insensitive characteristics resulted in an average on-chip loss of ~ 35 dB and polarization-dependent loss (PDL) of 3.2 dB. These devices are even now under intensive development with structure parameters optimized toward large-scale integration and furthermore scalability of optical switch capacity under low energy consumption [53].

4.5 Summary

Si photonics is now an inevitable device concept in the photonic network as key nano-photonic integrated devices, that is, optical switches and transceivers available by SOI CMOS-compatible process technologies. Their features of small size, low cost, high power efficiency, and scalability are reinforcing application fields to new areas of smart sensing (LiDAR, etc.) and neural network areas. In the early 2000s, Si photonics applications were pioneered by two innovative technologies; fabrication of nano-scale smooth Si photonic waveguides with small sidewall surface roughness by IBM, and realization of an integrated optical grating coupler connected to Si waveguides by Luxtera, both using a standard SOI CMOS process. Afterward, passive/active Si photonic applications were widely deployed: a heterogeneously integrated narrow-linewidth III-V/Si laser, chip-/wafer-bonded and hetero-epitaxial-grown hybrid lasers on the Ge/Si substrate, a new concept of fiber-to-chip grating couplers with sub-wavelength-pitch periodic passive nano-structures, etc. In the last decade, innovatively advanced traffic in the data-com market extremely enforced significant advancement of Si photonic interconnect transceivers: A "zero-change" Si photonic integrated transceiver was monolithically fabricated by the commercial 45-nm front-end SOI CMOS process. Micro-ring modulators (MRMs) in an existing sub-100-nm-thick crystalline Si layer enabled transmission of 40 Gb/s MRM NRZ signals with successful open eye diagrams. A fingertip-size Si photonic transceiver "optical I/O core" mounted on the FPGA was demonstrated for data centers. Miniaturized hybrid integration of a laser diode and optical pins for chip-fiber coupling enabled excellent transmission characteristics of 25 Gb/s $\times$ 12 channels, thus 300 Gb/s in total. In addition, a Si photonic transceiver with a polarization-independent bandpass filter and AWG with MMI waveguides was also developed for 5G mobile networks. Meanwhile, a variety of multi-port Si photonic switches for core networks and data centers are progressing continuously: For core networks, a multi-cast Si photonic TPA switch with polarization-independent Si rib waveguides and 152 TO MZ modulators was developed for CDC–ROADM networks. ~ 10 μs switching with an 8 $\times$ 48 TPA sub-system was successfully demonstrated for practical application. A new type of the TPA using a switching matrix with a MRR-based optical crossbar topology was proposed also. For data centers, an integrated 32 $\times$ 32 dynamic Si photonic path switch with a non-blocking PILOSS architecture was demonstrated. 2 $\times$ 2 MZ switches with total 1024 elements enabled transmission with 43-Gb/s digital-coherent QPSK signals. For broader bandwidth and faster switching speed, further investigations on TO versus EO materials/mechanisms and MZI versus MMR device configurations are enthusiastically continuing. Si photonics applications will continue to progress toward new next-generation systems together with innovating state-of-the-art integration of electronic and photonic technologies.

References

1. Murata K, Saida T (2011) Optical device technologies for future network evolution. NTT Tech Rev 9(3):1–6
2. "What is the All-Photonics Network?", Special features "IOWN", NTT R&D Website. https://www.rd.ntt/e/iown/0002.html#:~:text=The%20All%2DPhotonics%20Network%20introduces,with%20current%20electronics%20based%20technology
3. Jalali B, Fathpour S (2006) Silicon photonics. J Light Wave Technol 24:4600–4615
4. Fang Z, Zhao CZ (2012) Recent Progress in silicon photonics: a review. ISRN Opt 2012:1–27
5. Bowers JE, Komljenovic T, Davenport M, Hulme J, Liu AY et al (2017) Recent advances in silicon photonic integrated circuits. Proc SPIE 9774(977402):1–18
6. Pavesi L (2005) Will silicon be the photonics material of the third millennium? J Phys Condens Matter 15:1169–1196
7. Vlasov YA, McNab SJ (2004) Losses in single-mode silicon-on-insulator strip waveguides and bends. Opt Express 12:1622–1631
8. Wood M, Sun P, Reano RM (2012) Compact cantilever couplers for low-loss fiber coupling to silicon photonic integrated circuits. Opt Express 20:164–172
9. Cheben P, Xu D-X, Janz S, Densmore A (2006) Subwavelength waveguide grating for mode conversion and light coupling in integrated optics. Opt Express 14:4695–4702
10. Lalanne P, Hugonin J-P (1998) High-order effective-medium theory of subwavelength gratings in classical mounting: application to volume holograms. J Opt Soc Am A 15:1843–1851
11. Xu X, Subbaraman H, Covey J, Kwong D, Hosseini A, Chen RT (2012) Complementary metal–oxide–semiconductor compatible high efficiency subwavelength grating couplers for silicon integrated photonics. Appl Phys Lett 101:031109
12. Subbaraman H, Xu X, Hosseini A, Zhang X, Zhang Y, Kwong D, Chen RT (2015) Recent advances in silicon-based passive and active optical interconnects. Opt Express 23:2487–2511
13. Reed GT, Mashanovich G, Gardes FY, Thomson DJ (2010) Silicon optical modulators. Nat Photonics 4:518–526
14. Ferrara J, Yang W, Zhu L, Qiao P, Chang-Hasnain CJ (2015) Heterogeneously integrated long-wavelength VCSEL using silicon high contrast grating on an SOI substrate. Opt Express 23:2512–2523
15. Tanoto H, Yoon SF, Lew KL, Loke WK, Dohrman C et al (2009) Electroluminescence and structural characteristics of $InAs/In_{0.1}Ga_{0.9}As$ quantum dots grown on graded $Si_{1-x}Ge_x/Si$ substrate. Appl Phys Lett 95:141905
16. Liu H, Wang T, Jiang Q, Hogg R, Tutu F et al (2011) Long-wavelength InAs/GaAs quantum dot laser diode monolithically grown on Ge substrate. Nat Photonics 5:416–419
17. Komljenovic T, Srinivasan S, Norberg E, Davenport M, Fish G, Bowers JE (2015) Widely tunable narrow-linewidth monolithically integrated external-cavity semiconductor lasers. IEEE J Sel Top Quantum Electron 21:1501909
18. Silicon Photonics 2023 report, Yole Group. www.yolegroup.com
19. https://www.eetimes.com/globalfoundries-cuts-5-of-workforce/#. Source: EE Times and original source by IC Insights
20. Thomson D, Zilkie A, Bowers JE, Komljenovic T, Reed GT et al (2016) Roadmap on silicon photonics. J Opt 18. 073003
21. Silicon Photonics 2024 Report, Yole Group. www.yolegroup.com
22. Gunn G (2006) CMOS photonics for high-speed interconnects. IEEE Microelectron 26:58–66
23. Dhiman A (2013) Silicon photonics: a review. IOSR J Appl Phys 3:67–79
24. http://www.intel.com/pressroom/archive/releases/2010/20100727comp_sm.htm

25. Patterson D, De Sousa I, Achard L-M (2017) The future of packaging with silicon photonics. Chip Scale Rev
26. Lee YHD, Lipson M (2013) Back-end deposited silicon photonics for monolithic integration on CMOS. IEEE J Sel Top Quant Electron 19:8200207
27. V. Stojanović, Ram RJ, Popović M, Lin S, Moazeni S et al (2018) Monolithic silicon-photonic platforms in state-of-the-art CMOS SOI processes [Invited]. Opt Express 26:13106–13121
28. Sun C, Wade M, Georgas M, Lin S, Alloatti L et al (2016) A 45 nm CMOS-SOI Monolithic photonics platform with bit-statistics-based resonant micro-ring thermal tuning. IEEE J Solid-State Circuits 51:893–907
29. Moazeni S, Lin S, Wade M, Alloatti L, Ram R et al (2017) A 40Gb/s PAM-4 Transmitter based on a ring-resonator optical DAC in 45nm SOI CMOS. IEEE J Solid-State Circ 52:3503–3516
30. Nakamura T, Yashiki K, Mizutani K, Nedachi T, Fujikata J et al (2019) Fingertip-size optical module, Optical I/O Core, and its application in FPGA. IEICE Trans Electron E102-C(4):333–339
31. http://www.aiocore.com/
32. Yaegashi H (2017) Development of ultra-compact optical transceivers for IoT network utilizing silicon photonics technology. OKI Tech Rev 84(1):1–4 (Issue 229)
33. Okayama H, Onawa Y, Shimura D, Takahashi H, Yaegashi H, Sasaki H (2016) Low loss 100 GHz spacing Si arrayed waveguide grating using minimal terrace at slab–array interface. Electron Lett 52(18):1545–1546
34. Basch EB, Egorov R, Gringeri S, Elby S (2006) Architectural tradeoffs for reconfigurable dense wavelength-division multiplexing systems. IEEE J Sel Top Quant Electron 12:615–626
35. Sakamaki Y, Kawai T, Fukutoku M (2014) Next-generation optical switch technologies for realizing ROADM with more flexible functions. NTT Tech Rev 12:1–5
36. Gringeri S, Basch B, Shukla V, Egorov R, Xia TJ (2010) Flexible architectures for optical transport nodes and networks. IEEE Commun Mag 48:40–50
37. Cheng Q, Bahadori M, Glick M, Rumley S, Bergman K (2018) Recent advances in optical technologies for data centers: a review. Optica 5:1354–1370
38. Hinton HS (2013) An introduction to photonic switching fabrics. Springer Science & Business Media, Berlin
39. Yanagimachi S, Nakamura S, Takeshita H, Tajima A, Kato T, Hino T, Fukuchi K (2014) 8×48 transponder aggregator subsystem using silicon switch modules for flexible photonic network. In: Asia communications and photonics conference 2014, OSA Technical Digest (online), AF4B.3 (2014). https://doi.org/10.1364/ACPC.2014.AF4B.3
40. Nakamura S, Yanagimachi S, Takeshita H, Tajima A, Hino T, Fukuchi K (2016) Optical switches based on silicon photonics for ROADM application. IEEE J Sel Top Quantum Electron 22:3600609
41. Testa F, Oton CJ, Kopp C, Lee J-M, Ortuño R, Enne R, Tondini S, Chiaretti G, Bianchi A, Pintus P, Kim M-S, Fowler D, Ayúcar JA, Hofbauer M, Mancinelli M, Fournier M, Preve GB, Zecevic N, Manganelli CL, Castellan C, Parès G, Lemonnier O, Gambini F, Labeye P, Romagnoli M, Pavesi L, Zimmermann H, Pasquale FD, Stracca S (2016) Design and implementation of an integrated reconfigurable silicon photonics switch matrix in IRIS Project. IEEE J Sel Top Quant Electron 22:3600314
42. Tanizawa K, Suzuki K, Toyama M, Ohtsuka M, Yokoyama N et al (2015) Ultra-compact 32 × 32 strictly non-blocking Si-wire optical switch with fan-out LGA interposer. Opt Express 23:17599–17606
43. Lu L, Zhao S, Zhou L, Li D, Li Z et al (2016) 16 × 16 non-blocking silicon optical switch based on electro-optic Mach-Zehnder interferometers. Opt Express 24:9295–9307
44. Qiao L, Tang W, Chu T (2017) 32 × 32 silicon electro-optic switch with built-in monitors and balanced-status units. Sci Rep 7:42306
45. Dupuis N, Rylyakov AV, Schow CL, Kuchta DM, Baks CW et al (2016) Ultralow crosstalk nanosecond-scale nested 2 × 2 Mach-Zehnder silicon photonic switch. Opt Lett 41:3002–3005
46. Xu Q, Schmidt B, Pradhan S, Lipson M (2005) Micrometre-scale silicon electro-optic modulator. Nature 435(7040):325–327

47. Sherwood-Droz N, Wang H, Chen L, Lee BG, Biberman A, Lipson M (2008) Optical 4x4 hitless silicon router for optical networks-on-chip (NoC). Opt Express 16:15915–15922
48. Khope ASP, Hirokawa T, Netherton AM, Saeidi M, Xia Y et al (2017) On-chip wavelength locking for photonic switches. Opt Lett 42:4934–4937
49. Das Mahapatra P, Stabile R, Rohit A, Williams KA (2014) Optical Crosspoint matrix using broadband resonant switches. IEEE J Sel Top Quant Electron 20(4):1–10
50. Nikolova D, Calhoun DM, Liu Y, Rumley S, Novack A et al (2017) Modular architecture for fully non-blocking silicon photonic switch fabric. Microsyst Nano-eng 3:16071
51. Konoike R, Suzuki K, Inoue T, Matsumoto T, Kurahashi T, Ayahito U, Takabayashi K, Akiyama S, Sekiguchi S, Namiki S, Kawashima H, Ikeda K (2019) SOA-integrated silicon photonics switch and its lossless multistage transmission of high-capacity WDM signals. J Lightwave Technol 37:123–130
52. Suzuki K, Konoike R, Hasegawa J, Suda S, Matsuura H, Ikeda K, Namiki S, Kawashima H (2019) Low-insertion-loss and power-efficient 32 × 32 silicon photonics switch with extremely high-Δ silica PLC connector. J Lightwave Technol 37:116–122
53. Suzuki K, Konoike R, Yokoyama N, Seki M, Ohtsuka M, Saitoh S, Suda S, Matsuura H, Yamada K, Namiki S, Kawashima H, Ikeda K (2020) Nonduplicate polarization diversity 32 × 32 silicon photonics switch based on a SiN/Si double-layer platform. J Lightwave Technol 38:226–232
54. Dai D, Bowers JE (2011) Novel concept for ultra-compact polarization splitter-rotator based on silicon nanowires. Opt Express 19(11):10940–10949
55. Asakawa K, Sugimoto Y, Nakamura S (2020) Silicon photonics for telecom and data-com applications. Opto-Electron Adv 3:200011-1–200011-26

Chapter 5
Applications of Silicon Photonic Waveguides (II) LiDAR and Optical Neural Network

Abstract Recent progresses of Si photonic applications in new areas are introduced here. One is integrated solid state Si LiDAR (light detection and ranging) with optical beam steering arrays for autonomous cars, free-space communication links, etc. After surveying principles of key optical phased array techniques and frequency-modulated continuous wave (FMCW) operation, various FMCW–LiDAR systems developed to date are surveyed together with reviewing several beam steering elements such as high-performance optical phased arrays, focal plane arrays, and slow-light gratings. The other progress is optical neural network (ONN) using an integrated Si photonic chip for deep learning of image recognition and 1-D to 3-D dataset classification. First, basic elements of the ONN with input-/hidden-/output-layers, matrix–vector multiplication, and nonlinear activation function are reviewed. Then, novel various ONN architectures such as real/complex-valued networks, nonlinear mapping-based photonic classifier, and space-efficient diffractive ONN for Fourier transform are reviewed. Latest experiments on inference and training tasks such as logic gate realization, nonlinear dataset classification, and handwriting recognition are also shown.

5.1 Introduction

In recent years, Si photonic waveguides originally developed for nano-photonic devices in the photonic network and data center are now widely extending to innovative fields such as LiDAR and free-space communications using steering and shaping light beams in the free space. Originally, LiDAR illuminates a static/moving target with coherent light, detects reflected light with a detector, and measures a distance to the target and its velocity like radar in a radio wave/microwave range. Since the reflected light includes signals of return time and wavelength depending on the target position, LiDAR is capable of digital 3-D depictions of the target at a higher resolution than in the radar system. LiDAR available today uses a mechanical system such as gyro (gimbal) for beam steering with high peak-power laser. Such a mechanical system has drawbacks such as limited scan rate, low reliability, and high system

K. Asakawa et al., *Nano-photonics for Advanced Networks*,
https://doi.org/10.1007/978-3-031-80637-7_5

cost, while an integrated solid-state Si LiDAR has advantages of small footprint, low cost, and high operation speed. Future-promising applications of Si LiDAR extend to remote sensing in terrestrial, airborne, and mobile areas as well as point-to-point free-space communication links.

Meanwhile, another topic on the new Si photonic application is Si ONN for deep learning. In the modern society, all the information in many equipment and systems in data centers, autonomous vehicles, various consumer devices, etc., are extremely digitally transformed and artificial intelligence (AI) processing systems are tremendously operated. Current AI processing systems with CMOS-based enormous electronic digital processers involve serious problems of high power consumption and large operation latency. Consequently, innovative AI accelerators with low power consumption, low operation latency, and high data-processing speed without digital processors are now under development. Artificial neural network (ANN) is one of the most promising computing network model based on signal processing in the brain, later improved dramatically by means of many tasks including machine learning. Nevertheless, the current computing hardware based on the electronic AI accelerator is based on von Neumann computing approaches, so not fit for implementation of neural networks. In contrast, a Si photonic ANN accelerator, i.e., Si ONN chip is attracting much attention on account of many excellent features of high speed, low power consumption, ultra-low latency, high I/O bandwidth, and highly parallel optical processing using all the light characteristics of wavelength, amplitude, phase, and polarization. In this chapter, current states and future prospects of the integrated solid state Si LiDAR and Si ONN are reviewed.

5.2 Si Photonic LiDAR for Remote Sensing

5.2.1 Integrated Optical Beam Steering Arrays

Optical Phased Array

An optical phased array (OPA) can steer an optical beam in free space with no assistance of mechanical motion, so is an essential element for LiDAR and free-space communication network. Structures of the OPA are quite similar to a phased array antenna in radio wave and microwave systems [1, 2]. Figure 5.1a is a schematic showing a concept of the OPA with a phased array and a grating array [3]. A laser beam is incident on the input waveguide and divided into several beams through MMI trees, reaching the phased and grating arrays. As shown in Fig. 5.1b, the grating array steers a light beam in the far-field in free space in the longitudinal (θ) direction by tuning a wavelength of the laser beam, while the phased grating steers the light beam in the lateral (ψ) direction, as shown in Fig. 5.1c, by inducing refractive-index change through thermal optics via applied voltage on the electrodes located near the phased array. The shape and width of the far-field beam are determined by design

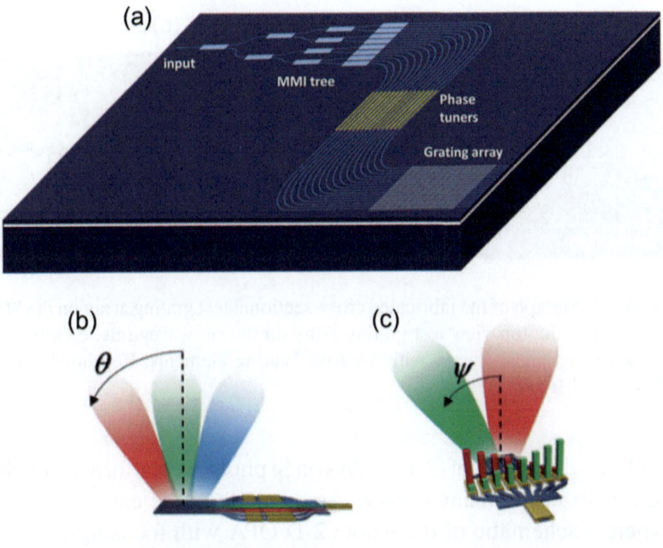

Fig. 5.1 **a** Schematic of elements in the optical phased array with a phased and grating arrays, radiating a laser beam to form far-fields. **b** Grating array for steering a laser beam in free space in the longitudinal (θ)) direction by tuning the laser beam. **c** Phase array for steering the laser beam in the lateral (ψ) direction by electrically inducing the refractive index through thermal optics via electrodes near the array. Reprinted with permission from Ref. [3]. © 2011 OSA

of these emitter arrays: A widely spaced emitter array induces a beam with side-lobes, while a densely spaced emitter array induces a broad beam. In addition, an array with increased number of emitters induces a sharp beam. Ref. [3] is a highly recommendable review article on the OPA technology including a wide viewpoint of photonic integrated circuits for optical beam steering.

Figure 5.2 shows photographs of the phased and grating array elements with a 16-channel device fabricated on the SOI wafer [3]. Figure 5.2a and b shows SEM images of the cross-sectional and top-view grating arrays on the rib waveguides, respectively, while Fig. 5.2c shows an optical microscope photograph of the phased array with resistive heating elements located near the phased array. By using the fabricated test device, the maximum θ-axis angle in the longitudinal direction and maximum ψ-axis angle in the lateral direction were measured as 39° and 33°, respectively. Refer to [3] for detail.

Integrated 2-D OPA

Initial work on the OPAs started with one-dimensional (1-D) arrays using indium phosphide [4] and Si photonic [5] technologies. In [5], a 12-channel edge emitting array realized over 32° beam steering using an unequally spaced thermo-optic phase tuning array. Due to design of non-uniform array spacing, unwanted side-lobes were suppressed.

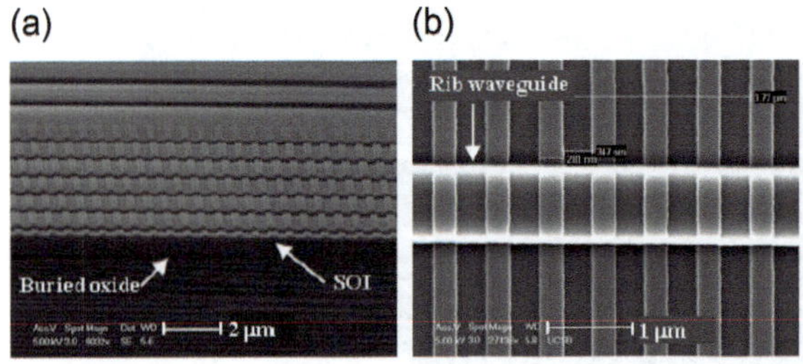

Fig. 5.2 **a** SEM photograph of the fabricated cross-sectional test grating array on the SOI substrate. **b** SEM photograph of the top-view test grating array on the rib waveguide. **c** Optical microscope image of the fabricated phased array with resistive heating elements. Reprinted with permission from Ref. [3]. © 2011 OSA

Integrated two-dimensional (2-D) OPAs on Si photonic platform using SOI CMOS process were realized by many people. An example in the early work is shown in Fig. 5.3, where a schematic of the whole 2-D OPA with focusing grating couplers fabricated on SOI and a top-view SEM image of the demonstrated 4 $\times$ 4 grating couplers are shown in the left and right panels, respectively [6]. Light impinging on an input waveguide is split via MMI (multi-mode interferometer) splitters into each row of an array of focusing grating couplers and radiated out-of-plane into air. The device was formed on a SOI substrate with a 220-nm thick Si top layer and a 2-μm thick buried oxide layer using standard CMOS processes. The waveguides including MMI splitters and bends were etched in 220-nm depth for loss reduction, while the grating couplers were etched in 70-nm depth for a high coupling efficiency, via two-step etching. The principle of the grating coupler is referred to in Sect. 1.4.6 and Fig. 4.4 in Sect. 4.2.1. The 2-D phased array antenna presented here was investigated to integrate Si photonic elements for free-space optical communication links, rather than LiDAR to be discussed later in detail. Beam steering in one direction was realized by wavelength tuning, resulting in a steering rate of 0.24°/nm and antenna loss less than 3 dB. Addition of phase tuners on the waveguide has been proposed to improve performances including the steering speed.

Hybrid-Integrated Silicon OPA

For the past decade, optical beam steering technologies for LiDAR were considerably advanced by introducing hybrid-integrated high-performance Si OPAs with individually tunable emitters [7]. The hybrid integration means integration between the Si photonics and III–V-based active devices like a laser, photodetector, and optical modulator. A typical example of the structure in hybrid-integrated Si photonics is shown later (Fig. 5.5): A pre-amplified tunable laser is divided into eight channels in this case. After amplified and phase-tuned separately, each channel is guided to an array of grating couplers. Propagating through the grating array, the laser beam is emitted out-of-plane at the emission angles in the θ direction determined by the wavelength and in the ψ direction by the relative phase. Demonstration of 2-D beam steering using the OPA with a Si 16-channel grating array resulted in the total field

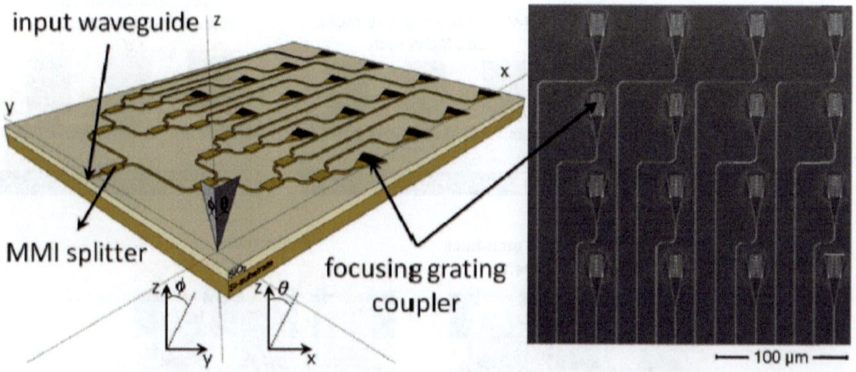

Fig. 5.3 Illustration of a 2-D OPA with focusing grating couplers on SOI, showing a whole schematic view (left) and a top-view SEM image of the 4 × 4 grating couplers (right). Light impinging on an input waveguide is divided through a MMI splitter into each row of an array of the focusing grating couplers and radiated out-of-plane into air. Reprinted with permission from Ref. [6]. © 2010 OSA

of view of 12° (ϕ), beam width of 1.8° × 0.6°, and full-window background peak suppression of 7 dB.

A large-scale 2-D Si OPAs with individually tunable emitters were developed by many groups. For instance, integration of 2-D OPAs with 64 × 64 vertical grating couplers as optical nano-antennas with a footprint of 576 µm × 576 µm was demonstrated by combining an off-chip laser source [8]. All the vertical grating couplers were designed uniformly and operated with precise power-balance, resulting in a non-Gaussian, or sinc-function pattern in the far-field. The non-Gaussian far-field pattern with undesired side-lobes was improved later to be a clean Gaussian pattern by apodizing a near-field of 8 × 8 array of emitters [9].

Among several implementations on optical beam steering with 2-D Si OPAs for LiDAR [10, 11], heterogeneous Si photonics sensing for LiDAR, based on the hybrid-integrated III–V/Si platform using 300 mm SOI CMOS facilities, is an extremely unique approach for practical applications [12]. The origin of the heterogeneous III–V/Si photonics was reported first in 2006 [13]. Figure 5.4a–e show a schematic process flow of a heterogeneously integrated III–V/Si array. Each process is as follows; (a) fabrication of SOI waveguide pattern, (b) bonding of III–V die, removal of III–V substrate, mesa etching of III–V and etching of multiple quantum-well (MQW), (c) SiO_2 passivation and n/p contact metallization, (d) benzo-cyclobutene (BCB) planarization, BCB etch-back and probe via open, (e) probe metallization.

The final panel (e) corresponds to a cross-sectional III–V/Si phase shifter located in one arm of the MZI modulator that acts as an important active OPA component: Most Si phase shifters use thermo-optic (TO) effects having both a merit of low optical loss in the Si waveguides and demerits of low bandwidth and thermal

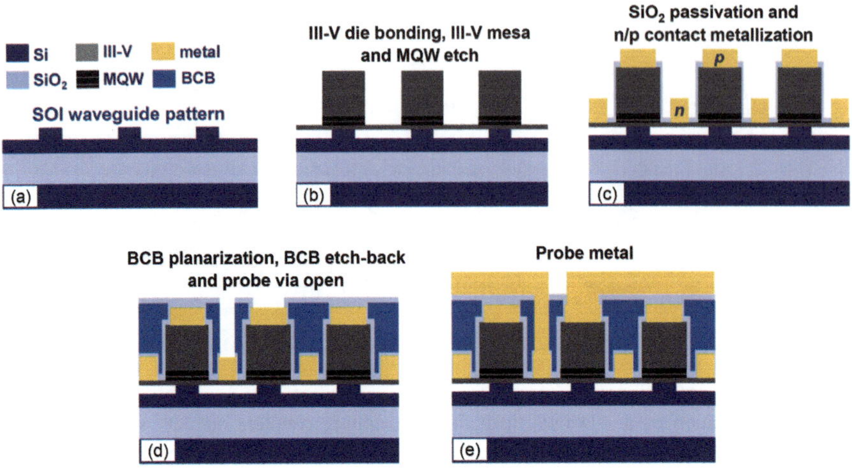

Fig. 5.4 Schematic of process flow showing an integrated array with III–V and Si heterogeneous materials. **a** SOI waveguide platform. **b** III–V die bonding, III–V mesa and MQW etch. **c** SiO_2 passivation and contact metallization. **d** BCB planarization/etch-back and prove via open. **e** Prove metal. Reprinted with permission from Ref. [12]. Panels (a) to (e) were extracted from original nine panels (a) to (i) in Fig. 1 in Ref. [12]. © 2019 OSA

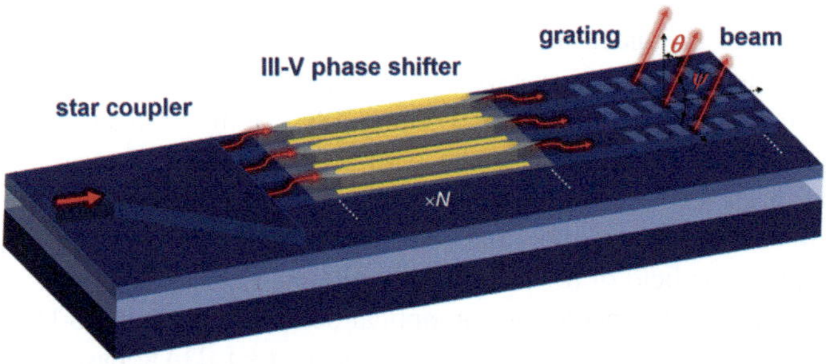

Fig. 5.5 Schematic of an OPA with a star coupler, III–V compound phase shifter, and grating array for beam steering. Reprinted with permission from Ref. [12]. © 2019 OSA

crosstalk. However, the hybrid-integrated III–V/Si OPA is capable of using a high-performance phase shifting effect, i.e., III–V-proper EO effects such as Franz–Keldysh effect and band filling effect. A gain region of an electrically tunable III–V laser is heterogeneously die-bonded (process (b)) prior to SiO_2 passivation and n/p contact metallization (process (c)).

Figure 5.5 shows a schematic of a III–V/Si OPA system [12]. The system is designed with a 1 × N star coupler, 5 mm III–V/Si phase shifter array, and 10 mmwaveguide grating array, where $N = 32$ and 240. The OPA and grating array are both in 2–4-μm pitches with III–V mesa of 1-μm width. Optical beam is steered in the ψ direction by phase tuning with the OPA, resulting in the 51° steering range with

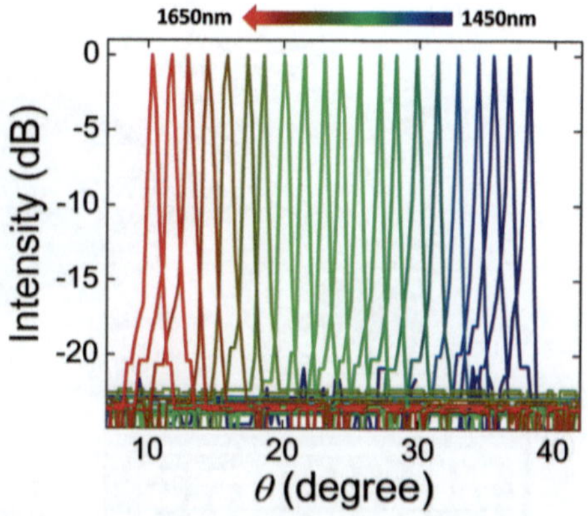

Fig. 5.6 Optical beam intensity versus θ ($\psi = 0$) for wavelength tuning (1450–1650 nm) in the OPAs. Reprinted with permission from Ref. [12]. © 2019 OSA

a beam width of $\delta\psi = 1.75°$ in FWHM, while the beam is steered in the θ direction by wavelength tuning (1450–1650 nm), resulting in 28° steering range with the beam width of $\delta\theta = 0.02°$. Figure 5.6 shows a 3-D plot for the beams by tuning both the OPA phase and wavelength (1520–1580 nm). These results are promising for OPAs in fully integrated LiDAR system.

Monolithic OPA Transceiver

Another unique approach of OPAs for practical LiDAR is a monolithically integrated OPA transceiver, as shown in Fig. 5.7 [14]. In Fig. 5.7a, an 8 × 8 active OPA with a radiating grating cell, phase control line, and amplitude control line is two-dimensionally arranged. An input laser incident in the input GC (grating coupler) impinges on the transmitter/receiver (T/R) optical switches and is delivered to the Si main bus waveguides, leading to the active OPA zone. The optical switch is configured as a Mach–Zehnder interferometer (MZI) with two-directional couplers at input and output ports. In the transmitter mode, each element of the OPA is controlled independently, thus emitting light with desired patterns. In the receiver mode, an optical fiber, placed apart from the chip surface with a tilted angle, irradiates optical signal on the whole OPA. The received optical signal is combined into the Si main bus waveguide, then led to the T/R optical switch, and transferred to a receiving fiber via an output GC off the chip. Residual optical beams in both modes, generated through the OPA elements without coupling, are terminated at the photodetectors to avoid the unwanted reflections. Capability of independent amplitude and phase control in each OPA is an essential function for producing arbitrary radiation pattern.

Figure 5.7b shows a microphotograph of the whole Si OPA chip (2 mm × 3 mm). All optical active elements are controlled by electronics. Figure 5.7c and d shows magnified views of the OPA core and active element, respectively. The active elements include a phase shifter, attenuator, and grating coupler as antenna. The OPA chip was manufactured by a SOI CMOS process. All optical as well as electronic elements were formed in a Si layer on a 1-μm thick SiO_2 in the SOI wafer.

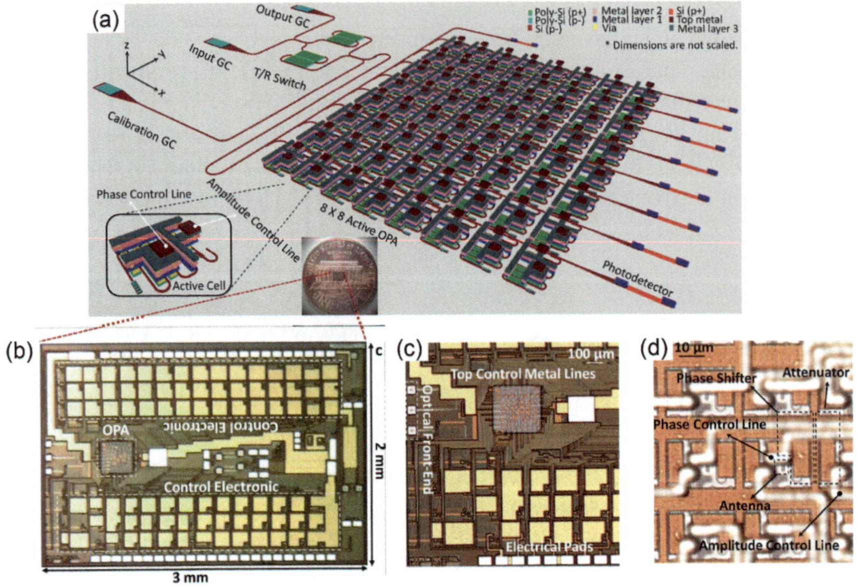

Fig. 5.7 **a** Schematic of a monolithic OPA transceiver chip composed of 8 × 8 elements with variable attenuators and variable phase shifters, and input/output optical grating couplers. **b** Microphotograph of the OPA transceiver chip with optical and electrical components. **c** Magnified OPA core including optical front end, top control metal lines, and electrical pads. **d** Magnified OPA elements including phase shifters, attenuators, and phase/amplitude control lines. Reprinted with permission from Ref. [14]. © 2015 OSA

Passive optical elements were realized in the 145-nm thick crystalline Si layer, usually assigned to MOSFETs.

Figure 5.8a and b are schematics of a TO variable attenuator and phase shifter with OPA elements, respectively [14]. The TO attenuator is arranged as an unbalanced MZI with polysilicon heaters (green) between two arms for amplitude control, while the TO phase shifter is arranged with a waveguide forming a meander line and provided with polysilicon heaters (green) to increase the power efficiency. An input/output optical grating coupler with a small footprint of 1 μm × 3.55 μm is designed as exhibiting a radiation efficiency of 50%.

As mentioned above, active elements of the OPA in the transmitter mode are controlled independently, thus emitting light beams with desired patterns. For experimental verification to the design of several patterns, near-and far-field patterns of the emitted light beams were measured, as shown in Fig. 5.9 [14]. 2-D patterns in the left two columns show designed phase-and amplitude-profiles. 2-D patterns in the center column show measured near-fields, while different patterns in the right two columns show simulated and measured far-fields. Radiation patterns from (a) to (e) correspond to five sets of differently designed phase-and amplitude-profiles, as follows:

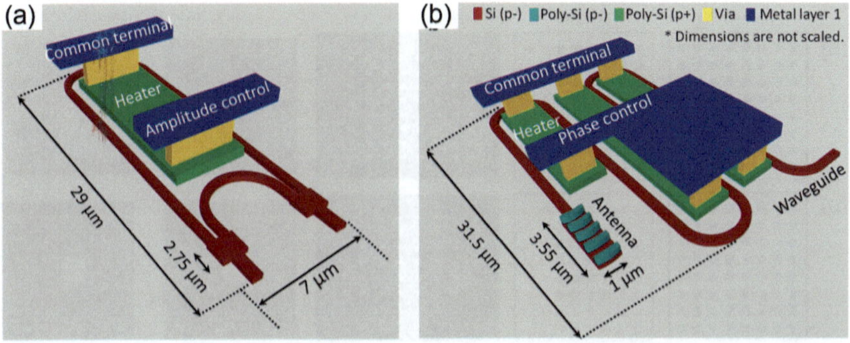

Fig. 5.8 **a** Schematic of a TO variable attenuator with a polysilicon heater formed between two arms of an unbalanced MZI. **b** Schematic of a TO variable phase shifter with a polysilicon heater formed between meandered waveguides. The input/output optical grating coupler as an optical antenna is designed with 50% radiation efficiency. Reprinted with permission from Ref. [14]. © 2015 OSA

(a) Uniform phase-and amplitude-profiles.
(b) Uniform phase-and Gaussian amplitude-profiles.
(c) Horizontally periodic phase-and Gaussian amplitude-profiles.
(d) Vertically periodic phase-and Gaussian amplitude-profiles.
(e) Horizontally straight line-profiles both in phase and amplitude.

A measured far-field in (a) involves first side-lobes as simulated theoretically, while a measured far-field in (b) involves a wider main-lobe with no deep side-lobe. The main beams for the designs in (c) and (d) are horizontally and vertically steered by 1.6° to the edge of each interference order, respectively. The results are obtained from shifting by π-phase between columns and rows in (c) and (d), respectively. A far-field pattern with a wide straight line and side-lobes for the design in (e) is created by straight line phase-and amplitude-profiles. Aforementioned demonstration of the electrically controlled OPA transceiver has a large potential to provide a promising optical beam steering technique for future various sensing, imaging, ranging, etc.

5.2.2 FMCW–LiDAR with High-Performance OPA

Principle of Frequency-Modulated Continuous Wave (FMCW)

Integrated OPAs were applied to demonstration of LiDAR system. Linear frequency-modulated continuous wave (FMCW) was widely used for operating LiDAR. Here, we first describe the principle of the FMCW [15] used in a LiDAR operating by OPAs with a transmitter and receiver, as shown in Fig. 5.11a [16]. In a homodyne configuration in the LiDAR system, a coherent beam emitted from the swept laser source is irradiated to the target as a transmitted beam (TX) and a part of the TX

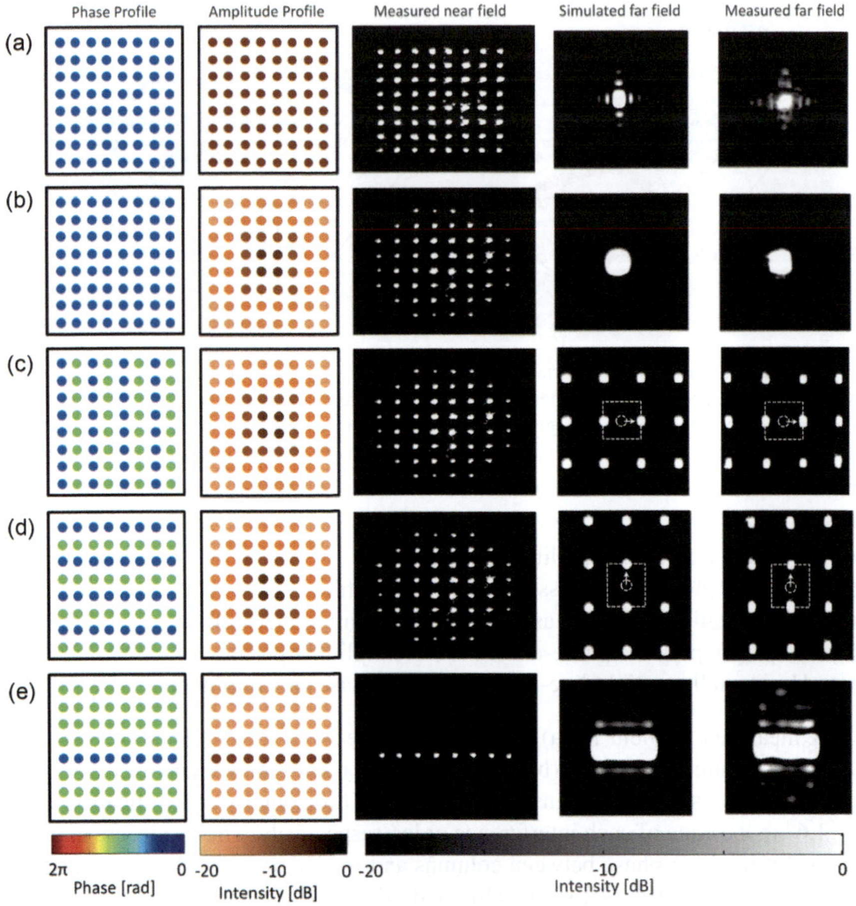

Fig. 5.9 Setting of phase-and amplitude-profiles in the 8 × 8 array of OPA transmitter mode, and measured near-field and simulated and measured far-fields. **a** to **e** Respective near-and far-fields by uniform phase-and amplitude-profiles in (**a**), by uniform phase-and Gaussian amplitude-profiles in (**b**), by horizontally periodic phase-and Gaussian amplitude-profiles in (**c**), by vertically periodic phase-and Gaussian amplitude-profiles in (d), and by proper phase-and amplitude-profiles in (**e**). Reprinted with permission from Ref. [14]. © 2015 OSA

divided by a coupler functions as a local oscillator (LO) for the optical receiver. After the field of the LO is mixed with the received beam (RX) delayed in time at the detector, a singular beat frequency, i.e., intermediate frequency (IF), f_{if}, is generated. Figure 5.10a shows time dependency of the TX (red), RX (green) and IF (red) frequencies for the linearly chirped frequency modulation with a triangular shape in a homodyne configuration corresponding to the case of the stationary target. Then f_{if} is described as:

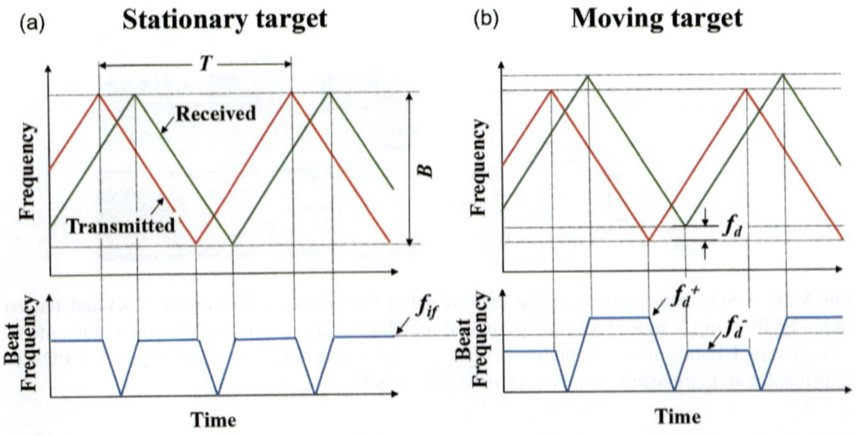

Fig. 5.10 (Upper panels) Time varying transmitted (red) and received (green) frequencies in linear FMCW system and (lower panels) resulting time varying IF of the homodyne signal, both for respective stationary (**a**) and moving (**b**) targets

$$f_{if} = \frac{4RB}{Tc}. \tag{5.1}$$

Here, R and B are a source-target distance and modulation bandwidth, respectively, while T and c are a waveform period and light velocity in air, respectively. When the target is moving, f_{if} is modified by a Doppler frequency (f_d) shift based on the moving target, resulting in two beat frequencies, i.e., $f_{if}^+ = f_{if} + f_d$ for the frequency waveform up-ramp and $f_{if}^- = f_{if} - f_d$ for the frequency waveform down-ramp, as shown in Fig. 5.10b. Then, f_{if} and f_d are given by:

$$f_{if} = \frac{f_{if}^+ + f_{if}^-}{2}, \tag{5.2}$$

$$f_d = \frac{f_{if}^+ - f_{if}^-}{2}. \tag{5.3}$$

The velocity component v of the target is derived as:

$$v = f_d \lambda, \tag{5.4}$$

where λ is the TX beam wavelength. By measuring beat frequencies, the range and velocity can be derived simultaneously. Regarding directions of a velocity vector, it is assumed that the target approaches toward the detector for simplicity.

Pioneering Demonstration of FMCW–LiDAR

The first demonstration of FMCW-based LiDAR was performed using a compact laser diode, Si photonic-integrated OPAs and the FMCW system [16]. Figure 5.11a

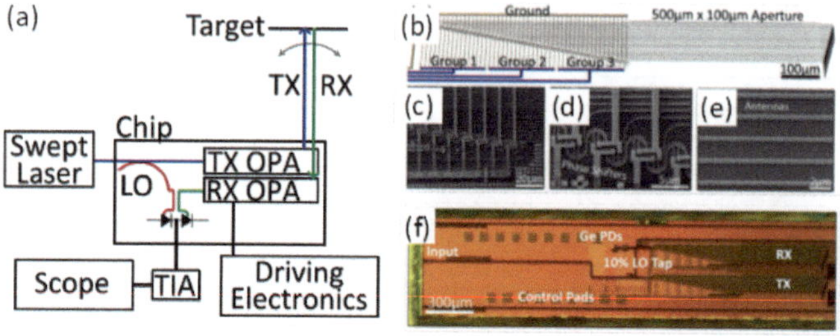

Fig. 5.11 **a** Schematic system of the LiDAR using OPAs with a transmitter (TX) and receiver (RX). **b** OPA chip. **c** to **e** SEM photographs of the phase shifter, their magnification, and antenna, respectively. **f** Photo-image of the whole OPA chip with TX, RX, LO tap, input port, Ge PDs, etc. Reprinted with permission from Ref. [16]. © 2017 OSA

shows a system of the LiDAR using OPAs with a transmitter (TX) and receiver (RX), both integrated on a Si chip monolithically. Figure 5.11b shows an OPA chip, while Fig. 5.11c–e shows SEM photographs of the phase shifter, their magnification, and grating antenna, respectively. Figure 5.11f shows a photo-image of a Si LiDAR device with a size of 6 mm × 0.5 mm. The phased array is composed of 50 grating antennas with a 2-μm pitch and a total length of 500 μm. Note that the antenna grating was fabricated by vertical etching of a Si waveguide with periodic perturbations on the side wall, as shown by the gradually periodical pattern on each edge line in Fig. 5.11e. The antenna radiates a beam with a spot size of 0.80° (full-width at half maximum) in the array dimension. A radiated beam is steered by thermal tuning in an array area, while it is steered by wavelength tuning in an antenna area, as described in Sect. 5.2.1. As a result, a steering angle with a range of 46° is realized by thermal tuning at 1.2 W power consumption, while 36° is achieved by wavelength tuning ranging at 1454 to 1641 nm. In evaluation of steering with the LiDAR, three targets were located at distances; 0.23, 0.32, and 0.45 m from the LiDAR in different directions. By steering the beam, each target could be measured separately with success. Unfortunately, the maximum angle range was restricted by an aperture size of the phased array in this work. This restriction is mitigated by scaling the LiDAR chip. As a whole, the above demonstration deserves the first demonstration of the coherent on-chip LiDAR using active OPAs.

Long-Range and High-Performance LiDAR

Long-range and high-performance on-chip coherent LiDAR was first realized by the same group in [16]. In this work, a group of phase shifters was combined to a common electrical line and wave front in each emitted beam was not controlled independently [16]. In the later work [17], however, 1-D OPA with an optical distribution network, phase shifters, and grating coupler, etc., was demonstrated with the new design for individually controllable phase shifters. Furthermore, a grating antenna with a long length (520 nm) provides good emission directivity and a narrow beam width. Since the phase shifter is based on the electro-optic (EO) effect, maximum static power consumption is only 20 μW, i.e., several orders of magnitude less than that of the widely used TO phase shifter. The OPA involves total 512 elements and an aperture area of ~ 10 mm^2.

Demonstration of the coherent LiDAR with OPAs was carried out by two methods; 2-D and 3-D LiDAR [17]. 2-D LiDAR was performed using two passive OPAs with no phase shifter, fiber-based couplers/detectors, and an external tunable laser. 3-D LiDAR was performed using two active OPAs with phase shifters. The two OPAs were controlled separately at different places, and set to raster scan a target repeatedly. Figure 5.12a shows a map of an outdoor test target with a measured distance of 185 m, while Fig. 5.12b shows a photograph of real-time visualization, both obtained by 2-D LiDAR. Characteristic objects are viewed in the parked cars, fence posts, and a pedestrian, all of which are distinguished by different colors depending on the range of the target. In addition, an IR camera took a photograph for real-time visualization of detections measured by the 3-D LiDAR, showing target and raster-scan patterns. The result of 3-D detection clearly indicated a person at ~ 7 m with prominent shapes such as arms and legs, two front and back walls at ~ 8 and ~ 12 m.

Besides coherent 2-D and 3-D LiDAR with large-scale OPAs, the work in [17] reports lens-free point-to-point, free-space optical communication links up to 50 m, using transducer (transmitter and receiver) OPAs. Beam steerable integrated transmitters enable a lens-free optical data-links, while integrated receivers with a large aperture are capable of coupling received light to an on-chip single-mode waveguide for further processing. Refer to [17] for details. These results suggest that the integrated OPA transducer has a potential for high data-rate lens-free point-to-point-based free-space optical communication as well as coherent LiDAR operation.

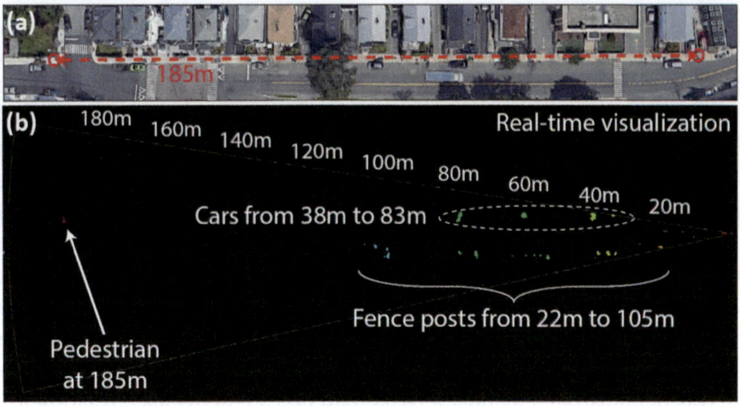

Fig. 5.12 **a** Aerial photograph showing a test target scanned with 2-D LiDAR. **b** Photograph of real-time visualization obtained by 2-D LiDAR showing detections at a 185 m range. Reprinted with permission from Ref. [17]. © 2019 IEEE

5.2.3 LiDAR with Focal Plane Arrays

Architecture of 3-D Imaging

A focal plane array (FPA) is an image sensing matrix with many small thin-film sensor elements in the focal plane of the image detector. A digital CMOS image sensor using a lens to focus light for image formation is considered to be the CMOS-based 2-D imaging FPA. Recently, a CMOS-based Si photonic 3-D imaging sensor for a FPA-based solid state-integrated photonic LiDAR has been developed [18]. Figure 5.13a shows a 3-D imaging architecture of the solid state LiDAR with two FPAs; a transmitter FPA for irradiating a small spot of the target (the cat here) through a collimating lens system and a receiver FPA for detecting scattered light from the target through the common lens system. In the transmitting line, a laser beam emitted from the external tunable laser is split at the power splitter into two; one to the transmitter FPA and the other toward the receiver FPA as a local oscillator (LO) light. The latter forms a coherent receiver FPA with the compact optical heterodyne detectors, resultantly, the FMCW–LiDAR scheme. Light scattered from the target is combined to the LO light in a heterodyne receiver, resulting in a beat frequency, as shown in Fig. 5.10. In Fig. 5.13b, the transmitter FPA with 16 gratings is controlled by a 1×16 electronic switching tree for off-plane beam steering, while the receiver FPA with 32×16 (512) pixels is controlled by a 1×8 electronic switching tree.

Figure 5.14 shows a schematic of an expanded receiver FPA unit. Each pixel is composed of a grating coupler as enlarged clearly in the left panel. Light scattered from the target and received through the grating coupler are coupled with the LO light on a 50/50 directional coupler and led to a balanced PIN photodiodes to cause a heterodyne signal as a result of the FMCW–LiDAR.

Characterization of 3-D Imaging LiDAR

Representative characteristics of the aforementioned 3-D imaging LiDAR with FPAs are shown in Fig. 5.15 [18]. Figure 5.15a shows typical signals with a Doppler shift of the linear chirp FMCW measurement obtained from homodyne and heterodyne receiver pixels. The stationary target in the upper panel represents a single beat frequency for both up-chirp and down-chirp, while a moving target in the lower panel represents two beat frequencies depending on up-and down-chirps. The results derive the distance and velocity of the target, as indicated in Fig. 5.10. Measurement precision (FWHM) difference on the depth noises between a target at a 18-m distance (85% in reflectance) and target at a 75-m distance (30% in reflectance) resulted in standard deviations of 1.8 mm and 3.1 mm, respectively. Next, 3-D imaging LiDAR was demonstrated against a basketball at a 17-m distance rotating at 1 rpm around a vertical axis, as shown by the photograph in Fig. 5.15c. Figure 5.15b shows histograms of its velocities at the left edge (upper) and right edge (lower) of the ball, resulting in a standard deviation of 1.0 mm/s. Highlight of the LiDAR operation is shown in Fig. 5.15d, where point clouds of the basketball are three-dimensionally arranged with their velocities distinguished by different colors. Furthermore, horizontal line cut of the velocity across the middle of the basketball exhibits almost

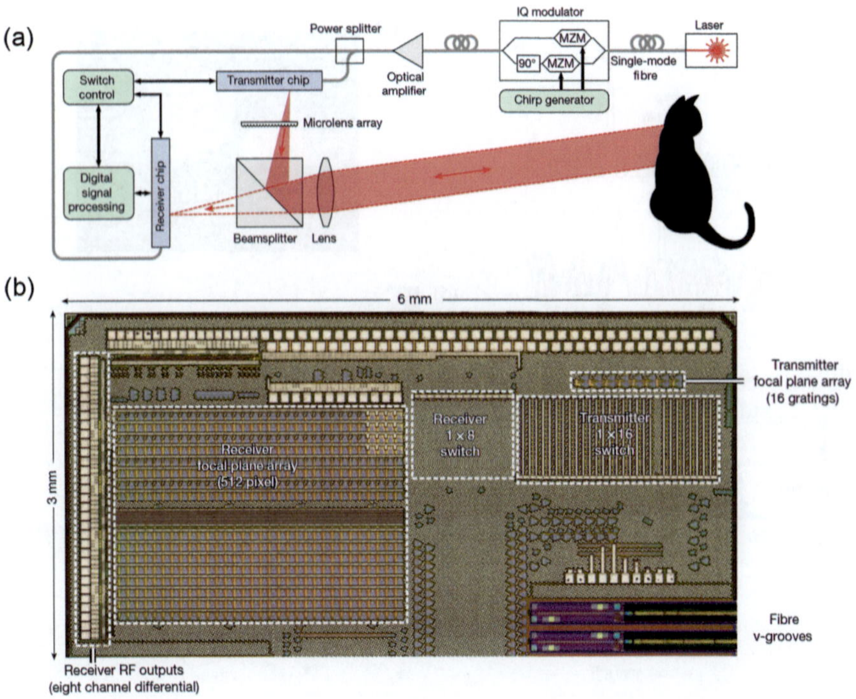

Fig. 5.13 **a** 3-D imaging construction with two FPAs; a transmitter illuminating a small target, and a receiver detecting scattered light from the target. **b** Optical image of the FPA chip including a transmitter switch and receiver FPA. Reprinted with permission from Ref. [18]. © 2021 Nature

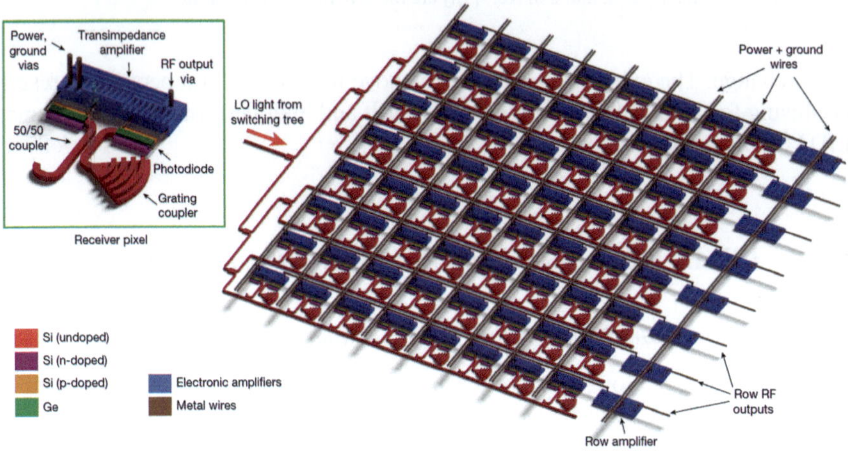

Fig. 5.14 Schematic of a receiver FPA design with matrix of receiver pixels including grating coupler, 50/50 coupler, photodiode, trans-impedance amplifier, etc. Reprinted with permission from Ref. [18]. © 2021 Nature

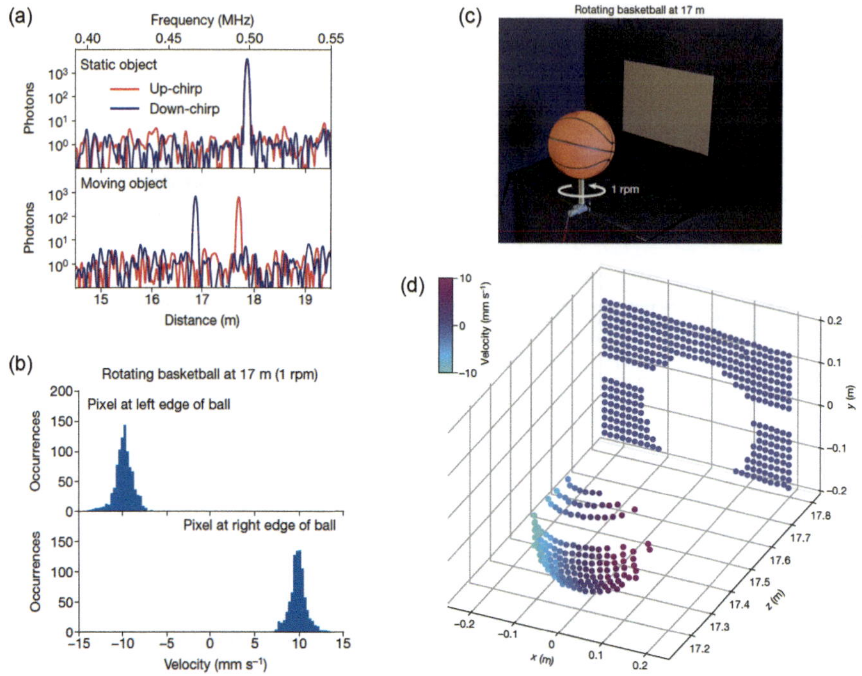

Fig. 5.15 Typical characteristics of 3-D imaging LiDAR with FPAs. **a** Typical Doppler-splitting signals of the linear chirp FMCW. **b** Velocity histograms for a basketball rotating at 1 rpm. **c** Photograph of a setup for measurement of basketball. **d** Point clouds of the basketball rotating at 1 rpm, 3-D arranged with their velocities distinguished by different colors. Reprinted with permission from Ref. [18]. Only Fig. 4a, c, d, and e in Ref. [18] are cited here. © 2021 Nature

linear change. The experiment mentioned above includes several persuasive results promising for practical applications to low-cost and high-performance 3-D imaging LiDAR with a compact size.

5.2.4 LiDAR with Slow-Light Grating and Prism Lens

Architecture for Slow-Light Grating Beam Scanner

Recently, a unique slow-light grating beam scanner for FMCW–LiDAR with Si photonic crystal (PC) waveguides was proposed and demonstrated [19, 20]. Fundamental structures and properties of PC waveguides are surveyed in Chap. 2, while slow-light in the PC exhibits a much smaller group velocity v_g, i.e., much larger group index n_g than those in vacuum, as discussed in Sect. 6.2 (Slow-Light Photonic Crystal Device).

Figure 5.16a shows a schematic of a LiDAR device with a slow-light waveguide-modified optical grating fabricated on SOI Si substrate [19]. The slow-light waveguide is realized by a lattice-shifted PC waveguide (LSPCW); waveguide with the third rows of holes from the line defect (waveguide core region) shifted along the waveguide of a triangular lattice [21]. The detailed slow-light grating is formed by inserting two kinds of grating periodicities in the waveguide; a hole-diameter-modulated double periodicity within the PC slab layer [22, 23] and a shallow (depth of ~ 10 nm) diffraction-grating groove on the surface for high emission, as noted (I) on the right panel in 5.16a. With this structure, the grating converts slow light to out-of-plane light beam irradiated in the θ direction given by the slow-light wavelength. Note that an angle of the beam radiating from the slow-light grating is extremely sensitive to the wavelength originated from the slow-light effect (small group velocity, or large group index). Figure 5.16a also shows that the irradiated beam accompanies a fan-shaped far-field pattern (FFP) in the lateral (ϕ) direction. Figure 5.16b shows a particularly designed prism lens for 2-D beam steering; steering in the θ direction with tuning the light wavelength and steering in the ϕ direction with shifting the LSPCW array, as shown in Fig. 5.16c and d. On account of large angular dispersion in slow light, slight variation of the wavelength causes substantial change of steering angle. To avoid this phenomenon, different beam steering without tuning wavelength was proposed: Tuning a group index n_g (a kind of phase shifting) of the waveguide by thermal change (TO effect) was experimented for a fixed wavelength [24]. Two kinds of integrated heating structures were used; a p-i-p doped Si waveguide and a waveguide with TiN heaters. The former resulted in a large beam steering range (21°) at a relatively low heating-power (1.3 W) with a high-speed response (100 kHz), while the latter showed the low heating efficiency.

Implementation of Wide-Beam-Steering LiDAR

The slow-light grating LiDAR device was fabricated using a SOI wafer with 200-mm diameter and CMOS process [19]. A chip size is 5.5 mm × 4.0 mm. An array of

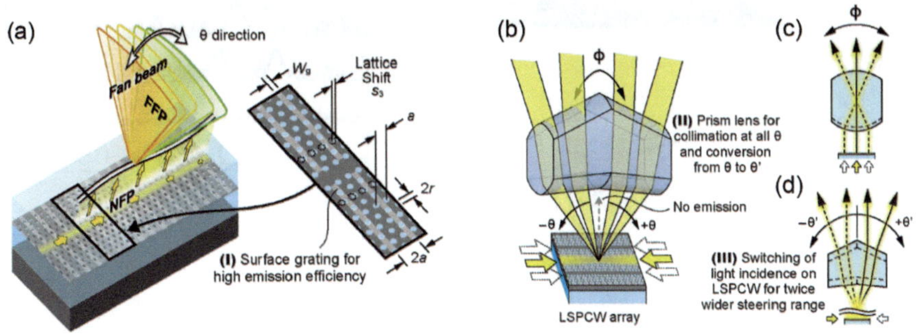

Fig. 5.16 **a** Schematic of the Si LiDAR using slow-light in the LSPCW with an optical grating on SOI substrate. **b** Particularly designed prism lens for 2-D beam steering. **c** and **d** Beam steering in the ϕ direction with shifting the LSPCW array and θ direction with wavelength tuning, respectively. Reprinted with permission from Ref. [19]. © 2020 OSA

32 LSPCWs was integrated in a 1.2 mm length and an 80 μm pitch. Selection of a LSPCW within the array was implemented by a six-stage TO–MZI optical switch circuitry. In the 1-D beam steering, a steering range of θ was 22° with a divergence $\delta\,\theta$ (FWHM) of 0.08°–0.15° for wavelength tuning of 23 nm. In the lateral direction (ϕ), the divergence $\delta\,\phi$ was 23°–40° and ~ 0.1° without and with the prism lens, respectively.

A slow-light grating FMCW–LiDAR experiment was implemented using a setup that shines the sidewall of a turntable (diameter: 330 mm) at a distance 4.08 m from the beam scanner, as shown in Fig. 5.17a [20]. The target was turning along the anticlockwise direction from the top view. The beam emitted from the slow-light grating was scanned in the θ direction from − 1.25° to + 1.25° with the wavelength sweep from 1535 to 1553 nm. Figure 5.17b shows measured beat frequency spectra for wavelengths changing with linearly chirped modulation, as discussed in Fig. 5.10 ($T = 40$ μs here). When the turntable is not moving, the beat spectrum is measured at 17.4 MHz. When the turntable is rotating at 45 rpm, the beat frequency is split into two; f_{up} and f_{down}, caused by the Doppler shift, at wavelengths other than 1554 nm ($\theta = 0.00°$) where the Doppler shift did not occur due to no velocity component of the turntable in the direction of the irradiated beam. From these measurements, the range and velocity of the turntable can be derived simultaneously. Figure 5.17c shows the off-line point cloud 3-D image of the range and velocity of the turntable, where the velocity is distinguished by the color. The point cloud image for the pedestal is plotted before that of the turntable.

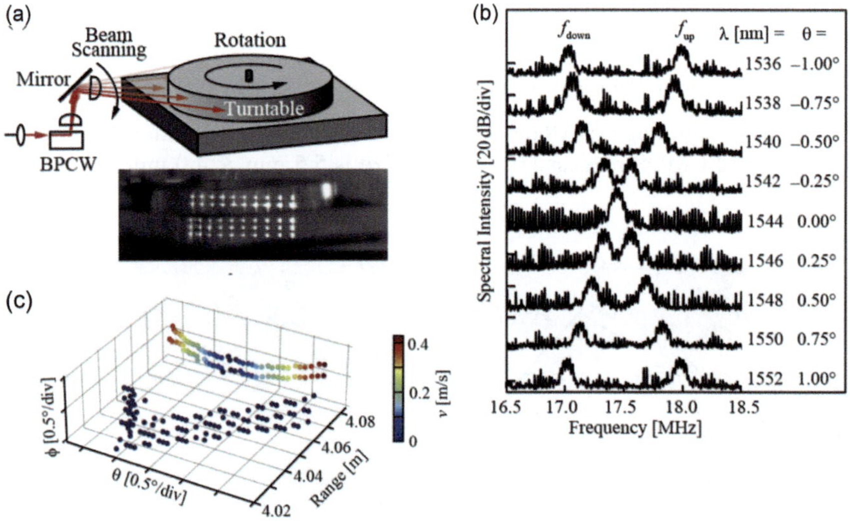

Fig. 5.17 **a** Experimental setup of slow-light grating FMCW–LiDAR that shines the sidewall of a turntable. **b** Measured beat frequency spectra for wavelengths changing with linearly chirped modulation. **c** Off-line point cloud 3-D image of the range and velocity of the turntable. Reprinted with permission from Ref. [20]. © 2021 OSA

Even after these experiments, continued improvement of the slow-light grating LiDAR device has been reported in Ref. [25], for instance.

Recently, a new type of integrated LiDAR system using a different material and beam steering principle has been reported; a GaAs/AlGaAs-based LiDAR chip with a photonic crystal surface-emitting laser (PCSEL). The beam-emitting principle is not based on the combined OPA and grating coupler mostly used in the Si LiDAR system mentioned above, but based on the electronically scanned beam emitted from the on-chip dually modulated PC laser. Fundamental LiDAR performances like measurements of the distance and velocity of poorly reflective targets in the field of view. However, since the structure is not included in the Si photonic platform, the PCSEL–LiDAR is introduced in Chap. 6 (Applications of Photonic Crystal Waveguides). Refer to Sect. 6.5.7 (Application of PCSEL to LiDAR Sensing Mobility) for detail.

5.3 Optical Neural Network for Deep Learning

5.3.1 Photonic Accelerator for "Beyond CMOS" Platform

Optical Neural Network (ONN)

As described in the introductory part, information equipment in data centers, autonomous vehicles, etc., strongly necessitate operation of an AI processing system. Yet, the AI processing system full of tremendous CMOS-based electronic digital processers causes problems of high power consumption and large operation latency. Accordingly, AI accelerators with low power consumption, low operation latency, and high data-processing speed are now urgent matters. Here, an accelerator is a particular processor located at a front end of the digital computer and performs a specific function with faster speed and energy efficiency larger than electronic processors such as CPU, GPU, and FPGA [26–29]. A Si photonics-based optical neural network (ONN) processor has now attracted much attention as a powerful candidate for such innovative AI accelerators, i.e., a "Beyond CMOS" processer aiming at resolving the bottleneck of the electronic digital processor [29].

Architecture for Matrix Multiplication-Enabling ONN

Artificial neural network (ANN) is one of the most promising computational network model provided by the signal processing architecture of the brain in the living body. Now a variety of machine learning tasks such as vowel and image recognition have significantly improved the ANN architectures. Nevertheless, turning to the current computing hardware, electronic AI accelerators supporting the hardware are based on the von Neumann architecture, so not fit for implementation of neural networks. Alternatively, Si photonic ANN accelerators have attracted much attention because of excellent capabilities with high speed, low power consumption, ultra-low

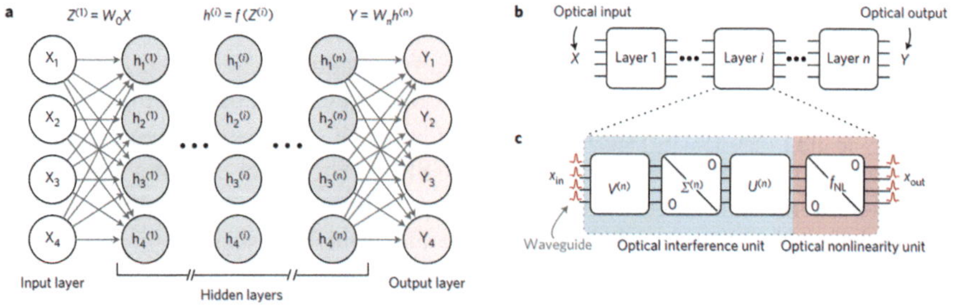

Fig. 5.18 **a** ANN with input artificial neurons connected to at least one hidden layer and an output layer. **b** Decomposition of the ANN into individual layers. **c** Detailed ANN architecture of the optical version, i.e., ONN with OIU and ONU. Reprinted with permission from Ref. [30]. © 2017 Springer Nature

latency, high I/O bandwidth, and highly parallel optical processing making full use of wavelength, phase, polarization, and amplitude of light [26–29].

A general architecture for ANN is briefly illustrated in Fig. 5.18 [30, 31]. Figure 5.18a shows ANN with input artificial neurons connected to at least one hidden layer and an output layer [30], while each layer is cascaded, as shown in Fig. 5.18b. Information propagates through each layer by a linear combination such as matrix multiplication and a nonlinear activation function. Note that, as shown in Fig. 5.18a, several artificial neurons are connected to form network by many complicated arrows with their proper weight parameters. For training ANN, training data are entered into the input layer, while the output is computed by forward propagation. With this process, proper weight parameters mentioned above are optimized during the course of computation mostly by backward propagation and gradient descent.

Figure 5.18c shows a detailed ANN architecture of the optical version, that is, each layer assuming ONN with two essential optical units: One is an optical interference unit (OIU) for optical matrix multiplication. The other is an optical nonlinearity unit (ONU) for nonlinear activation. A general theory on the matrix operation can be simply presented as follows [32]: A matrix (W) with real values can be decomposed as $W = U\Sigma V$. Here, U, Σ, and V are an $m \times m$ unitary matrix, an $m \times n$ rectangular diagonal matrix with non-negative real values on the diagonal, and a complex conjugate of $n \times n$ unitary matrix, respectively. Note that any unitary transformation (U and V) can be performed using an optical beam splitter and phase shifter, while Σ can also be performed using an optical amplifier or attenuator [32–34]. Meanwhile, the ONU is represented by a nonlinear activation function. In general, an activation function in the ANN allows for linear and nonlinear activation functions. An output layer allows for a linear (or identity activation) function, $f(x) = x$, whereas a hidden layer necessitates a nonlinear activation function (e.g., sigmoid function, tanh, and ReLU) for suppressing stimulated output values (otherwise expanding after each matrix multiplication) and enhancing the identification performance. The ONU in the integrated Si platform can be implemented using, e.g., saturable absorbers on

Si photonic waveguides [33]. A sinusoidal transmission curve in the Mach–Zehnder interferometer (MZI) is available also for such a purpose in the ONU, as shown later.

A general real-valued matrix $W = U\Sigma V$ with decomposed two unitary matrix switches based on MZIs is shown in Fig. 5.19 [27]. Each MZI, consisting of two directional couplers (DCs) for 50:50 division of the input signal and two phase (ϕ, θ) shifters (PSs) for input signal modulation, can modulate the input signal into two output signals with arbitrary ratios by adjusting ϕ and θ. Accordingly, it can be used as a 2×2 unitary transformation. During the course of unitary transformation, adjusted ϕ and θ are linked to the proper weight parameters in Fig. 5.18a. All MZIs are arranged in order as given topologies such as Reck's implementation (triangular shape) and Clements's implementation (rectangular shape). Clements's implementation has advantages of a smaller size, higher fidelity, and excellent tolerance to noise.

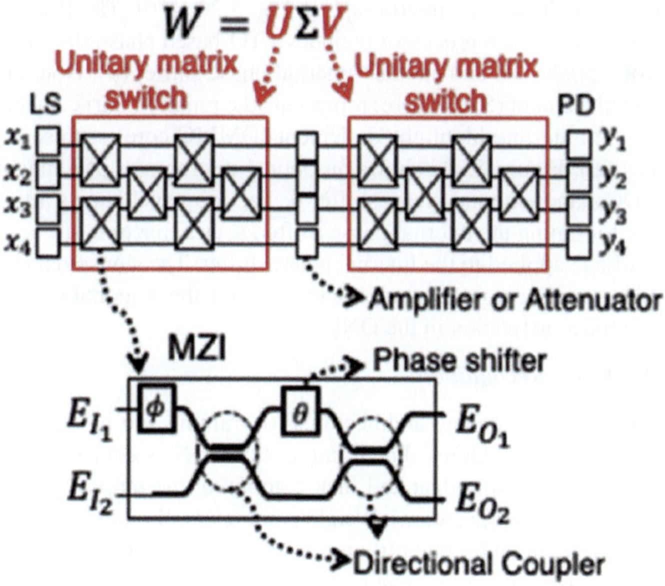

Fig. 5.19 General real-valued matrix $W = U\Sigma V$ with decomposed two unitary matrix switches based on MZIs. Reproduced from Fig. 7 of [27] (panel (c) only), released under a CC BY 4.0 license; https://doi.org/10.1063/1.5108912; Fig. 7 (c) of [27] itself adapted from [30]

5.3.2 *Programmable Nano-photonic Processor for On-Chip ONN*

Architecture for Integrated Optical Interference Unit

New ONN architectures for enhancement of computational speeds and power efficiencies over electronics were implemented for conventional inference tasks [30, 31]. Among them, a pioneering work on the Si photonics-based ONN with a programmable nano-photonic processor (PNP) is introduced here. Figure 5.20 shows an illustration of the demonstrated optical units using the PNP; OIU for optical matrix multiplication for vowel recognition [30]. Figure 5.20a shows a schematic of an experimented ONN architecture with two layers. Each layer consists of two OIUs (OIU_1 and OIU_2). Figure 5.20b shows experimental feedback and control loop for vowel recognition. A PNP for OIU was demonstrated using a Si photonic platform, as shown by the optical micrograph in Fig. 5.20c [30]. The PNP involves 56 programmable MZIs. Each processor includes a TO-based phase shifter (θ), leading to two sets of a 50:50 DC and another external phase shifter (ϕ). Note that the red-line-highlighted element (SU(4) core) represents the unitary matrix switches (U and V), whereas the blue-line-highlighted element (DMMC core) represents the diagonal matrices (Σ). Figure 5.20d shows an optical micrograph of a magnified MZI with 50:50 DCs, and external phase shifter ϕ. The top panel shows a transmission characteristic for tuning the internal phase shifter, θ, as a function of $v^2{}_{\Delta\theta}$, where $v_{\Delta\theta}$ indicates a voltage applied to the internal phase shifter. The sinusoidal transmission curve exhibits a nonlinear performance demanded for the diagonal matrices Σ, thus providing nonlinear activation in the ONU.

Inference for Vowel Recognition

In machine learning in general, architectures for training and inference tasks need different several pre-procedures: For training, the ONN is educated by providing the input layer with training data and then computing the output by the exclusive algorithm (via forward propagation). Weight parameters, W_ns, in the matrix, as indicated in Fig. 5.18a, are subsequently optimized using another exclusive algorithm (via backward propagation and gradient descent).

In the articled [30] proposing an example of ONN, inference for vowel recognition task was implemented using the developed PNP. The PNP was reprogrammed for realizing all required OIUs, while ONU using a saturable absorber was simulated on a computer. Then, the PNP was evaluated on a test set of the vowel recognition for implementing the ONN architecture. Figure 5.21a shows the resultant correlation matrix using the demonstrated ONN, which obtained 138/180 cases in correct identification (76.7% accuracy). The similar matrix using a conventional computer for comparison resulted in correctness of 165/180 (91.7%), as shown in Fig. 5.21b. The difference in correctness between them was analyzed as the difference in computational resolution. The current resolution of the ONN was limited by several factors such as the noise and dynamic range of the photo-detection, noise of phase-encoding, thermal crosstalk between phase shifters, etc.

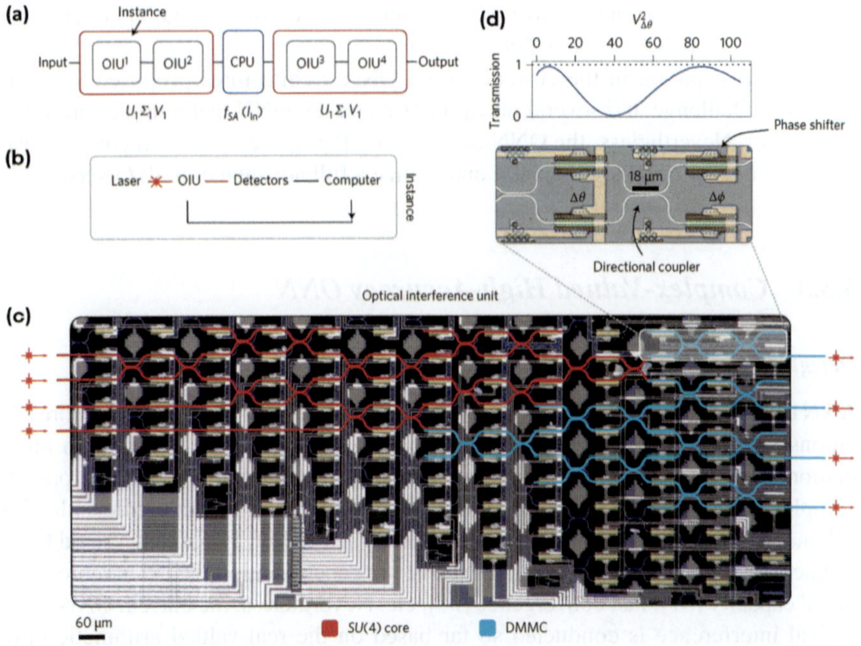

Fig. 5.20 (a) Schematic of an experimented ONN architecture with two layers involving OIU_1 and OIU_2 per each. (b) Demonstrated feedback and control loop. (c) Optical micrograph of fabricated OIU for realizing matrix multiplication (in red) and attenuation (in blue). (d) Optical micrograph of a MZI with phase shifters and directional couplers (lower) and a transmission curve showing a nonlinear sinusoidal performance (upper). Reprinted with permission from Ref. [30]. © 2017 Springer Nature

Fig. 5.21 **a** Correlation matrix using the demonstrated ONN resulting in 138/180 cases in correctness. **b** Similar matrix using a conventional computer for comparison resulted in correctness of 165/180. Reprinted with permission from Ref. [30]. © 2017 Springer Nature

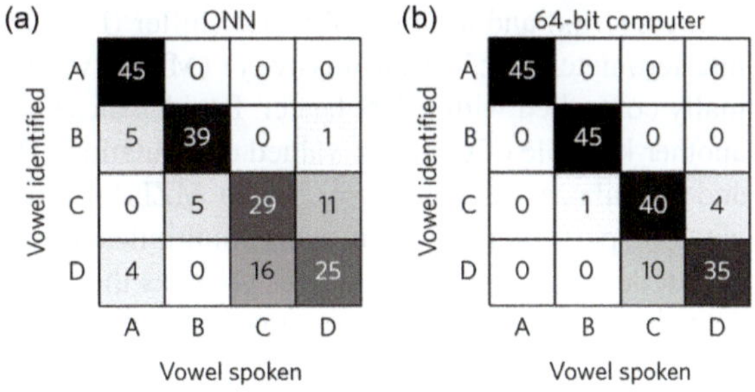

The above experiment was implemented for inference of vowel recognition only. However, training of the vowel recognition was carried out also by simulation on on-chip training of the unitary matrix using a new algorithm; forward propagation, and the finite difference method instead of conventional backward propagation and gradient descent algorithm with an electronic computer. As described before, processing big data using a conventional computer has growing challenges at a high speed with low power and low latency.

Power dissipation in the currently used ONN architectures proposed here still remains a challenge and several new proposals on the ONN technologies are introduced later. Nevertheless, the ONN using the PNP proposed here will pave the way to reduce energy costs significantly and realize a full advantage of all ONNs.

5.3.3 Complex-Valued High-Accuracy ONN

Design of Complex-Valued Neural Network

ANN using a conventional electronic computer is operated with real-valued representations and incapable of complex-valued operations. In contrast, ONN with information encoded in both amplitude and phase of the photonic device can operate complex-valued arithmetic using optical interference, thus providing considerably enhanced operation speed and energy efficiency [35]. Such a complex-valued ONN can perform excellent advantages such as strong generalization [36], rich representational capacity [37], fast convergence [38], etc. Nevertheless, the current ONN using optical interference is conducted so far based on the real-valued arithmetic only, missing the benefit of the complex-valued ANN.

Recently, complex-valued high-accuracy ONN was implemented [35]. Figure 5.22 shows the whole aspect of the complex-valued coherent ONN corresponding to the diagram in Fig. 5.18a. Figure 5.22a shows a schematic of a complex-valued optical neural chip (ONC) with functions of network input operation, weight multiplication, and coherent detection, all involving Mach–Zehnder interferometers (MZIs) integrated on a Si platform chip. The MZI consists of a set of two beam splitters (BSs) and an internal phase shifter (PS). The BS consists of a multimode interferometer (50:50 transmissivity) (MMI), whereas the internal PS (θ) is thermally controlled with a TiN heater. Besides these elements, the external PS (ϕ) has another key role of complex-valued computation. The chain of MZIs in red divides and modulates the input laser, while MZIs in blue, constructing a 6×6 neural network, perform the weight-matrix multiplication and accumulation, leading to the output detectors. A MZI in green separates the reference light needed for coherent detection. Remaining MZIs in gray are used for coherent detection which detects phase difference between the reference and signal light. All light signals including reference light are generated from an external laser ($\lambda = 1550$ nm).

Initial input signals are prepared with both optical amplitudes and phases as complex-valued encodes. The input preparation is followed by the light signals propagating through the 6×6 ONN indicated by the MZIs in blue. During light propagation through the network, the input states S_{in} is transformed into the output states S_{out} according to the equation: $S_{\text{out}} = U(N)\, S_{\text{in}}$, where $U(N)$ is a 6×6 unitary weight matrix ($N = 6$ in the figure) represented by product of a multiple rotation matrix T and a diagonal matrix D, as $U(N) = TD$. The matrix T is related to internal

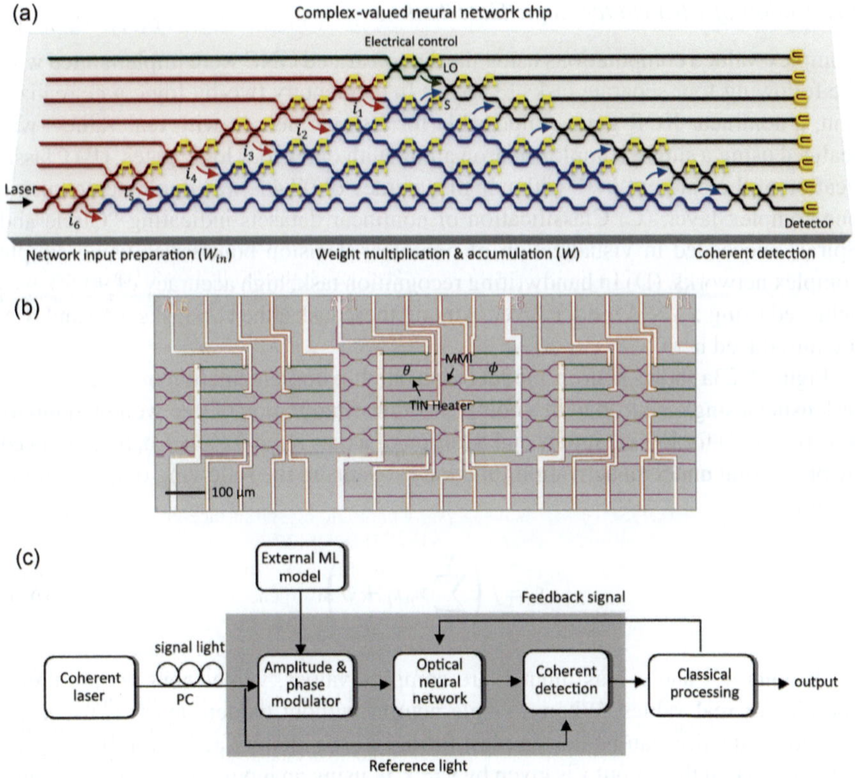

Fig. 5.22 Structure of complex-valued coherent ONC. **a** Schematic of ONC with functions of network input operation, weight multiplication, and coherent detection including MZIs on a Si platform chip. **b** Micrograph of the ONC constructing the MZI network with MMIs, PSs, and TiN heaters. **c** Schematic of a signal light flow of the ONC system. Reprinted with permission from Ref. [35]. © 2021 Nature Comm.)" to "Adapted from [35] (amalgamation of Figs. 1 and 2), released under a CC BY 4.0 license; https://doi.org/10.1038/s41467-020-20719-7

and external phase shifters, PS (θ) and PS (ϕ), as:

$$T = ie^{i\frac{\theta}{2}}\begin{bmatrix} e^{i\phi}\sin\frac{\theta}{2} & e^{i\phi}\cos\frac{\theta}{2} \\ \cos\frac{\theta}{2} & -\sin\frac{\theta}{2} \end{bmatrix} \tag{5.5}$$

Multiplication of the weight matrix is implemented under this formulation. Values of θ and ϕ are dictated from the external machine learning (ML) model, as shown later. Figure 5.22b shows a micrograph of the ONC constructing a MZI network with MMIs, internal PSs, external PSs, and TiN heaters. Figure 5.22c shows a schematic of a signal light flow, where an amplitude and phase of the incident laser are modulated by the external ML model that characterizes the targeting task, while the detected output light processed in a conventional digital processor is sent back to the ONN through the feedback loop.

Realization of Elementary Logic Gate Task

Complex-valued computations using the demonstrated ONC were implemented with the following four separate tasks [35]: (A) In elementary two-bit logic gate realization, a nonlinear XOR gate—impossible for a single neuron with real value—was realized using a single complex neuron along with other four logic gates. (B) Classification of the dataset "*Iris*" with a high accuracy of 97.4% was implemented using one complex layer. (C) Classification of nonlinear datasets indicating "Circle and Spiral" succeeded in visualization of nonlinear decision boundaries using simple complex networks. (D) In handwriting recognition task, high accuracy of 90.5% was achieved using a 4 × 4 hidden layer. Among these tasks, the two tasks: (A) and (C), are introduced here.

Figure 5.23a shows neuron-model based on-chip implementation of the logic gate task using a single neuron with a complex value. Summation of two weighted inputs $w_i x_i$ ($i = 1, 2$) for logical inputs and addition of a bias input b ($b = 1$ here), followed by processing under an activation function f, results in the following output y of the neuron, as:

$$y = f\left(\sum_{i}^{N} w_i x_i + b\right). \tag{5.6}$$

Weights w_is and a bias input b are complex values, while input x_is are either complex or real values. When a single neuron without hidden layers is used, the activation function can be the simplest form, i.e., the identity activation function ($f(x) = x$), so that the output y is given by $y = \boldsymbol{x}\,\boldsymbol{W}$ using an input vector $\boldsymbol{x}$ and a weight vector $\boldsymbol{W}$. Note that $\boldsymbol{W}$ is updated after each iteration of the workflow, as shown in Fig. 5.23c. Weight matrix is decomposed into values in the phase shifters, θ_i and ϕ_i ($i = 1, 2, 3$) on chip. In this way, training processes of NAND and XOR gates resulted in Fig. 5.23b and c, respectively. Processing was iterated by 10 times and each logic gate was recorded. Quadrants in blue represent "logical 0", while quadrants in pink represent "logical 1". Note that, in the NAND gate in Fig. 5.23b, three logical inputs, [0 0], [0 1], and [1 0], converge to the expected zone of "logical 1" from a random starting point, while the logical input [1 1] only converges to the expected zone of "logical 0" on account of the neuron with complex values. Similarly, training processes of the XOR gate are also realized expectedly, as shown in Fig. 5.23c.

Classification of Nonlinear Datasets Circle

In classifying nonlinear datasets of circle, nonlinear decision boundaries for the shape of the circle were visualized by using a simple complex network [35]. Figure 5.24a shows visualized datasets of two classes of circles (outer blue and inner red ones), while Fig. 5.24b shows model construction of their classification using a mono layer (complex vs real value design for comparison) using two inputs with real values and two neurons. Intensities of the chip outputs are detected by PDs. Figure 5.24c and d shows simulated decision boundaries of the circle performed by the real and complex models, respectively. Zones in purple and orange belong to the outer circle (blue) and

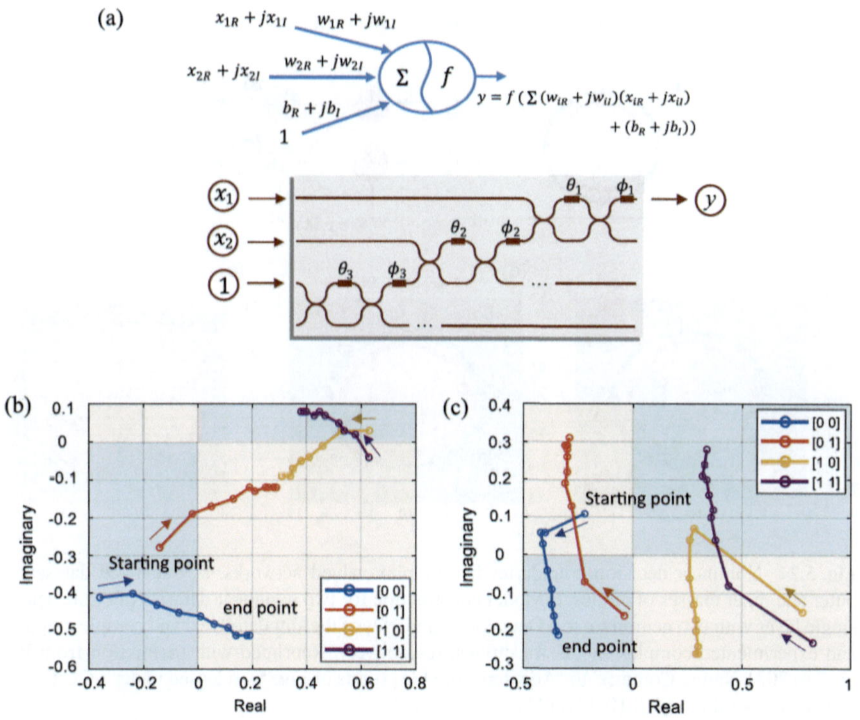

Fig. 5.23 Single complex-valued neuron for logic gate task. **a** Neuron-model based on-chip operation of the logic gate task using a single neuron with a complex value. **b** and **c** Training processes of NAND and XOR gates, respectively. Reprinted with permission from Ref. [35]. © 2021 Nature Comm.)" to "Reproduced from [35], released under a CC BY 4.0 license; https://doi.org/10.1038/s41467-020-20719-7

inner circle (red), respectively. Figure 5.24e shows an experimented complex model, where the black and white wires are expected and experimented decision boundaries, respectively. Each accuracy after classification is noted in the below right corner in all panels. Note that the decision boundaries by the real model are straight lines, resulting in failure in classification of the two circles. In contrast, those by the complex model provide the nonlinear curves with perfect coincidence to the entangled shape of the datasets. Resultantly, the complex model achieved a simulated accuracy of 100%, far exceeding the real model with the simulated accuracy of 55%. The complex-valued model implemented here has proved many significant advantages over conventional real-valued models in several ONN tasks in language and image recognition.

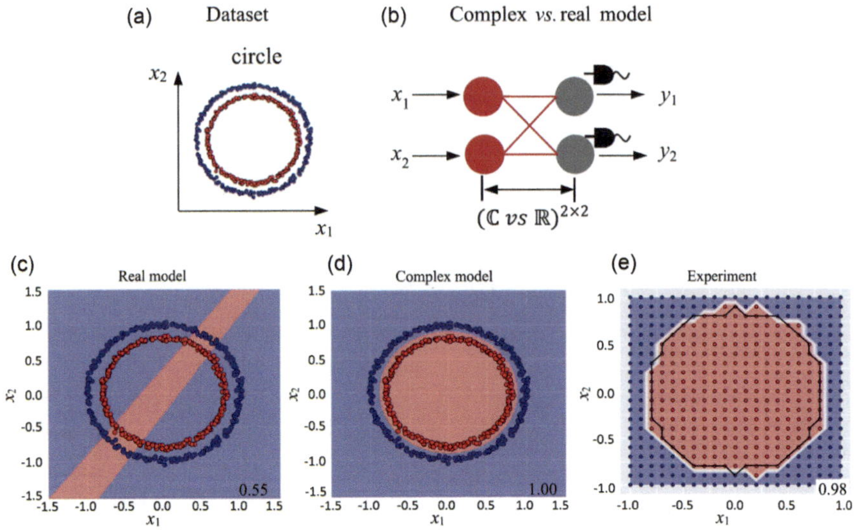

Fig. 5.24 Nonlinear decision boundaries by complex-valued networks. **a** Visualized datasets of outer and inner classes of circles. **b** Model construction for two nonlinear datasets of circle using a single layer with two neurons. **c** to **e** Decision boundaries of the simulated real and complex models, and experimented complex model for circles, respectively. Reprinted with permission from Ref. [35]. © 2021 Nature Comm.)" to "Adapted from [35] (parts of Fig. 5), released under a CC BY 4.0 license; https://doi.org/10.1038/s41467-020-20719-7

5.3.4 New Model for Nonlinear Mapping-Based Classification

Architecture for Projection-Based Photonic Classifier (PPC)

In the previous sections, Si photonic-integrated neuro computing based on the ANN algorithm has been executed to certificate the advantages of photonics; high calculation speed, high energy efficiency, and low latency. As mentioned in Fig. 5.18c, the ANN-based ONN algorithm essentially necessitates nonlinear activation functions. Nevertheless, the current ONN circuit makes it difficult to monolithically integrate photonic nonlinear activation functions, so that it has allowed a hybrid architecture to transform optical signals to electrical signals and leave an ANN operation of the nonlinear function with an electronic-circuit-based digital calculation. Future ONNs need to be developed for realizing full advantage of all-optical neural networks including capability of photonic nonlinear response.

Recently, a new photonic machine learning model based on a projection-based photonic classifier (PPC) capable of the photonic nonlinear response was proposed using a Si photonic-integrated Mach–Zehnder interferometer (MZI) circuit [39]. For executing the nonlinear response, the proposed model adopts a support vector machine (SVM)-like algorithm [40]. The new algorithm determines a nonlinear

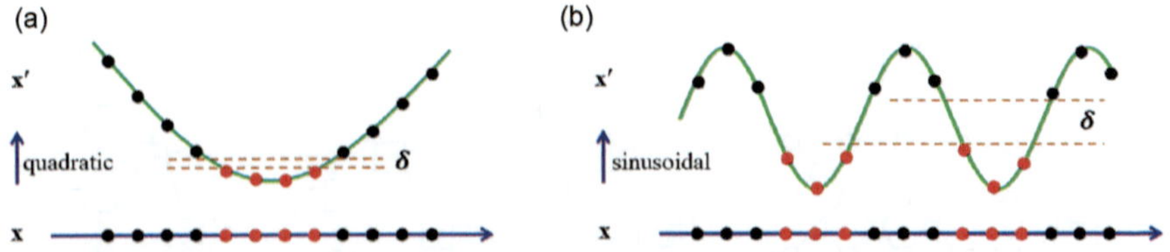

Fig. 5.25 **a** Schematic quadratic mapping function in conventional SVM. **b** Sinusoidal mapping function provided by the transmission vs voltage relation in the MZI device in the proposed PPC algorithm. Reproduced from [39 (supplementary information)], released under a CC BY 4.0 license; https://doi.org/10.1038/s41467-022-30906-3

mapping function by arranging the data-input scheme in the MZI mesh, thus realizing projection-enabled nonlinear classification in the higher-dimensional complex-amplitude space. Originally, the SVM is a powerful machine learning algorithm for solving problems such as classification and regression, i.e., the approach to determine the optimum hyperplane which separates binary-class data into two different classes by maximizing the margin of the support vectors (nearest data to the hyperplane). Figure 5.25a shows a schematic of a quadratic mapping function in the conventional SVM. Entangled two-class (black and red) data, inseparable into two in the low-dimensional space (x; on a linear function here), are easily classified into two by the hyperplane with δ separation after nonlinear mapping into the higher-dimensional space (x'; quadratic function here). In the proposed PPC with MZI devices, the quadratic function in the conventional SVM is replaced by the sinusoidal function provided by the transmission vs voltage relation in the MZI device, as shown in Fig. 5.25b.

Figure 5.26a shows a topology of the proposed PPC with MZIs in square shapes and external phase shifters (PhSs) in orange lines [39]. The top panel shows a schematic of the device topology with a delta-shaped area of data input for providing mapping functions $G(x)$, a column of PS, and a network of MZIs for implementing vector–matrix multiplication (VMM). Red and blue arrows show optical inputs/outputs and electronic data inputs, respectively. Figure 5.26b, c, and d shows topologies for classifying exclusive OR (XOR), Iris, and three-class nonlinear datasets (Circle, Moon, and Spiral), respectively, implemented using this system. XOR and nonlinear datasets involve two input parameters, namely, $x = (x_1, x_2)$, where $x_\mathrm{i} =$ 0 or 1 for XOR and $x_\mathrm{i} \in [0, 1)$ for nonlinear datasets with the exception that $x = (x_1, x_2, x_3)$ where $x_\mathrm{i} \in [0, 1)$ for the spiral dataset. The Iris dataset has four input parameters, i.e., $x = (x_1, x_2, x_3, x_4)$. Figure 5.26e shows a microscopic picture of the fabricated Si PPC chip, while Fig. 5.26f shows an enlarged MZI unit. An arm of the MZI involves a heater for driving an internal PS with a TO effect, thus giving a phase bias θ_jk as its learning parameter. Both ends of the MZI are connected to the directional couplers (DCs) for 50:50 switching.

Input parameters of XOR, Iris, and nonlinear datasets mentioned above produce their specific eight $G(\mathrm{x})$ equations, G_0 to G_7, with phase bias parameters θ_{01} to θ_{24}, as shown in Fig. 5.26b–d. Derivation of these equations are based on the MZI equation

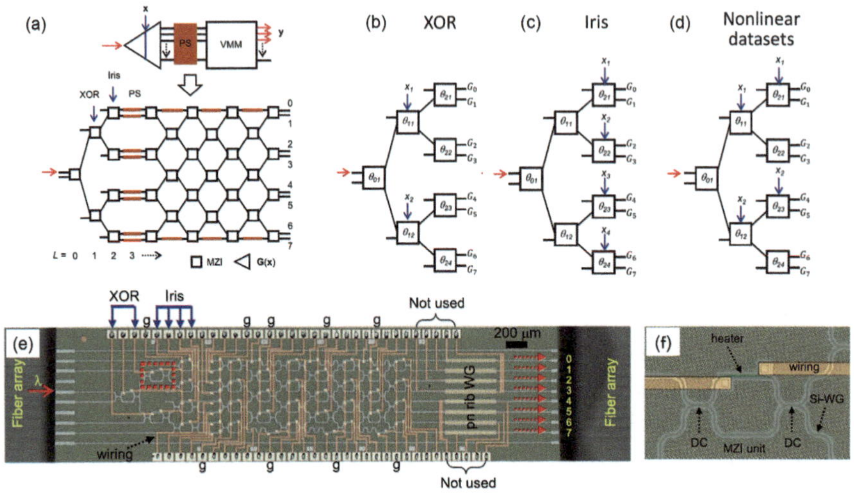

Fig. 5.26 **a** Topology of the PPC with MZIs and external PSs. The top panel shows a schematic of the device topology with a data-input area for forming mapping functions G(x)s, an array of PSs, and a neural network area for implementing VMM. **b** to **d** Topologies for classifying XOR, Iris, and nonlinear datasets (Circle, Moon, and Spiral). **e** and **f** Microscopic pictures of Si PPC chip and enlarged MZI unit, respectively. The MZI is operated with a TO effect of the internal PS. Both MZI ends are connected to the DCs for 50:50 switching. Reproduced from [39] (composite of Figs. 1 and 2a and b), released under a CC BY 4.0 license; https://doi.org/10.1038/s41467-022-30906-3

similar to Eq. (5.5) in Sect. 5.3.3, where ϕ corresponding to the outer phase shifter is zero in G(x)s. Input signals converted from electrical to optical domains via mapping functions G_i are changed to the output signals in the form of output complex vector $\mathbf{y} = \mathbf{A}G_i$, where **A** is a matrix product of the VMM and the PS (diagonal matrix) mentioned above. For training the device, as discussed later, the loss function of the mean squared error (MSE) with a factor of $(\mathbf{y}_t - \mathbf{y})^2$ plays an important role, where $\mathbf{y}_t$ is the target vector. Minimizing the loss is to maximize the power of the output **y**.

The proposed PPC has a large advantage that the algorithm of the PPC is based on the nonlinear SVM with no need of a nonlinear activation function and the device operator with a nonlinear projection operator $\boldsymbol{G}(\mathbf{x})$ and a linear matrix operator **A**. This means that the PPC can be realized only by passive Si photonic platform, different from the ANN-based ONN which needs on-chip nonlinear activation functions.

In the next, an application of the PPC to on-chip training for Iris classification is introduced.

Bacterial Foraging Training for Iris Classification

In the experiment, on-chip training using bacterial foraging optimization (BFO) [41] was implemented in the Si photonic circuit without a pre-trained network model on the computer as was often the case in the conventional ANN frame. The PPC device was applied to training and inference of Boolean logics (AND, OR, NAND, XOR,

XOR-AND, and OR-NAND) separation, Iris classification, and classification of three-class nonlinear datasets (Circle, Moon, and Spiral) [39, 41]. Figure 5.27 shows experimental results of the Iris classification. Input electronic data enter through four terminals (x_1, x_2, x_3, x_4) into the MZI network (Fig. 5.26c), transformed into the optical signals, and are projected to the high-dimensional complex-amplitude space via a nonlinear response in the MZI, thus operating the multiplication and accumulation. This operation successfully provides a boundary of the classification. The final result after classification is clarified by the position of the output port showing the maximum optical power. Parameters in the MZI are assigned by machine learning using the BFO algorithm. Figure 5.27a and b show measured power maps for the train and test sets after BFO training, respectively, using the benchmark; Iris flower classification. This benchmark identifies three species of the iris flowers (Setosa, Versicolor, and Virginica) from four datasets (lengths and sizes of the petals and sepals). The sample #1 ~ 30, #31 ~ 60, and #61 ~ 90 in the horizontal axis are assigned to the three iris species; Setosa, Versicolor, and Virginica, respectively, while the output ports, 1, 3, 5 out of eight ports in the vertical axis are assigned to these three iris species. The classification is correct if the output port giving the maximum optical power coincides with the port assigned to each iris species. Before training, the sample number and the optical power at the output port don't correlate and the result is unclassified yet. After training with 90 samples, however, classification is correct with a 94.44% accuracy rate. Similarly, the test experiment revealed a 96.67% accuracy rate, as shown in Fig. 5.27, where red arrows indicate the wrong recognition. Operation speed and power consumption in the experiment were less than one thousandth (~ 100 ps) and one tenth (360 mW) of digital electronic circuit operation, respectively. The proposed PPC work has a large potential for the photonic circuit capable of on-chip nonlinear classification.

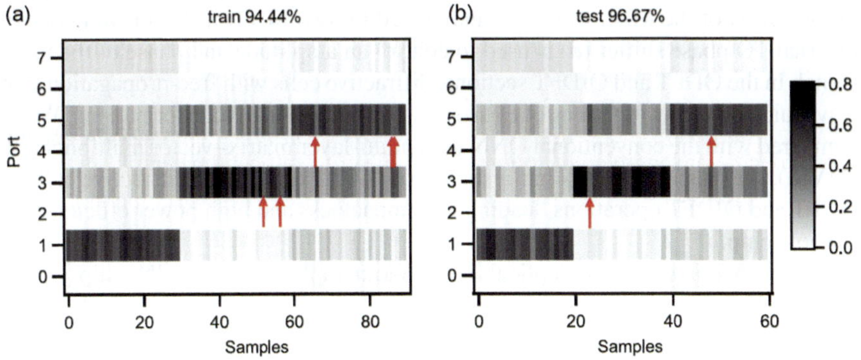

Fig. 5.27 **a** and **b** show measured power maps of the train set and test set, respectively after BFO training in Iris classification. Red arrows indicate the wrong recognition. (Reproduced from [39] (Fig. 5d), released under a CC BY 4.0 license; https://doi.org/10.1038/s41467-022-30906-3)

5.3.5 Space-Efficient Integrated Diffractive ONN

Architecture for Fourier Transforms in Diffractive Cell

As shown in Fig. 5.18a, ONN is an ANN architecture where each layer is usually connected to any other layers in the multi-layer networks. Meanwhile, a convolutional neural network (CNN) is another ANN commonly used to analyze tasks such as image/video recognition and natural language processing [42]. For fast processing, CNNs use convolution operation between input series (such as test image) and target series (such as target image) in their layers instead of general matrix multiplication.

Recently, a CNN-based new architecture using Fourier transform in a diffractive cell was proposed. The proposed idea is an integrated diffractive neural network (IDNN) chip using Si photonic-integrated circuits [43]. Figure 5.28a shows a concept of the multi-layer IDNN. Input coherent light is encoded with preprocessed signals by way of amplitude and phase modulation and then input into the multi-layer IDNN. As shown in Fig. 5.28b, input signals propagate through matrix multiplications and lead to a nonlinear activation function. Each layer surrounded by a yellow-line includes three sections; operation of optical discrete Fourier transform (ODFT) (W^{l}_{ODFT}), amplitude and phase modulation with complex values after ODFT to achieve Hadamard product, and operation of optical inverse discrete Fourier transform (OIDFT) (W^{l}_{OIDFT}). A nonlinear activation function is generated by detecting intensity of complex outputs. Figure 5.28c shows a whole diagram of the IDNN chip with four sections; (i) input preparation showing network of an optical waveguide tree with MZIs, MMIs, and phase shifters, (ii) ODFT operation showing a slab waveguide for free propagation of optical beams connected to input/output waveguides, (iii) complex modulation showing an optical waveguide network similar to (i), and (iv) OIDFT operation showing an optical free-propagation slab waveguide similar to (ii). In the input operation, two internal TO phase shifters (θ) are located on two arms of the MZI, which is sandwiched by two MMIs as beam splitters. An external TO phase shifter (ϕ) is used to control an amplitude and phase of the input signal. In the ODFT and OIDFT sections, diffractive cells with free-propagation slab waveguides serve as functions of Fourier and inverse Fourier transformations. When compared with the conventional ONN with multi-layer matrix–vector multiplication (MVM), the proposed IDNN has a superiority of the ultra-small diffractive cell for ODFT and OIDFT operations, leading to compactness and high power efficiency as a whole chip.

Figure 5.29a shows an optical microphotograph of the IDNN chip, while Fig. 5.29b shows a magnified photo of the compact diffractive cell involving an optical free-propagation planer waveguide for core operation of the Fourier transform. Figure 5.29c shows a simulated distribution of optical beams in the free-propagation waveguide, where an electric field E_0 (x) of the input light propagating along two input central waveguides is transformed into E_{i} (k) in an output k plane in the diffraction cell with arc radius R after ODFT operation. Then, simulated output intensity distribution as a function of k and normalized intensities at ten channels are shown in Fig. 5.29d and e, respectively. Meanwhile, Fig. 5.29f shows experimentally

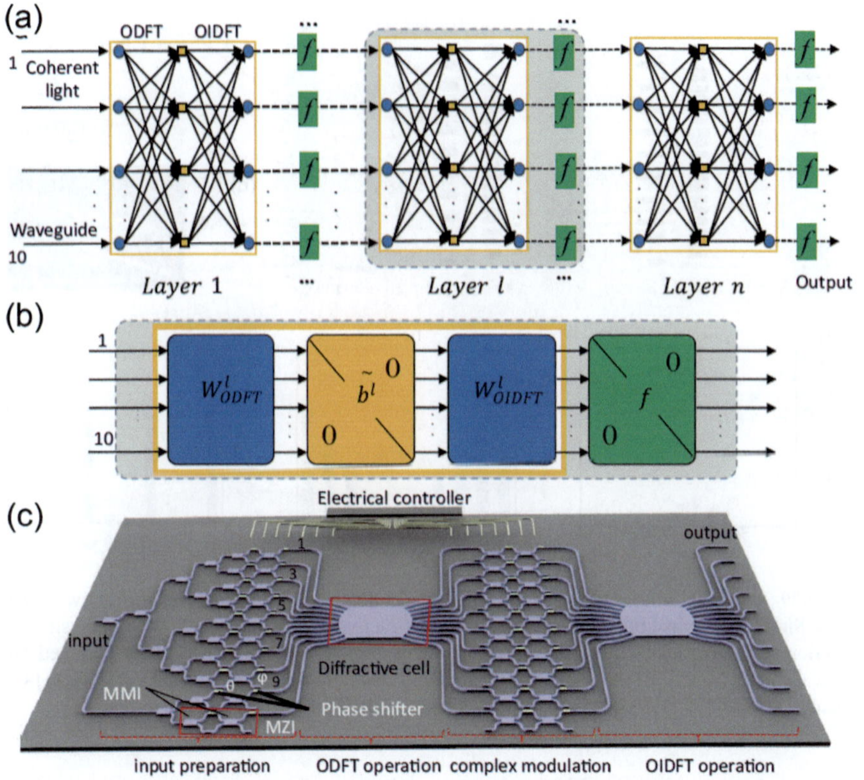

Fig. 5.28 **a** Theoretical framework of the IDNNs. Each layer contains processers for ODFT operation, amplitude/phase modulation, and OIDFT operation. **b** Three layers each with functions of ODFT (W^l_{ODFT}), amplitude/phase modulation after ODFT, and OIDFT (W^l_{OIDFT}). **c** Whole diagram of the IDNN chip with four sections including above three sections and an input preparation section. Reproduced from [43], released under a CC BY 4.0 license; https://doi.org/10.1038/s41467-022-28702-0

retrieved 1-D input sequence [1001001000] signals at ten output ports after operation of OIDFT. Note that the input signals are well-reproduced by operations of ODFT and OIDFT.

Recognition of 1-D Sequence and 2-D Digit Images

The proposed IDNN mentioned above performed the following tasks; recognition of 1-D sequence and 2-D three-digit images, classification of Iris flower, classification of handwritten digit, etc. [43]. As an example, the demonstrated recognition of 1-D sequence and 2-D three-digit images are shown in detail using convolution implemented by the proposed IDNN. Figure 5.30a shows a schematic of an image recognition process using the IDNN. Assuming two functions; $f(x)$ for input series and $g(x)$ for target series, a mathematical relation between the Fourier transform and convolution is expressed as:

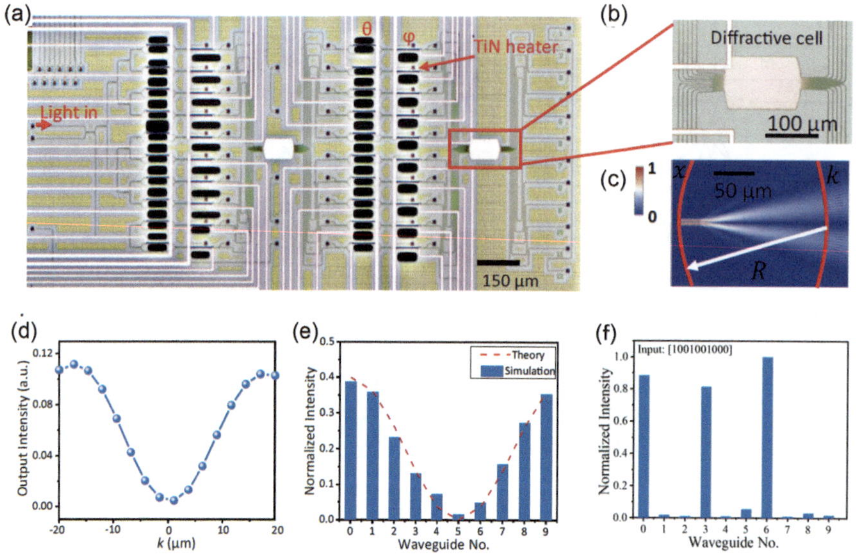

Fig. 5.29 **a** Optical microscope image of the whole IDNN chip. **b** Magnified diffractive cell in (**a**). **c** Simulated optical beam distribution in the diffractive cell. **d** Simulated output intensity as a function of *k*. **e** Normalized intensity at ten channels after ODFT. **f** Experimentally retrieved 1-D input sequence [1001001000] signals from all output channels after OIDFT. Reproduced from [43], released under a CC BY 4.0 license; https://doi.org/10.1038/s41467-022-28702-0

$$\mathscr{F}\{f(x) * g(x)\} = F(w) \circ G(w), \tag{5.7}$$

where *, ○, and $\mathscr{F}$ indicate convolution, Hadamard product, and Fourier transformation, respectively, while $F(\omega)$ and $G(\omega)$ indicate each Fourier transformation against $f(x)$ and $g(x)$, respectively. Furthermore, by using two vectors, $S(s_1, s_2, \ldots s_n)$ and $T(t_1, t_2, \ldots t_n)$, Hadamard product W is expressed as:

$$W = S \circ T = W(s_1 t_1, s_2 t_2, \ldots s_n t_n). \tag{5.8}$$

Turning back to Fig. 5.30a, convolution of input image ($n \times n$) and target image ($m \times m$) is performed by Hadamard product between $F(\omega)$ and $G(\omega)$. Resultant matching between the test image and target image is determined by the height of the convolution peak. For instance, in application to image recognition of 1-D 5-digit sequence, coincidence of a sample sequence [10101] and target sequence [10101] was experimentally obtained, as shown in Fig. 5.30b. Figure 5.30c shows another experimented recognition of the 2-D image, three-digit images (1, 2, 9) each with a 5 × 5 matrix. As a consequence, it is natural that the diagonal line shows strong correlation peaks among nine different digit images.

In an experiment on classification of Iris flower, 30 test images were correctly evaluated with only one incorrect classification using the one-layer IDNN chip [43]. An achieved classification accuracy was 96.7%, which was the same as the result in

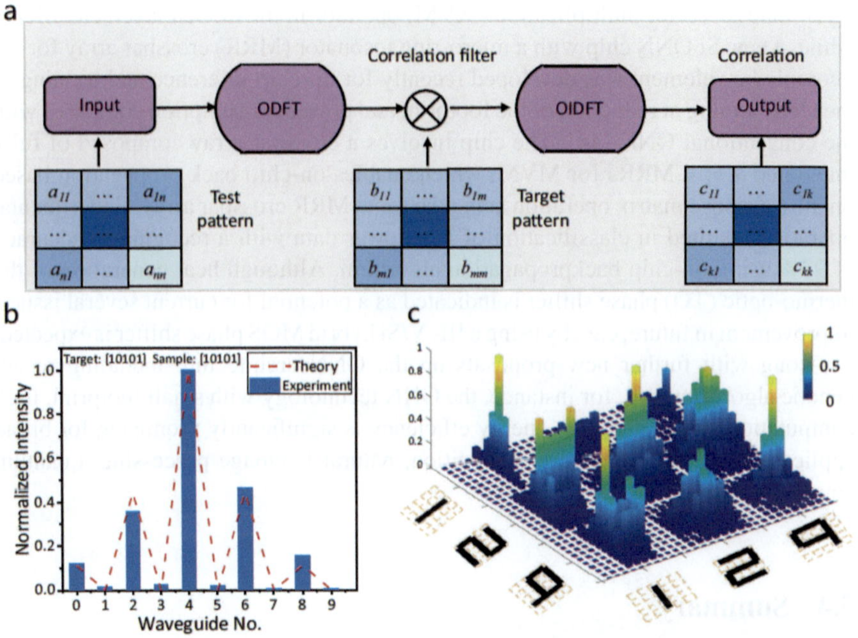

Fig. 5.30 **a** Schematic of an image recognition process. **b** Normalized output waveguide intensity with the same 1-D sequence [10101] for both the sample and target datasets. **c** Experimental results for 2-D digit image recognition using three-digit images (1, 2, 9). Reproduced from [43], released under a CC BY 4.0 license; https://doi.org/10.1038/s41467-022-28702-0

the conventional ONN with one layer. In the higher-level experiments on classification of the handwritten digit and fashion product, more complicated datasets were used to verify operation of the IDNN chip, namely the dataset "MNIST" for a typical handwritten digit and dataset "Fashion MNIST" for a typical clothing image. As a result, the classification tasks were successfully achieved with smaller footprint and lower energy consumption than the fully connected conventional ONN, accompanied also by comparable classification accuracies. Refer to [43] with supplementary information for more details on the testing tasks and their computational algorithms.

5.3.6 Ever-Growing Si Photonic Neural Networks

For the last decade, Si photonics-based integrated ONN technologies have significantly evolved from viewpoints of both the neural network architecture and the key element of photonic devices, as shown in the previous sections.

Referring to the device key element, for instance, a set of MZIs and phase shifters serves as conventional key elements of input optical signal encoder in the input layer, matrix–vector multiplication (MVM) operator in the hidden layer, etc. Meanwhile, a new Si ONN chip with a micro-ring resonator (MRR) crossbar array for the photonic key element was developed recently for on-chip inference and training of the ONN aiming at reduction of the footprint and power consumption compared with the conventional ONN [44]. The chip involves a crossbar array composed of fully integrated 4×4 MRRs for MVM, which enables on-chip back propagation based on the transpose matrix operation using the same MRR crossbar array. Experimental operation resulted in classification of Iris flower data with a recognition accuracy of 93% using on-chip backpropagation algorithm. Although heat generation of the thermo-optic (TO) phase shifter is indicated as a potential for current several issues, improvement in future, e.g., by using a III–V/Si hybrid MOS phase shifter is expected.

Along with further new proposals on the ONN architecture including recent genetic algorithm [45], for instance, the ONN technology with small footprint, high computation speed, and high energy efficiency is significantly promising for broad application abilities in pattern recognition, natural language processing, quantum computing, etc.

5.4 Summary

Two latest applications of Si photonics were introduced here: one is an integrated solid state Si LiDAR system with an optical beam steering device for remote sensing, autonomous car, etc. The other is a Si photonics-based ONN system with a matrix–vector multiplication architecture for deep learning of image recognition, dataset classification, etc.

For introducing a Si LiDAR system, a principle of a frequency-modulated continuous wave (FMCW) method was first reviewed for signal processing of radiated and reflected coherent light beams. Integrated optical phased arrays (OPAs) and waveguide grating arrays for beam steering were also reviewed. As examples of LiDAR, steering angles of $51^\circ \times 28^\circ$ was realized by a 32-channel III–V/Si OPA system with full 2-D beam steering. Outdoor targets including parked cars, fence posts, etc., were detected by on-chip 2-D FMCW–LiDAR using passive 512 OPAs without phase shifters. A 3-D LiDAR using active OPAs with phase shifters and an IR camera image recognized a person standing at ~ 7 m distance with his notable arms and legs, and two walls separated between 8 and 12 m. Unique steering and sensing devices for LiDAR were developed also using other techniques: One is a 3-D imaging LiDAR system using transmitter and receiver focal plane arrays (FPAs) with CMOS-based 2-D integrated image sensing devices. LiDAR operation for a rotating basketball succeeded in demonstration of point clouds distinguishing different point velocities with different colors. Another is a 3-D LiDAR system with a Si photonic crystal slow-light grating beam scanner. Slow light exhibits large sensitivity to tuning wavelength for beam steering. By using a rotating turntable with a diameter of 330 mm at a distance ~ 4 m, beat frequency spectra in FMCW were successfully measured with off-line point cloud 3-D image for both range and velocity of the turntable.

Meanwhile, Si ONN is based on an ANN model originated from signal processing in the brain. Here, a neural network architecture for input-, hidden-, and output-layers implementing matrix–vector multiplication and nonlinear activation function was reviewed first. Next, four types of ONN systems based on such architectures were introduced: The first type is a pioneering Si photonics-based ONN system with a programmable nano-photonic processor. The system includes an optical interference unit for matrix multiplication and an optical nonlinear unit for nonlinear activation. Training and inference tasks for vowel recognition was demonstrated. The second type is a high-accuracy ONN system with complex-valued input-/output-/hidden-layer networks having many integrated MZIs for input operation, weight multiplication, and coherent detection. Integrated MZIs are composed of two sets of 50:50 MMI beam splitters, an internal phase shifter (PS) in one arm, and an external PS for complex-valued computation. High-accuracy (> 90%) tasks were implemented for realization of two-bit logic gate including nonlinear XOR gate, classification of an Iris dataset, nonlinear circle datasets, etc. The third type is a projection-based new photonic classifier with a Si MZI circuit. The system demonstrated nonlinear mapping using the support vector machine (SVM)-like algorithm, thus realizing projection-enabled nonlinear Iris flower classification with high-accuracy rate of 94.44%. The fourth type is a space-efficient integrated diffractive ONN system with a Fourier transform architecture, i.e., convolutional neural network (CNN) without a fully connected multi-layer network like a usual ONN. The CNN algorithm conducts fast operation between input series (such as test image) and target series (such as target image) instead of general matrix multiplication. The CNN technique was applied to image recognition of 1-D five-digit sequence and 2-D three-digit images (1, 2, 9), Iris flower classification with accuracy of ~ 97%, etc. Si photonics applications will ever continue to progress toward new next-generation systems together with innovating state-of-the-art integration of electronics and photonics.

References

1. Marino RM, Davis WRJ (2005) A foliage-penetrating 3D imaging laser radar system. Lincoln Lab J 15:23–36
2. McManamon PF, Bos PJ, Escuti MK, Heikenfeld J, Serati et al (2009) A review of phased array steering for narrow-band electro-optical system. Proc IEEE 97:1078–1096
3. Doylend JK, Heck MJR, Bovington JT, Peters JD, Coldren LA, Bowers JE (2011) Two-dimensional free-space beam steering with an optical phased array on silicon-on-insulator. Opt Express 19(13):21595–21604
4. Flamand G, De Mesel K, Moerman I, Dhoedt B, Hunziker W, Kalmar A, Baets R, Van Daele P, Leeb W. (2000) InP-based PIC for an optical phased-array antenna at 1.06 m. IEEE Photonics Technol Lett 12:876–878
5. Kwong D, Hosseini A, Zhang Y, Chen RT (2011) 1 × 12 unequally spaced waveguide array for actively tuned optical phased array on a silicon nano-membrane. Appl Phys Lett 99:051104
6. Van Acoleyen K, Rogier H, Baets R (2010) Two-dimensional optical phased array antenna on silicon-on-insulator. Opt Exp 18(13):13655–13660

7. Doylend JK, Heck MJR, Bovington JT, Peters JD, Davenport ML, Coldren LA, Bowers JE (2013) Hybrid silicon free-space source with integrated beam steering. Proc of SPIE 8629:862911–862921
8. Sun J, Timurdogan E, Yaacobi A, Hosseini ES, Watts MR (2013) Large scale nano-photonic phased array. Nature 493:195–199
9. Sun J, Yaacobi A, Cole DB, Leake G, Coolbaugh D, Watts MR (2014) Two-dimensional apodized silicon photonic phased arrays. Opt Lett 39:367–370
10. Kwong D, Hosseini A, Covey J, Zhang Y, Xu X, Subbaraman H, Chen RT (2014) On-chip silicon optical phased array for two-dimensional beam steering. Opt Lett 39(4):941–944
11. Hulme JC, Doylend JK, Heck MJR, Peters JD, Davenport ML, Bovington JT, Coldren LA, Bowers JE (2015) Fully integrated hybrid silicon two dimensional beam scanner. Opt Express 23(5):5861–5874
12. Xie W, Komljenovic T, Huang J, Tran M, Davenport M, Torres A, Pintus P, Bowers J (2019) Heterogeneous silicon photonics sensing for autonomous cars [Invited]. Opt Express 27(3):3642
13. Fang W, Park H, Cohen O, Jones R, Paniccia MJ, Bowers JE (2006) Electrically pumped hybrid AlGaInAs-silicon evanescent laser. Opt Express 14(20):9203–9210
14. Abediasl H, Hashemi H (2015) Monolithic optical phased-array transceiver in a standard SOI CMOS process. Opt Express 23(5):6509
15. Pierrottet D, Amzajerdian F, Petway L, Barnes B, Lockard G, Rubio M (2008) MRS Proc 1076:1076
16. Poulton CV, Yaacobi A, Cole DB, Byrd MJ, Raval M, Vermeulen D, Watts MR (2017) Coherent solid-state LIDAR with silicon photonic optical phased arrays. Opt Lett 42(20):4091–4094
17. Poulton CV, Byrd MJ, Russo P, Timurdogan E, Khandaker M, Vermeulen D, Watts MR (2019) Long-range LiDAR and free-space data communication with high-performance optical phased arrays. IEEE J Sel Top Quantum Electron 25:1–8
18. Rogers C, Piggott AY, Thomson DJ, Wiser RF, Opris IE, Fortune SA, Compston AJ, Gondarenko A, Meng F, Chen X, Reed GT Nicolaescu R (2021) A universal 3D imaging sensor on a silicon photonics platform. Nature 590:11
19. Ito H, Kusunoki Y, Maeda J, Akiyama D, Kodama N, Abe H, Tetsuya R, Baba T (2020) Wide beam steering by slow-light waveguide gratings and a prism lens. Optica 7(1):47–52
20. Suyama S, Ito H, Kurahashi R, Abe H, Baba T (2021) Doppler velocimeter and vibrometer FMCW LiDAR with Si photonic crystal beam scanner. Opt Express 29(19):30727
21. Hamachi Y, Kubo S, Baba T (2009) Slow light with low dispersion and nonlinear enhancement in a lattice-shifted photonic crystal waveguide. Opt Lett 34(7):1072–1074
22. Kondo K, Tatebe T, Hachuda S, Abe H, Koyama F, Baba T (2017) Fan beam steering device using a photonic crystal slow-light waveguide with surface diffraction grating. Opt Lett 42:4990–4993
23. Abe H, Takeuchi M, Takeuchi G, Ito H, Yokokawa T, Kondo K, Furukado Y, Baba T (2018) Two-dimensional beam-steering device using a doubly periodic Si photonic-crystal waveguide. Opt Express 26:9389–9397
24. Takeuchi G, Terada Y, Takeuchi M, Abe H, Ito H, Baba T (2018) Thermally controlled Si photonic crystal slow light waveguide beam steering device. Opt Express 26(9):11529–11537
25. Baba T,. Tamanuki T, Ito H, Kamata M, Tetsuya R, Suyama S, Abe H, Kurahashi R (2022) Silicon photonics FMCW LiDAR chip with a slow-light grating beam scanner. IEEE J Selected Top Quantum Electron 28(5)
26. Jalali B, Mahjoubfar A (2015) Tailoring wideband signals with a photonic hardware accelerator. Proc IEEE 103:1071–1086
27. Kitayama K, Notomi M, Naruse M, Inoue K, Kawakami S, Uchida A (2019) Novel frontier of photonics for data processing—Photonic accelerator. APL Photon 4:090901
28. Reck M, Zeilinger A, Bernstein HJ, Bertani P (1994) Experimental realization of any discrete unitary operator. Phys Rev Lett 73:58–61
29. Clements WR, Humphreys PC, Metcalf BJ, Kolthammer WS, Walmsley IA (2016) Optimal design for universal multiport interferometers. Optica 3:1460–1465

30. Shen Y, Harris NC, Skirlo S, Prabhu M, Jones TB, M. Hochberg, X Sun, Zhao S, Larochelle H, Englund D, Soljačić M (2017) Deep learning with coherent nano-photonic circuits. Nature Photon 11:441–447. www.nature.com/naturephotonics
31. Shen Y, Harris NC, Skirlo S, Prabhu M, Jones TB, Hochberg M, Sun X, Zhao S, Larochelle H, Englund D, Soljacic M (2016) Deep learning with coherent nano-photonic circuits. arXiv: 1610.02365v1 [physics. optics] 7 Oct 2016
32. Miller DAB (2015) Perfect optics with imperfect components. Optica 2:747–750
33. Cheng Z, Tsang HK, Wang X, Xu K, Xu J-B (2014) In-plane optical absorption and free carrier absorption in graphene-on-silicon waveguides. IEEE J Sel Top Quantum Electron 20:43–48
34. Harris NC et al (2015) Bosonic transport simulations in a large-scale programmable nano-photonic processor. https://arxiv.org/abs/1507.03406
35. Gu ZM, Jiang XD, Thompson J, Cai H, Paesani S, Santagati R, Laing A, Zhang Y, Yung MH, Shi YZ, Muhammad FK, Lo GQ, Luo XS, Dong B, Kwong DL, Kwek1 LC,Liu AQ (2021) An optical neural chip for implementing complex valued neural network. Nat Commun 12(457):1–12
36. Hirose A, Yoshida S (2012) Generalization characteristics of complex-valued feedforward neural networks in relation to signal coherence. IEEE Trans Neural Netw Learn Syst 23:541–551
37. Reichert DP, Serre T (2013) Neuronal synchrony in complex-valued deep Networks, Preprint at https://arxiv.org/abs/1312.6115v5
38. Arjovsky M, Shah A, Bengio Y (2016) Unitary evolution recurrent neural networks. In: Langford J (ed). International conference on machine learning, pp 1120–1128
39. Cong G, Yamamoto N, Inoue T, Maegami Y, Ohno M, Kita S, Namiki S, Yamada K (2022) On-chip bacterial foraging training in silicon photonic circuits for projection-enabled nonlinear classification. Nat Commun 13:3261
40. Schölkopf B, Smola AJ (2022) Learning with Kernels: support vector machines, regularization, optimization, and Beyond Ch. 1 and 2. MIT Press, London
41. Cong G, Yamamoto N, Inoue T, Okano M, Maegami Y, Ohno M, Yamada K (2019) Arbitrary reconfiguration of universal silicon photonic circuits by bacteria foraging algorithm to achieve reconfigurable photonic digital-to-analog conversion. Opt Express 27:24914–24922
42. Valueva MV, Nagornov NN, Lyakhov PA, Valuev GV, Chervyakov NI (2020) Application of the residue number system to reduce hardware costs of the convolutional neural network implementation. Math Comput Simul. Elsevier BV. 177:232–243
43. Zhu HH, Zou J, Zhang H, Shi YZ, Luo SB, Wang N, Cai H, Wan LX, Wang B, Jiang XD, Thompson J, Luo XS, Zhou XH, Xiao LM, Huang W, Patrick L, Gu M, Kwek LC, Liu AQ (2022) Space-efficient optical computing with an integrated chip diffractive neural network. Nat Commun 13(1044):1–9. https://doi.org/10.1038/s41467-022-28702-0
44. Ohno S, Tang R, Toprasertpong K, Takagi S, Takenaka M (2022) Si microring resonator crossbar array for on-chip inference and training of the optical neural network. ACS Photon 9:2614–2622
45. Zhang H, Thompson J, Gu M, Jiang XD, Cai H, Liu PY, Shi Y, Zhang Y, Karim MF, Lo GQ, Luo X, Dong B, Kwek LC, Liu AQ (2021) Efficient on-chip training of optical neural networks using genetic algorithm. ACS Photon 8(6):1662–1672

Chapter 6
Applications of Photonic Crystal Waveguides

Abstract Applications of PC waveguides to advanced photonic networks are overviewed here by selecting four typical ones. Their origins are introduced in Chap. 2 as selected topics on the PC waveguides, i.e., slow light in the defect-mode PC band, fabrication of low-loss air-bridge PC waveguides, extremely high-Q 2-D PC nano-cavity, and coherent surface emission at the Γ-point in the 2-D PC. The first slow-light PC application to signal processing includes a tunable delay/dispersion device, symbol-rate variable differential quadrature phase-shift keying (DQPSK) receiver, etc. The second ultra-fast all-optical switch application includes multi-functional key configuration of symmetrical MZI-based 2-D PC device modified with large optical nonlinearity of quantum dots (QDs) embedded in the PC. The third optical nonlinearity-based bi-stable device application includes a high-Q nano-cavity in the 2-D PC waveguide for integrated optical DRAMs. The final innovative 2-D PC surface-emitting laser (PCSEL) application includes various excellent characteristics; unprecedented tens-of-watt class high power operation using 3 mm-emission area with single mode and high beam-quality. Unique applications to compact and high-resolution 3-D LiDAR are also reported using new types of PCSELs such as double-hole lattice and dually modulated PCSELs.

6.1 Introduction

A novel concept of the photonic crystal (PC) providing ultimate optical energy confinement into high index-contrast nano-scale periodical media was proposed in 1987. Since then, unique PC-specific structures such as nano-scale cavity with missing air-holes and line-defect PC waveguide with various shapes were reported tremendously from the 1990 to the early 2000s. Among many review articles dealing with such PC-related optical devices, Benisty overviewed advancement of PC devices in wide perspective from brief history and basic concept of the PC to PC circuitry, passive/active components in 2-D PCs, and 3-D PCs [1]. Among them, PC circuitry includes waveguide dispersion, waveguide loss, waveguide bends, and micro-cavity.

K. Asakawa et al., *Nano-photonics for Advanced Networks*,
https://doi.org/10.1007/978-3-031-80637-7_6

Passive 2-D PC components include super-prisms/collimators and negative refraction [2], optical delay lines [3], add-drop filters [4], PC-guided mode coupling [5], while active 2-D PC components include surface-/edge-emitting lasers and nonlinear PC devices. Some of them are introduced in Chap. 2 as selected topics on the PC waveguides, that is, slow light in the dispersive waveguide, low-loss 2-D waveguide, high-Q nano-cavity, and 2-D surface emission.

In this chapter, applications of PC waveguides to advanced photonic networks are overviewed by selecting four topical PC waveguides as more advanced ones than those in Ref. [1]. The first application is a slow-light PC for signal processing devices such as tunable delay/dispersion device. As mentioned in Chap. 4 (Fig. 4.22), a node device connecting multi-ring networks is inevitable in the WDM optical communication system. In a photonic packet router as such a node device, an optical buffer is indispensable for control of delay in optical signal and management of timing in sending optical packets [6]. Yet, current tunable delay lines in practical use are difficult to realize a compact on-chip optical buffer because they are forced to depend on mechanically movable optics for changing an optical route in a slow (~ millisecond) response time. Slow light in PC increases light-matter interaction (or compresses optical energy) within the 2-D PC waveguide, resulting in enhancing gain, phase shift, nonlinearities, etc., in the waveguide devices. The signal processing device such as tunable delay using the slow-light PC is based on these physical mechanism. In this chapter, the slow-light PC provides a compact delay line with faster continuous tunable ability and other related tunable devices using a Si photonic platform.

As the second PC application, a line-defect 2-D PC waveguide with embedded semiconductor quantum dots (QDs) provides an all-optical switch with ultra-fast (~ ps) switching time and ultra-low (~ fJ) switching energy based on large optical nonlinearity of the QDs [7]. A key switch element is a MZI structure with GaAs/AlGaAs-based 2-D PC waveguides and InAs-based optical nonlinear QDs. Analog and digital ultra-fast all-optical switches for future photonic network are proposed with two different configurations of switching elements; a symmetrical MZI analog switch (PC-SMZ) and a flip-flop MZI digital switch (PC-FF). Analog switching was demonstrated at repetition frequencies of 20–40 GHz, while digital flip-flop switching operation was confirmed as a proof-of-principle experiment by using discrete optical components instead of the completed PC-FF. See Sect. 6.3.2 for detail.

The third PC application is an integrated optical DRAM using 2-D PC high-Q nano-cavities [8]. The optical DRAM is the first implementation of integrated wavelength addressable optical bit memories in a large scale. A noticeable point lies in configuration of wavelength-dependent PC high-Q nano-cavities connected to a coupled common bus waveguide on a chip. Using the PC high-Q nano-cavity, operation of a very small bi-stable optical memory under low energy consumption was implemented based on enhanced optical nonlinearity in the nano-cavities. In the later section, 28-bit and 128-bit optical DRAM operations using InGaAsP/InP BH-PC laser substrate and Si PC substrate are introduced, respectively. Note that initial bias-light, write pulse, reset pulse, and read pulse impinging into the PC waveguide produce bit information in the bias-light output, thus establishing large-scale bi-stable switches/memories.

The final PC application is a robust challenge of a PC surface-emitting laser (PCSEL) capable of band-edge (Γ-point) emission in the 2-D PC structure with variously modified lattices. Outstanding performances established recently involve single-mode RT-CW operation with more than tens watt-class output power and less than 0.2° beam divergence at a large emission area (mm-class) [9–11]. Conventional vertical-cavity surface-emitting lasers (VCSELs) are widely applied in practical use such as computer mouse, laser printer, optical communication, etc. Yet, single-mode output powers in the VCSEL are limited to mW classes due to small emission areas, while higher powers with transverse multi-modes degrade beam qualities seriously due to mixing of higher-order modes. In contrast, PCSELs have solved such problems by various innovative ideas and now verified many excellent features. One of the greatest innovations is a design with a double-lattice configuration, which has a theoretical possibility of 100 W–kW-class optical power in a cm-class lasing area under the most advanced design method [12]. These results are overviewed in Sect. 6.5.5 [13, 14]. Another modified PCSEL is now under a new application to a LiDAR system, which is also introduced in Sect. 6.5.7 [15]. Works on these PCSELs represent an important milestone for innovation in the high-power laser processing market. They also open the door for various applications in remote sensing, laser medicine, nonlinear optics, etc.

6.2 Slow-Light Photonic Crystal Device

6.2.1 Potential Application of Slow-Light Devices

Slow light in the specially designed 2-D PC lattice can be applied to digital signal processing that manipulates optical pulse delays controlled by the dispersive media in the photonic network. Recent applications of these slow-light effects can be categorized into tunable delay devices and enhanced light-matter interaction devices, as shown in Fig. 6.1 [6].

The first category shows applications to the tunable delay devices: optical buffer memories have long been expected in the photonic network, yet, had problems to be cleared because optical memories practically need over 1-μs for a buffer time. If we consider a 2-D PC waveguide with a group index n_g of 40, a 1-μs buffer time is equivalent to a propagation length of 7.5 m; an almost impractical length. Here, n_g is a reciprocal of the group velocity, meaning that a low/zero group velocity corresponds to a large/infinity group index. Instead, more realistic applications are retiming and (de)multiplexing of pulses, which do not need such long delays. For other applications such as dispersion compensation, arrayed waveguides, pump-probe, and correlation measurements, slow light contributes to arbitrary delay tuning by demonstrating its ability of group-velocity-dispersion (GVD) control.

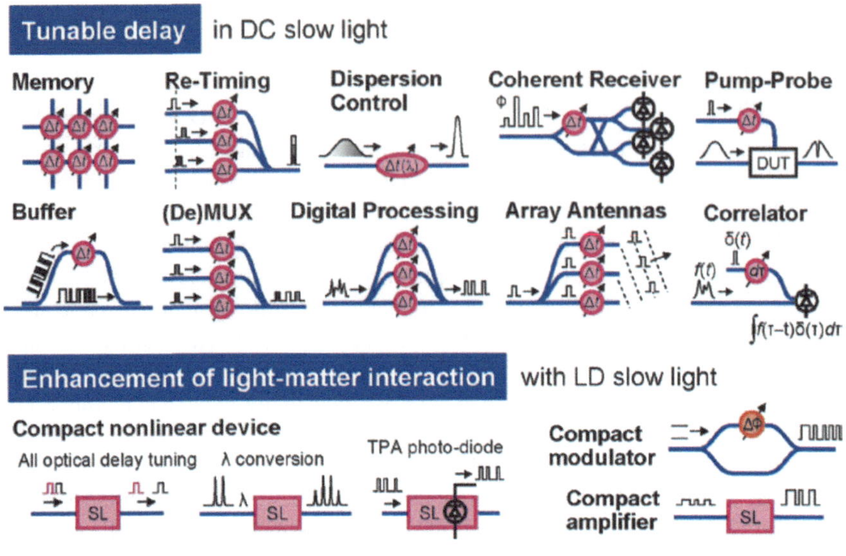

Fig. 6.1 Typical applications of slow light. **a** Applications based on tunable delays. **b** Applications based on enhanced light-matter interaction. Reprinted with permission from Ref. [6]. © 2013 IEICE

On the other hand, the second category shows applications to enhanced light-matter interaction devices as a result of an enlarged light-propagation period and enhanced peak intensity of pulses caused by the slow-light effect. Capable applications within this category are compact nonlinear devices, compact modulators, and amplifiers caused by two-photon absorption, self-phase modulation, etc. In the next section and later, several promising applications are introduced in more detail.

6.2.2 Dispersion Control of Slow-Light Waveguide

Dispersion-Compensated and Low-dispersion Slow Light

Slow light emerging near the band edge induces a narrow bandwidth, while slow light caused by the distortion of the photonic band leads to higher-order dispersions. Based on these properties, two different slow-light origins can be considered. Figure 6.2a and b shows slow-light behaviors based on dispersion compensation and low dispersion, respectively [16]. The dispersion-compensated slow light appears in the photonic band where opposite GVD characteristics exist at the opposite site of a flat area. The photonic band shifts in position owing to a chirped waveguide with varied structural parameters along the waveguide. Each wavelength component of an incident short optical pulse is slowed down and delayed, resulting in total compensation of the GVD after passing through the opposite GVD area. Meanwhile, the slow light based on the low dispersion appears in the nearly standard PC waveguide

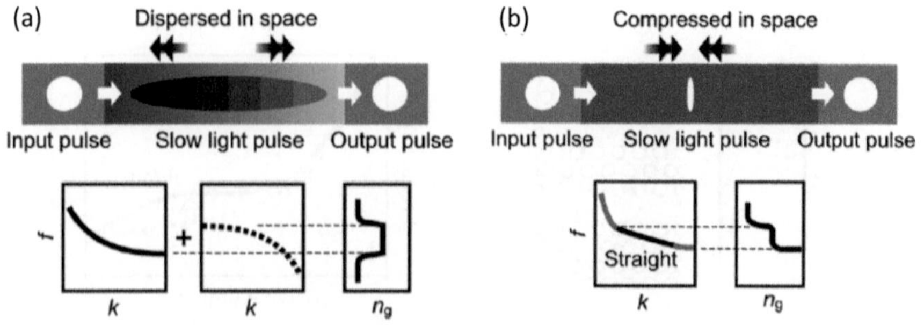

Fig. 6.2 Schematics explaining two types of slow light. **a** Slow light with dispersion compensation. **b** Slow light with low dispersion. (Upper) Models for behaviors of slow light. (Lower) Schematic relations between PC mode spectra and n_g spectra. Reprinted with permission from Ref. [16]. © 2009 The Japan Academy

with slightly modified lattice configuration, where a tilted part with a small angle appears in the photonic band, as shown at the bottom panel in Fig. 6.2b. In general, the PC band is caused by anti-crossing of a PC mode with confinement induced by line-defects and a PC mode with confinement induced by a bandgap. A straight slow light band is caused by an appropriate balance between these two effects. In the followings, two peculiar PC slow light waveguides are introduced in more detail.

Chirped Photonic Crystal Coupled Waveguide

One of the pioneering works on realizing a slow light waveguide is a proposal of a chirped PC coupled waveguide (PCCW), where a symmetric directional coupler (DC) is used for the coupled waveguide [17]. As shown in Fig. 6.3a, a symmetric PC DC is based on two identical PC waveguides with a line-defect part isolated by spacing with several air-hole lines (three lines here). In the PC DC, two photonic bands; even and odd modes (coupled modes), are considered, as indicated in the left panel. See Fig. 1.10 for detailed DC structure. Figure 6.3b shows a schematic of even (dotted curve) and odd (solid curve) modes. Focusing again on the air-hole spacing region in (a) with an increased air-filling factor, the even mode has a larger influence on the effect of decrease in an effective refractive index in the spacing region than the odd mode due to a larger-intensity mode field on the spacing region. Thus the even mode raises its own band into the high-frequency direction, as shown in (b), consequently producing a flat zone with the zero v_g in the even mode; the slow-light zone. The proposed PC DC for the slow light is based on this idea. Figure 6.3c shows a schematic of the designed slow-light PC pattern with a symmetric DC. The triangle lattice consists of a lattice constant a, background index n of 2.963, and basic air-hole diameter $2r$ of $0.66a$, while the center air hole is enlarged as $2r = 0.89a$. For detailed investigation, the band shape varied largely with an air-hole shift $d = 0$–$0.25a$ and the positive slope (negative υ_g) resulted in zero at $d = 0.25a$. Figure 6.3d shows calculated even (thick) and odd (thin) mode bands, where the target state of $\upsilon_g = 0$ is found to exist in the even mode.

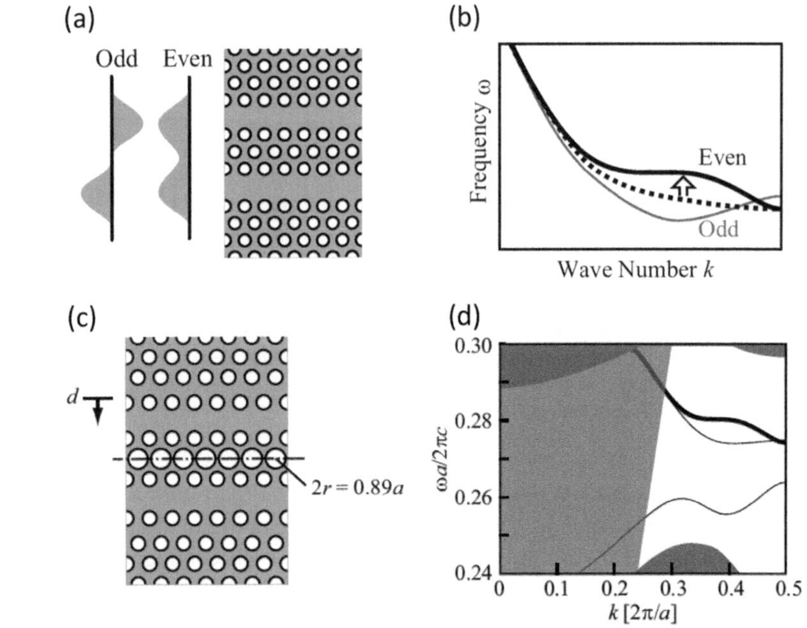

Fig. 6.3 **a** (Right) Schematic of lattice pattern in PC DC. (Left) Even and odd coupled-mode profiles. **b** PC bands in corresponding even mode with a flat zone and odd mode. **c** Typical lattice pattern in proposed PCCW. **d** PC bands in (**c**), where air-hole shift d is 0.25a. Thick and thin lines correspond to even and other modes, respectively. Reprinted with permission from Ref. [17]. © 2005 OSA

For developing the new PC slab slow-light device, the proposed PCCW was investigated by adopting an index-chirped DC in the structure shown in Fig. 6.3c [17]. Wavelength λ and lattice constant a were 1.55 μm and 0.42 μm, respectively. For designing the PCCW with a chirped length of 50 μm, n was linearly varied as 2.963–2.870. On a finite-difference time-domain (FDTD) analysis, a TE-polarized pulse with a 260-fs width in the even mode was incident on the chirped DC, reached a zero-υ_g zone (slow-light zone), stopped temporarily, then propagating again to reach the output edge of the DC. The observed delay time ΔT was 3.5 ps, i.e., an estimated average group velocity of c/20.9 (average group index of 20.9). On another FDTD analysis using a differently chirped DC band with two extreme points, the pulse was found to behave a unique motion. The incident light pulse, repeating back- and forth-propagation between the end and start points, finally passed through the device toward the output end. These slow-light behaviors of the optical pulse including the "back and forth" motion are extremely expected to produce novel applications such as tunable delay and enhancement tool of light-matter interaction, as shown in Fig. 6.1.

As explained above, the PCCW needs an even mode in the DC for making the most of the effective flat-band property. To this end, input/output structures optimized for the even mode have already been proposed: A pair of half-elliptical tapers are inserted between the PCCW input-end and the bend-waveguide-based input branch, while a circular funnel taper is attached to the PCCW output-end as an output confluence [17].

Lattice-Shifted Photonic Crystal Waveguide

The PCCW composed of the DC is a useful structure for producing dispersion-compensated slow light, as shown in Fig. 6.2a, because optical signals are first dispersed, then recovered by the dispersion compensation. Yet, the whole structure is not so simple for applications to an enhanced nonlinear effect device, for instance. Instead, a new type of the PC slow-light waveguide—a lattice-shifted PC waveguide (LSPCW) with a low group-velocity and low group-velocity dispersion (LVLD)—was proposed [18]. An example of the typical LSPCW is shown later by the inset in Fig. 6.4a. The LSPCW is structured by narrowing a width of the line defect, increasing a diameter $2r$ in the air-hole, and decreasing a diameter $2r'$ of the innermost air-hole. Importantly, the LSPCW band shows a linear and almost flat region. The LVLD areas with flat group-velocity regions are provided by arranging parameters $2r/a$ and $2r'/2r$. The LSPCW as well as PCCW will be applicable to optical buffer devices and nonlinearity enhancement of optical signals.

Optical Nonlinearity Enhancement in LSPCW

The LSPCW mentioned above is effective for spatially compressing optical pulses and enhancing longitudinal optical density, so useful for applications such as nonlinearity enhancement device, as shown in Fig. 6.4 [19]. The inset in Fig. 6.4a shows an example of the triangular-lattice LSPCW pattern, where parameters a and $2r$ mean a period and diameter of air-holes, respectively, while a parameter s means a shifted value of the third rows of air-holes along the line defect. Four low-dispersion

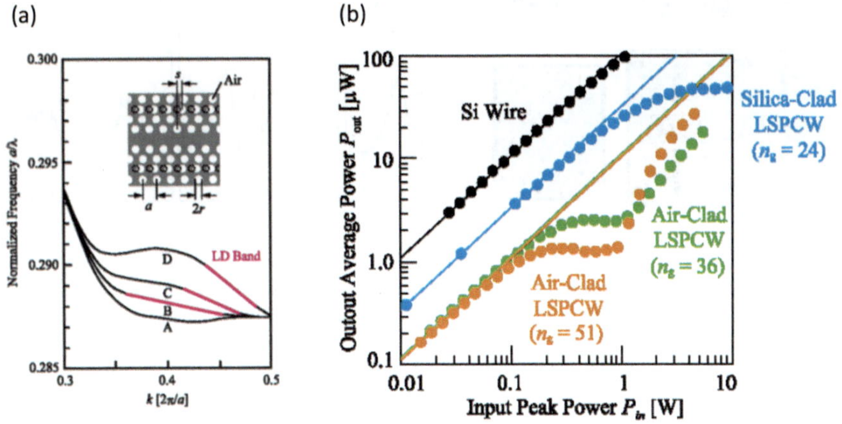

Fig. 6.4 **a** Photonic bands of the air-clad LSPCW waveguide with $2r/a = 0.658$. A to D correspond to s/a of 0, 0.13, 0.18 and 0.22, respectively. **b** Output power P_{out} versus input power P_{in} for four types of PC waveguides; Si wire, silica-clad LSPCW and air-clad LSPCW. Reprinted with permission from Ref. [19]. © 2011 OSA

photonic bands are distinguished by A, B, C and D, where $s/a = 0$, 0.13, 0.18, and 0.22, respectively. Note that the straight sections in the photonic band in red represent low-dispersion slow-light regions.

Nonlinear characteristics using the LSPCW were measured using an optical pulse width of 3.1 ps and input power P_{in} of less than 30 W. In general, a Si nano-wire waveguide increases an optical intensity within the waveguide by 1000 times, for instance, compared to that in a fiber. As a result, several optical nonlinear phenomena such as self-phase modulation (SPM), two-photon absorption (TwPA), etc., are generated easily [20]. For comparison with the Si wire waveguide, experiments on several LSPCWs were implemented, as follows: Fig. 6.4b shows output power intensity responses versus P_{in} for LSPCWs. For the Si wire waveguide, the output average power P_{out} saturates over $P_{in} \sim 1$ W mainly based on TPA. For air-clad LSPCWs (n_g = 35 and51), however, the output power P_{out} saturates at $P_{in} \sim 0.1$ W, whereby it was verified that optical nonlinearity with slow light in the LSPCW is enhanced by over 100 times when compared with Si wires.

These devices were fabricated using a CMOS-compatible process on SOI. A top Si-layer and silica-box-layer were 220 nm and 2 μm thick, respectively. Since a Si wire waveguide edge was efficiently coupled to an optical fiber via spot-size-converters (SSCs), input optical pulses were introduced to the Si waveguide without serious power loss. These technologies were successfully applied to the tunable delay devices, as shown later (Fig. 6.5).

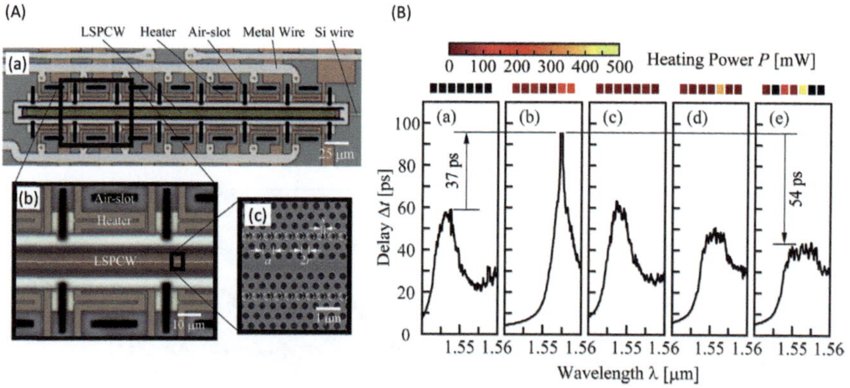

Fig. 6.5 **a** Demonstrated tunable delays and dispersion device with LSPCW. (a) Optical microscope photograph of the whole device configuration. (b) Photograph of LSPCW sandwiched with heaters. (c) Enlarged photograph of air-hole configuration in LSPCW. **b** Heating-pattern-dependent delay spectra. (a) Spectrum without heating. (b) to (e) Spectra with delay caused by various heating patterns (indicating with seven squares). Reprinted with permission from Ref. [21]. © 2012 AIP

6.2.3 Demonstration of Slow-Light Device

In the early-stage development of the PC slow-light waveguides, a wide range of theoretical proposals and experimental demonstrations were carried out [6, 16]. High-quality PC waveguides with PCCW and LSPCW configurations were realized using a CMOS-compatible process on SOI substrate and their applicability to optical pulse-delay devices such as integrated tunable delay device using the LSPCW, symbol-rate variable DQPSK receiver, and compact MZ modulator were certified. These applications are briefly overviewed in the followings.

Tunable Delay and Dispersion Device

A tunable delay and dispersion device with the PC slow-light waveguide and integrated multiple heater for compensating fabrication errors were demonstrated [21]. Figures 6.5a shows entire and magnified optical micrographs of the LSPCW device: (a) shows a whole LSPCW. (b) shows a magnified area around the heater. (c) shows a magnified area of a missing air-hole line-defect waveguide and surrounding lattice-shifted air-holes in the LSPCW. The PC slow-light waveguide is composed of triangular-lattice air-holes with an air-bridge-type Si slab. For realizing the LSPCW, the third-row lattice was shifted by $s = 100$ nm with wide slow-light band. For providing a tunable delay function, TiN heaters in seven pairs were arranged on each side of the LSPCW. In addition, air slots were prepared around the heaters for raising a heating efficiency. Figures 6.5b show several delay spectra with various heating patterns on top (color rectangles) indicating heating powers, namely (a) a spectrum with no heating, (b) a spectrum with a sharp peak delay owing to the adjusted heating, and (c)–(e) decreasing spectra-tops with maximum delay-reduction of 54-ps owing to further-change of heating patterns. Note that the sharp spectral peak in (b), degraded before heating (see (a)) due to fabrication errors of ≤ 5 nm in air-hole diameter, was ideally improved by the effect of the linear index-chirp. The device shown here is promising for tunable dispersion compensator.

Symbol-Rate Variable DQPSK Receiver

As indicated above, a low-v_g slow light is capable of inducing a large delay with a compact structure, thus promising for tunable-delay slow-light devices in optical signal processing, as shown in Fig. 6.1. In the explosively growing traffic in networks like data centers, electrical signals suffer from severe challenges that limit their performances, namely delay and loss generated in copper wires for electronics. Optical transceivers with low cost and high performance are capable of solving this problem. In a recent year, a differential quadrature phases-shift keying (DQPSK) receiver—an optical coherent receiver—has been proposed for one of such applications [22]. In Fig. 6.6a, (a) shows a schematic configuration of the Si DQPSK receiver, while (b) shows an optical microscope photograph of the demonstrated Si DQPSK receiver. In (a), input optical signals are divided into two identical variable optical attenuators (VOAs) with heater-controlled asymmetric Mach–Zehnder interferometers (MZIs). VOAs are needed for balancing input powers, leading to the

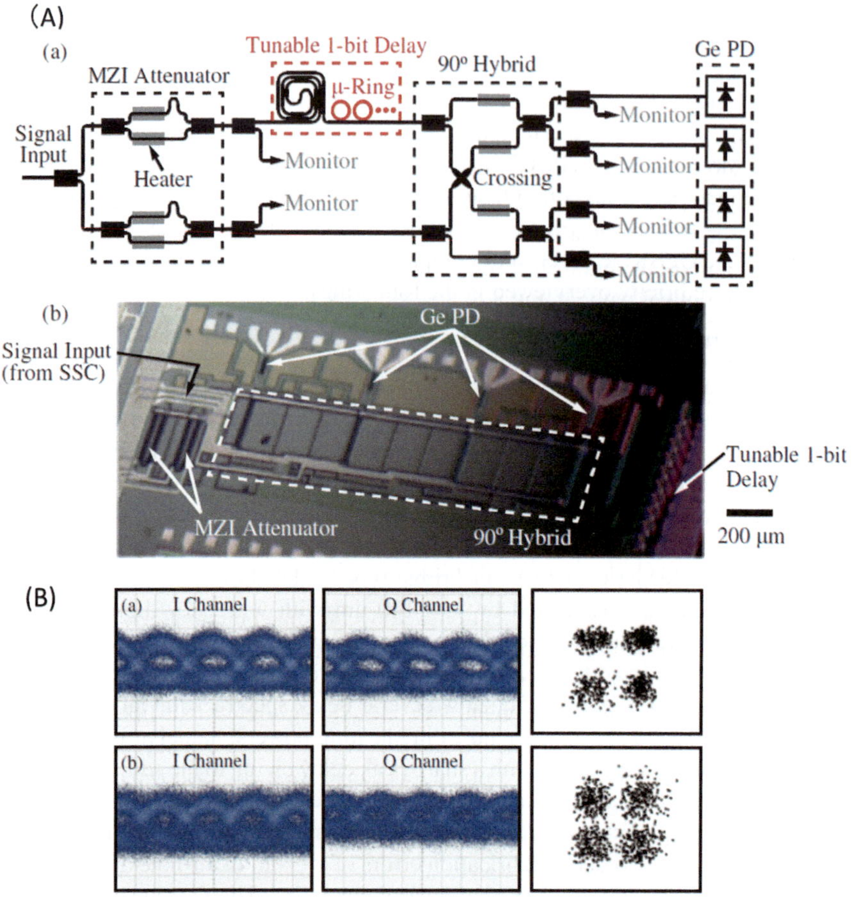

Fig. 6.6 **a** Structure of Si DQPSK receiver. (a) Schematic configuration of the whole receiver chip. (b) Photograph of the fabricated receiver chip. **b** Modulated DQPSK signals. (Left and center) Eye-patterns. (Right) Constellations. (a) Symbol rate of 7.4 Gbaud. (b) Symbol rate of 9.0 Gbaud. Reprinted with permission from Ref. [22]. © 2012 OSA

90° hybrid. One side of the VOA is connected to a tunable one-bit delay line with a race-track micro-ring array (MRA). The tunable delay line involves two parts; a fixed-delay part with a long Si waveguide and a tunable delay (in red). Note that the tunable delay involves a 10 race-track MRA as a slow-light device. Individually different heating power (3.0–0 mW) injected into each ring successfully achieves one-bit delay corresponding to each of three different DQPSK symbol rates (7.4–9.0 Gbaud), as shown later. Then, the signals are led to the 90° hybrid with a low-loss crossing and phase shifters, and to balanced Ge PDs. In the 90° hybrid component, input and output ports are arranged with 1×2 MMI splitters and 2×2 MMI couplers, while heater-controlled phase shifters are arranged in between. Furthermore, two identical circuits are connected to a mode-expansion-type crossing. Footprint in the monolithically integrated chip is as small as ~ 1.0 × 2.5 mm^2, while the Si waveguide with a 7.13-mm length causes a 100-ps delay ($n_g = 4.2$).

Tunable one-bit delay in the device was evaluated by the pattern-locked waveform of delayed non-return-to-zero (NRZ) pseudo-random binary sequence signals. Figure 6.6b shows eye-patterns and constellations for 7.4 Gbaud (a) and 9.0 Gbaud (b), corresponding to bit-rates of 14.8 and 18.0 Gb/s, respectively. Demodulated constellation signals in right panels are well-split into four parts although their convergences are not so good. These successful results are owing to good performances in the following key components: Input powers were balanced with VOAs, while symbol rates of the input signals were optimized with tunable one-bit delay elements and phase-adjusted in the 90° hybrid.

Compact MZ Modulator

A MZI optical modulator is a key optical component in the modern photonic network. Yet, compatibility of size reduction to a hundred μm or less and high performance in speed and bandwidth are still challenging issues for applications to optical transceivers, switches, etc. In spite that the PC waveguide is expected for improving these challenges, a fabrication-induced imperfection in the PC waveguide is likely to bring about random phase differences between the two Mach–Zehnder modulator (MZM) arms. To solve this problem, a new compact MZM with a LSPCW (LSPCW–MZM) for reducing an arm length was proposed [23]. For demonstration, two different configurations of the LSPCW–MZM were prepared; dual LSPCW–MZM and single LSPCW–MZM. Brief configuration in device component and performance in RF modulation for both devices is explained, as follows. The dual LSPCW–MZM is composed of symmetric MZIs with p-n doped LSPCW phase-shifters and thermo-optic (TO) phase shifters. The LSPCW itself is a SiO_2-clad PC waveguide and completely buried in SiO_2 after lattice air-holes are etched. On the other hand, the single LSPCW–MZM is composed of different elements in two arms. One arm involves an asymmetric MZI with a p-n doped LSPCW phase-shifter and a TO phase shifter, while the other arm involves a fixed-length delay line and a TO–MZI variable attenuator for matching group delay between two arms. The LSPCW lattice configuration is similar to that in (c) in Fig. 6.5a. In addition, all devices are fabricated using a standard CMOS process. So far, the double LSPCW–MZM achieved 10 Gb/s in operation and 12.5 nm in operating bandwidth using a 50 μm phase-shifter, while the single LSPCW–MZM achieved 40 Gb/s with a 90 μm phase-shifter, both with much short lengths owing to larger group indices. Moreover, the single LSPCW–MZM successfully realized eye-patterns for 10 and 40 Gb/s with a bit error rate (BER) of 10^{-5} order. It should be noted that 40 Gb/s modulation was first demonstrated using a compact MZM with length of sub-100 μm.

6.3 Ultra-fast All-Optical Photonic Crystal Switch

A semiconductor quantum dot (QD) is one of the electronic quantum-state family together with a quantum well and quantum wire. The semiconductor QD exhibits a sharp electronic density of state, quite different from the bulk semiconductor with a continuous density of state, whereby it is applicable to a low-threshold semiconductor laser with high temperature stability. Large third-order optical nonlinearity (χ^3) of the QD, represented by saturable absorption, is also a promising nature for application to an ultra-fast low-energy photonic switch. In this section, QDs are combined with 2-D PC waveguides to present a novel integrated ultrafast all-optical switch [7]. Here, the QD-combined PC switch was developed in two styles; an analog PC switch with a symmetric Mach–Zehnder interferometer (MZI) and optical nonlinear QDs (hereafter referred to simply as PC–SMZ) and a digital flip-flop PC switch with PC–SMZs combined in a feedback loop (similarly referred to as PC–FF). The PC–SMZ was confirmed to operate sequential switching with repetition frequencies at 40 GHz using a 25-ps-period pulse train input into the switch. The PC–FF was confirmed, as a preliminary work, to operate optical flip-flop operation using 40 ns durations of set/reset pulses with discrete optical components instead of the completed PC–FF. Configuration of each device and operational characteristics are overviewed together with a unique design method called topology optimization technique, as follows.

6.3.1 *Quantum Dot-Combined Analogue Photonic Crystal Switch*

Symmetric Mach–Zehnder Interferometer

The PC–SMZ is composed of a symmetric MZI and QDs embedded in the 2-D PC waveguide. QDs exhibit third-order optical nonlinearity (ONL) effects, i.e., absorption saturation in the switch. Figure 6.7a, b shows schematics of the principle on a

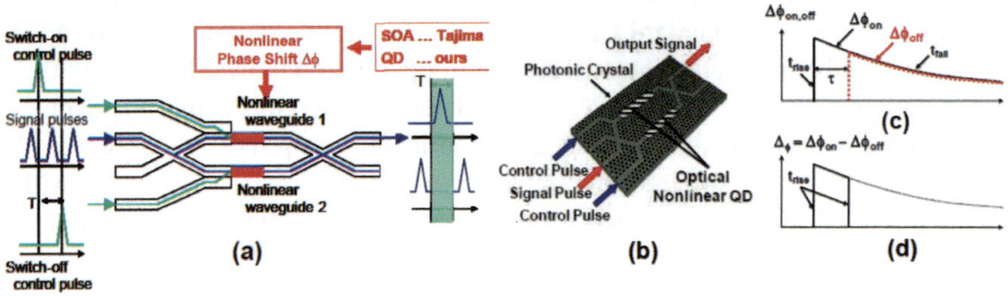

Fig. 6.7 **a** Schematic of operational principle in all-optical PC–SMZ switch. **b** Schematic of the PC–SMZ with line-defect waveguides. **c** Change of the phase shift by on/off CPs at an arbitrary time difference τ. **d** Output pulse with rapid rise and fall time caused by $\pi/2$- or π-phase-shift difference between two identical arms after slowly decaying components canceled to each other. Reprinted with permission from Ref. [7]. © 2012 IEICE

time-differential MZI all-optical switch and configuration of the waveguide in PC–SMZ, respectively [7]. In Fig. 6.7a, aswitch-on control pulse (CP) (on the left side) is incident in the upper ONL waveguide 1 and induces change in refractive index, $\delta_{n,}$ owing to a large optical nonlinearity of the ODs, thus causing a phase shift $\Delta\phi_{\text{on}}$ in the signal pulse (SP). Similarly, a switch-off CP in the lower ONL waveguide 2 generates a phase shift $\Delta\phi_{\text{off}}$ in the lower ONL waveguide 2. When the phase shift differs by $\pi/2$ or π between the upper and lower ONL arms, the SP is switched spatially. Time responses of δ_n, $\Delta\phi_{\text{on}}$ and $\Delta\phi_{\text{off}}$ are rapid in rise (sub-ps) but slow in fall (sub-ns), as shown in Fig. 6.7c. However, note that, since the δ_n is excited time-differentially, the δ_ns at components decaying slowly are canceled, as shown in Fig. 6.7d. Consequently, the PC–SMZ behaves like a time-differential phase modulator with a time response in a rectangular-like shape. Single-pulse switching with 2 ps in rise/fall times and 10–100 fJ in switching energy has ever been demonstrated under excitation by the single set of CPs.

Figure 6.8 shows scanning electron microscope (SEM) and optical microscope photographs of the PC–SMZ with a single line-defect PC slab waveguide in an air-bridge structure [7]. Figure 6.8a shows a cross-section of a molecular beam epitaxy (MBE)-grown GaAs core layer above AlGaAs on a GaAs substrate. Figure 6.8b shows an expanded cross-sectional GaAs core layer (250-nm thick), embedding four InAs-QD layers, as shown by arrows. An AlGaAs layer under the GaAs layer is removed by selective chemical etching to form an air-bridge 2-D PC slab waveguide structure. Figure 6.8c shows a tilted view of the PC–SMZ chip in total with a total length of 500 μm, while Fig. 6.8d shows a magnified cross-section of the air-bridge structure. A circle in red solid line shows a missing air-hole line-defect waveguide with 0.7-μm width and 0.25-μm thickness.

Optical Nonlinearity-Induced Phase-Shift in Quantum Dot

Optical nonlinear characteristics of the InAs-QDs in the GaAs layer are investigated by the Kramers–Kronig relation, as shown in Fig. 6.9a, where absorption and refractive-index curves in the QD-embedding waveguide are schematically shown. Solid- and dotted-line curves show spectra before and after pumping by an incident CP, respectively [24]. Note that deformation around the absorption peak by

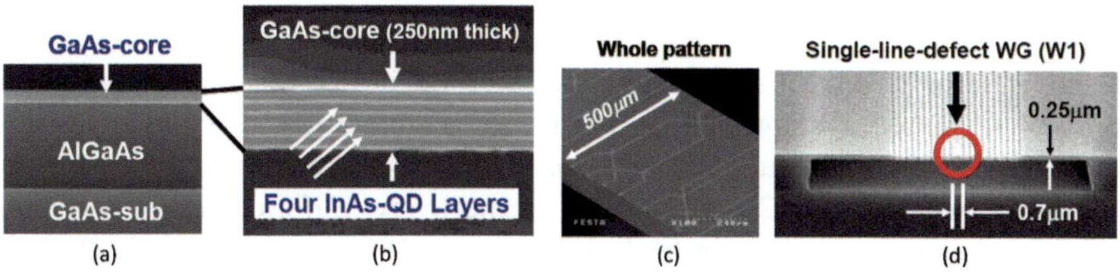

Fig. 6.8 SEM photographs of the PC–SMZ. **a** GaAs core layer on a sacrificial AlGaAs layer. **b** Four InAs layers with embedded QDs in the GaAs core layer with. **c** Whole PC–SMZ chip pattern with a total length of 500 μm **d** Magnified cross-section of the air-bridge waveguide. Reprinted with permission from Ref. [7]. © 2012 IEICE

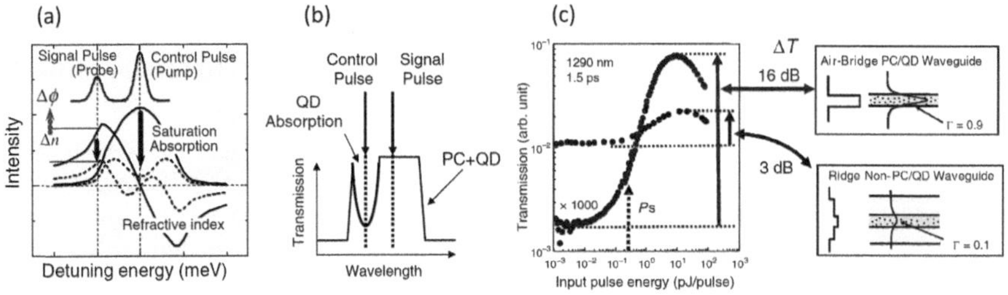

Fig. 6.9 **a** Schematic of Kramers–Kronig relation, showing absorption and refractive index spectra of optical nonlinearity in QDs before pumping by the CP (solid line) and their changes after pumping by the CP (broken line). **b** Relationship in detuning between the CP and SP in the transmission spectrum of the 2-D PC waveguide. **c** Large difference in transmission changes ΔTs between the air-bridge PC waveguide with buried QDs (right-top) and conventional PC ridge waveguide without QDs (right-bottom). Reprinted with permission from Ref. [24]. © 2006 IOP Publishing Ltd.

the pumping energy induces decrease in the refractive-index spectrum (Δn) at the detuned energy, resulting in the ONL phase shift $\Delta\varphi$. Figure 6.9b is a schematic of wavelength-dependent transmission spectrum in the 2-D PC hybrid waveguide with embedded QDs (PC + QD), that shows the detuning relationship between the CP and SP wavelengths. The CP wavelength is selected to the bottom of the spectrum, while the SP wavelength is selected around 1.3 μm for the GaAs PC waveguide with InAs QDs. Since the detuning wavelength between CP and SP is arranged to be 15–20 nm considering the width of the QD absorption peak, a band width of the 2-D PC waveguide should be arranged to be more than 20 nm, preferably 40–50 nm.

An optical nonlinearity of the QD was evaluated, as shown in Fig. 6.9c. The left panel shows measured transmittance versus energy in incident CP using an air-bridge PC waveguide with buried QDs and a conventional non-PC ridge waveguide. Their schematic cross-sections are shown in two right panels. Note that, taking the maximum transmission change of ΔT, the air-bridge PC waveguide with buried QDs (right-top) exhibits a much larger ΔT of 16 dB than that of 3 dB for the non-PC ridge waveguide (right-bottom). The former larger ΔT is resulted from strong confinement of an optical beam in the 2-D PC waveguide. On account of the large ΔT, the PC–SMZ was successfully operated with low-energy CP, as shown next.

40 GHz Switching Operation

Demonstration of optical switching in the PC–SMZ device was implemented with on- and off-pulses, as shown in Fig. 6.10 [7, 25]. Figure 6.10a shows an input pulse train with 25-ps period four pulses. Rise and fall times and energy of the input pulse were expected values; ~ 2 ps and 100 fJ, respectively. Unfortunately, a switching ratio (modulation) was 50% instead of the desired 100%, as a result of $\pi/2$ phase shift instead of the desired π phase shift. Nevertheless, a high repetition rate was successfully operated using the PC–SMZ. Figures 6.10b and c show output pulses with repetition frequencies at 40 GHz and 20 GHz, respectively [7]. Although a switching ratio was as low as 16%, ultra-fast switch operation was confirmed as a

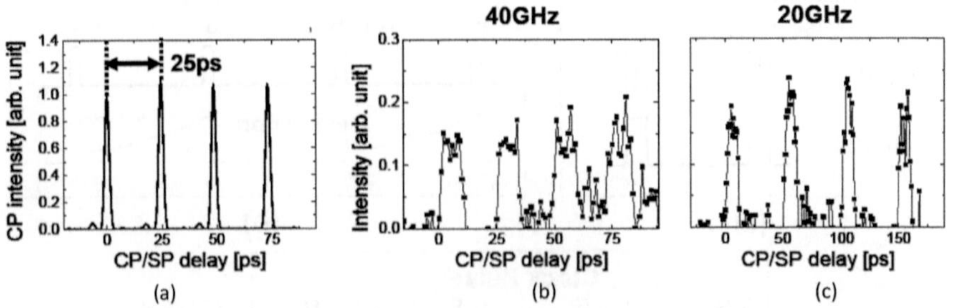

Fig. 6.10 Demonstration of optical switching in the PC–SMZ device. **a** Input CP intensity versus CP/SP delay with 25-ps period. **b** and **c** Output pulse intensity train with repetition frequencies at 40 GHz and 20 GHz, respectively. Reprinted with permission from Ref. [7]. © 2012 IEICE

proof of principle owing to two expected factors; large optical nonlinearity of the QDs in the PC waveguide and large optical-field confinement in the nano-scale line-defect air-bridge PC waveguide.

6.3.2 *Flip-Flop-Based Digital Photonic Crystal Switch*

In the modern reliable and flexible digital signal processing network, demand for an optical flip-flop that enables synchronization with a high-frequency clock is increasing recently. Yet, a conventional optical flip-flop often suffers from a limited switching speed between bi-stable states due to a long feedback loop needed for bi-stability. To this demand, a new type of an optical flip-flop with a high-speed was proposed modifying configuration of aforementioned PC–SMZs with a very compact feedback mechanism [7]. The new PC–SMZ switch is a digital flip-flop switching mode named PC–FF using two similar PC–SMZs for constructing a compact feedback loop [26]. Here, the principle of pseudo flip-flop operation with the PC-SMZ is described first. Then, the principle of feedback loop-induced flip-flop operation of the PC–FF and its preliminary demonstration are overviewed next, as follows:

Pseudo Flip-Plop Operation with PC–SMZ

As mentioned above, the PC–SMZ switch was implemented with on- and off-pulses with an expected phase shift ($\pi/2$ or π) for time-differential phase modulation. This mechanism is similar to the mechanism of the flip-flop switch, as shown in Fig. 6.11a [7]: In the flip-flop operation, two input pulses; S (set) and R (reset) pulses, generate an output pulse with a rectangular shape. These input pulses correspond to on- and off-pulses in the PC–SMZ switch. Now consider the PC-SMZ switch for the flip-flop operation: If the "switch-off" control pulse (reset pulse) comes delayed far from the "switch-on" control pulse (set pulse), the excited on-state is remarkably decayed originated from carrier relaxation of the QD, resulting in unfavorably incomplete bi-stable operation, as shown in Fig. 6.11b. The result is overcome, for instance,

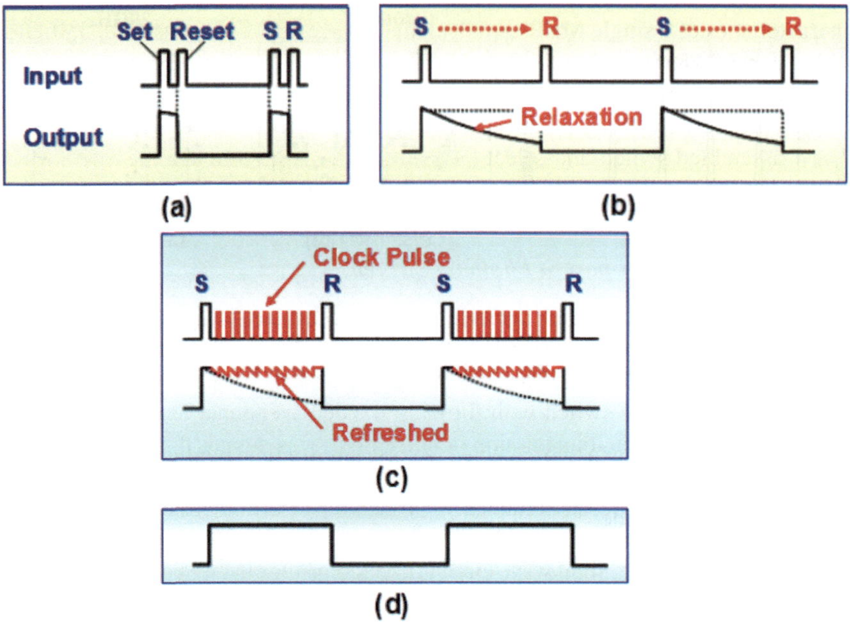

Fig. 6.11 **a** Schematic of switching with flip-flop operation using on- and off-pulses in the PC–SMZ device. (b) Remarkable decay of output (on-state) originated from carrier relaxation of the QD caused by the significantly delayed reset pulse in (**a**). **c** Approximate refreshment of the output (on-state) decay owing to introduction of a series of refreshment clock pulses. **d** Approximately reformed flip-flop operation. Reprinted with permission from Ref. [7]. © 2012 IEICE

by introducing a series of refreshment clock pulses between set and reset pulses, as shown in Fig. 6.11c, whereby desirable flip-flop operation is approximate reformed, as shown in Fig. 6.11d. This idea is realized by a modified waveguide configuration in the PC–SMZ, as shown next [7, 26].

Feedback Loop-Induced Flip-Flop Operation

Based on the principle of pseudo flip-flop behavior using the PC–SMZ, feedback loop induced flip-flop behavior as a new style of PC–SMZ, i.e., PC–FF, is introduced in Fig. 6.12a, together with its preliminary demonstration [26]: The whole PC–FF is composed of two identical PC–SMZs; SMZ-1 and SMZ-2, both connected with a feedback loop to perform successful bi-stable operation. Figure 6.12b shows a schematic of the PC–FF chip arranged on the GaAs air-bridge PC slab substrate. Two rectangular areas indicated by QDs show InAs-QDs buried in the GaAs core layer by a newly developed selective-area-growth technique [7]. In Fig. 6.12a again, output Q is controlled to 1 or 0 by set and reset pulses, while a portion of the output Q is split as feedback light A1 to excite the SMZ-2. Refresh clock pulses (or CW light) A2 are incident in the SMZ-2 to generate output A3. The A3 is incident into the upper arm (S) in the SMZ-1 as a feedback signal, involving nonlinear phase shifts

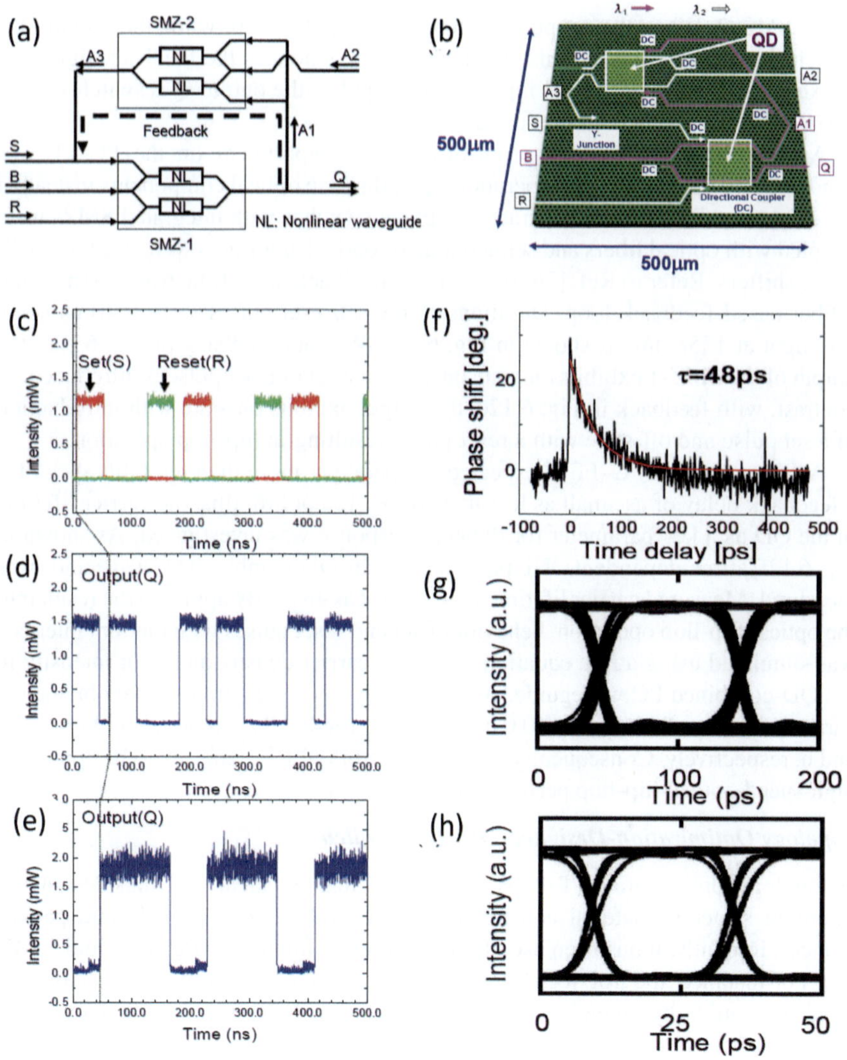

Fig. 6.12 **a** Configuration of an optical flip-flop operation using two PC-SMZs (SMZ-1 and SMZ-2). **b** Schematic of PC and QD configuration for PC-FF. **c**–**e** Experiment of bi-stability using imitated setup: **c** Set and reset pulse train. **d** Output without feedback exhibiting no flip-flop performance. **e** Output with feedback exhibiting successful flip-flop performance. **f** Measured time-dependent nonlinear phase shift. **g** and **h** Simulated output of PC–FF corresponding to 10 Gb/s with carrier lifetime of 100 ps and 40 Gb/s with 25 ps, respectively. Reprinted with permission from Ref. [26]. © 2008 OSA

in the SMZ-2. When the set pulse is zero, the output Q is 0, while the output Q is switched to 1 when the set pulse is incident, and generates the feedback signal A1 to excite the SMZ-2 again. On input of a reset pulse, the output Q is switched from 1 to 0, which remains until the set pulse is input.

As a preliminary work, a proof-of-principle experiment on the PC–FF was conducted by optical flip-flop operation using discrete optical components instead of the completed PC–FF: two separate all-optical switches with integrated SMZs were coupled with optical fibers and semiconductor optical amplifiers were used for ONL phase shifters. Refer to Ref. [26] for detail. A feedback delay time was ~ 40 ns using a fiber-based feedback loop. Durations of the set/reset pulses were ~ 40 ns using CW light at 1558 nm, as shown in Fig. 6.12c. Without feedback in Fig. 6.12d, the output of the SMZ-1 exhibits on-state only when a set or reset pulse is introduced. In contrast, with feedback in Fig. 6.12e, the output exhibits on-state with introduction of a set pulse and off-state with a reset pulse, resulting in flip-flop operation.

A QD-combined PC–FF is expected to operate at more than ten-Gb/s speed for a feedback delay of as small as less than 10 ps. To confirm this, the carrier lifetime in the QD as a key parameter for dynamic response was investigated. As shown in Fig. 6.12f, atime-dependent ONL phase shift in the QD-combined PC waveguide was measured. Measured carrier lifetime was as short as 48 ps. By applying the result into the optical flip-flop operation, behavior of set and reset pulses with random intervals was simulated using a rate equation regarding carrier density and light intensity in the QD-combined PC waveguide. As a result, eye-opened outputs were obtained at carrier lifetimes of 100 ps for 10 Gb/s and 25 ps for 40 Gb/s, as shown in Fig. 6.12g and h, respectively. Consequently, the completed PC-FF is expected to be applied to high-speed optical flip-flop performance.

Topology Optimization-Designed Low-Loss Switch

A topology optimization (ToP) is an efficient mathematical tool that designs an optimum structure/material construction of general devices with desired performances. Recently, it has been used for designing optimum 2-D PC waveguides [27, 28]. For instance, the ToP technology was used to maximize transmittance of a Z-bend waveguide by changing shapes of circular air-holes to non-circular shapes, whereby effective refractive indices of the area to design are optimized.

The ToP technology was used also for the PC–SMZ and PC–FF mentioned above to design low-loss elements such as 2-D PC intersection, bend, DC, and wavelength-selective Y-junction waveguides. A dramatic efficiency of this method was verified experimentally by applying it to design the 2-D PC Y-junction and bend, as shown in Fig. 6.13 [29]. Figure 6.13a shows a schematic of targeted design domains of a Y-junction and bend corners. As shown in Fig. 6.13b, transmittances through AC and BC channels are maximized at λ_1 and λ_2, respectively, while transmittances through unwanted AB and BA channels are minimized at λ_1 and λ_2, respectively, thus optimized completely at two different wavelengths. Figure 6.13c shows air-hole patterns at the bend corner (left) and Y-branch (right), each designed by the ToP method. Figure 6.13d shows calculated AC and BC transmission spectra with 20 dB isolation at 30-nm-apart wavelengths corresponding to those in Fig. 6.13b. Figure 6.13e shows measured results. Although measured isolations at 30-nm-apart wavelengths were 10–4 dB instead of calculated 20 dB. As a result, the ToP method was verified to be extremely effective for designing low-loss waveguides, significantly difficult by the conventional design method, for the PC-SMZ and PC-FF design [29].

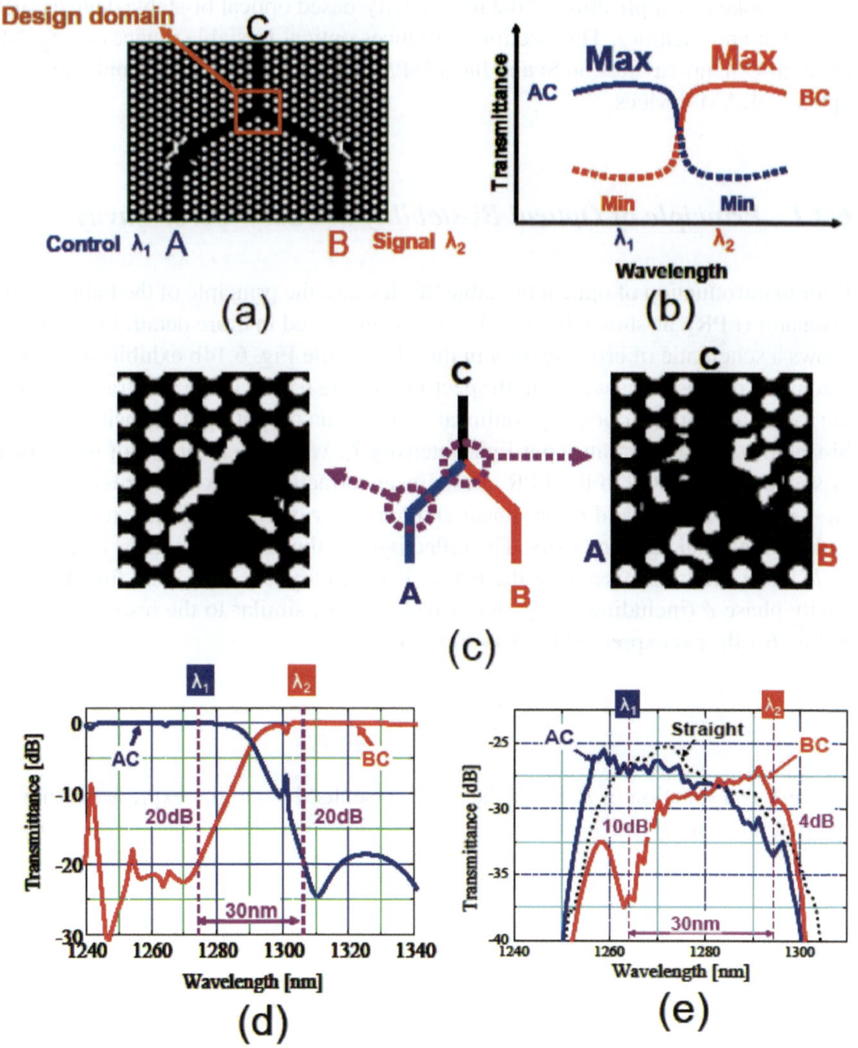

Fig. 6.13 **a** Schematic of targeting TO-design domains of a Y-junction and bend corners. **b** Schematic of asymmetric transmission spectra with the targeting TO-design domains. **c** SEM images of the TO-designed air-hole patterns at the corner (left) and Y-branch (right). **d** and **e** Calculated and measured transmission spectra for AC and BC channels, respectively, corresponding to those in (**b**). Reprinted with permission from Ref. [29]. © 2008 IOP

6.4 Nano-cavity-Based Photonic Crystal DRAM

In the PC nano-cavity, optical energy is strongly confined by the Bragg reflection based on a periodic air-rod lattice, as shown in Fig. 2.13 (Sect. 2.7.2). The enhanced optical energy induces optical nonlinearity in the cavity. Such a nonlinear effect in the PC evokes an application of the nano-cavity-based optical bi-stable devices such as switch and memory. This section introduces optical bi-stable devices using 2-D PC high-Q nano-cavities on Si and InGaAsP/InP laser substrates and applications to optical DRAM devices.

6.4.1 Principle of Optical Bi-stability in Fabry–Perot Cavity

Prior to introduction of optical bi-stable PC devices, the principle of the Fabry–Perot resonator (FPR), as shown in Sect. 1.4.3, is overviewed in more detail. Figure 6.14a shows a schematic of cross-section in the FPR, while Fig. 6.14b exhibits a series of resonance modes in the wavelength spectrum, where FSR is a free spectral range. For the resonator with an optically nonlinear material, an output light intensity I_{o} shows bi-stable states against an input light intensity I_{i}, which is an origin of the optical bi-stable switch [30]. In the FPR, a nonlinear refractive index is expressed as $n = n_0 + n_2 I_c$, where n_0 and n_2 are linear and nonlinear optical constants, respectively, and I_c is internal light intensity. The reflectivity of the resonator mirror is indicated as R. Then the transmission of the Fabry–Perot cavity, $\tau = I_{\mathrm{o}}/I_{\mathrm{i}}$, is a function of a cavity phase ϕ (including n_2, I_c, and wavelength λ), similar to the resonance curve in Fig. 6.14b. τ is expressed by Airy function as;

$$\tau_1 = \frac{I_o}{I_i} = \frac{1}{1 + F \sin^2 \phi/2}, \tag{6.1}$$

where F is a function of R. τ can also be represented by another expression with an internal intensity I_{c} and n_2 as;

$$\tau_2 = \frac{I_c}{I_i} \frac{(1 - R)}{(1 + R)}. \tag{6.2}$$

Equations (6.1) and (6.2) are drawn in Fig. 6.14c, where τ_1 is a resonance curve (black) involving a resonance peak, while τ_2 is a straight line (green) whose slope decreases as I_{i} increases. Note that, when I_{i} increases, cross-points A to F formed by τ_1 and τ_2 derive a relation between output and input intensities, I_{o} and I_{i}, as shown in Fig. 6.14d. Here, the input intensity I_{i} at position 3 on the I_{i} axis produces three cross-points X, Y, and Z. However condition Y disappears due to its unstable state. Consequently, X and Z provide a bi-stable state in the Fabry–Perot cavity. In particular, consider the case when an initial condition in the wavelength of the input light is detuned from the resonance peak wavelength by $\delta\lambda$. This means that, in

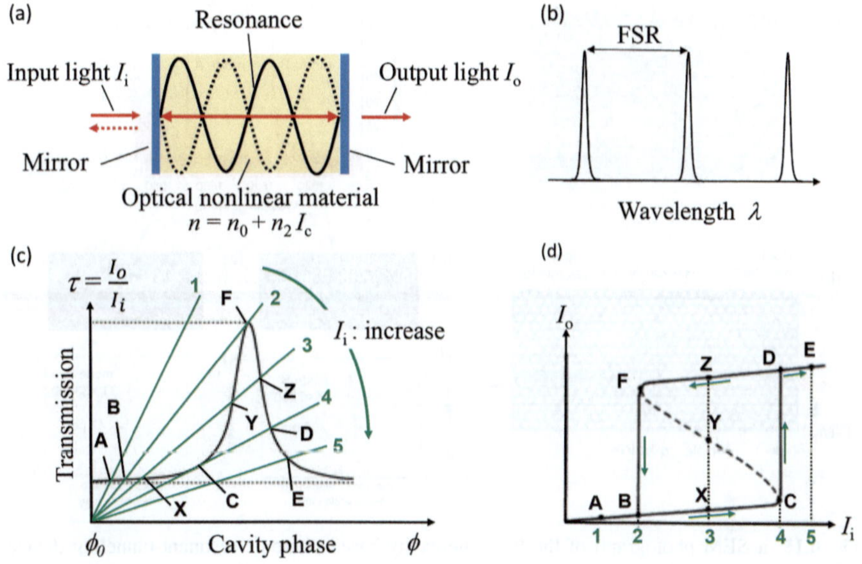

Fig. 6.14 **a** Schematic of cross-sectional view of FPR. **b** Series of resonance modes in a wavelength spectrum. FSR: free spectral range. **c** Graphic solution of cross-points between two sorts of transmission functions, $\tau_1(\phi)$ and $\tau_2(\phi)$. Sequence (green) 1–5 is equivalent to five different I_i, while alphabet A–F is equivalent to increasing and decreasing the input light intensity I_i. **d** Hysteresis loop indicating bi-stable relation between I_o versus I_i

Fig. 6.14c, the distance between the resonance curve in black (τ_1) and the straight line in green (τ_2) in the ϕ (cavity phase) axis direction varies. As a result, the hysteresis loop in Fig. 6.14d changes significantly. In a particular case, it appears or disappears depending on $\delta\lambda$. These behaviors are important for designing this kind of device. For instance, see Fig. 6.16, which shows bi-stable operation in the 2-D PC nano-cavity-based photonic resonant-tunneling device.

6.4.2 *Bi-stable Switch in Si High-Q Nano-cavity*

Si Resonant-Tunneling Nano-cavity

A 2-D PC nano-cavity-based photonic resonant-tunneling device was proposed, as shown in Fig. 6.15 [31]. Figure 6.15a shows a SEM image of a nano-cavity fabricated on a SOI (silicon-on-insulator) substrate, while Fig. 6.15b shows a schematic of a PC lattice pattern involving a high-Q nano-cavity resonator coupled evanescently to input and output PC waveguides. Input light intensity incident into the Si nano-cavity modifies refractive indices of the nano-cavity due to optical nonlinearity (two-photon absorption) and induces a resonant wavelength change, which causes a bi-stable switch operation. These behaviors are explained in more detail, as follows:

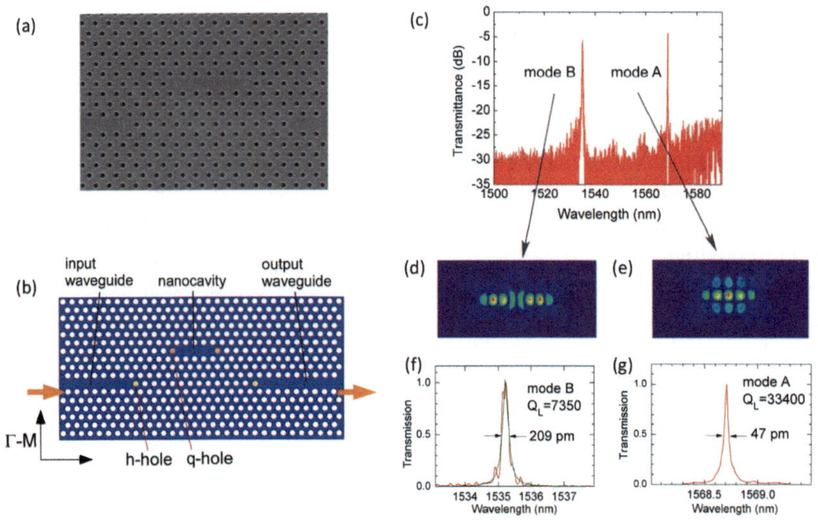

Fig. 6.15 **a** SEM photograph of the PC nano-cavity-based photonic resonant-tunneling device. **b** Schematic of a PC lattice pattern with a high-Q nano-cavity coupled to input/output PC waveguides. **c** Transmission spectrum in a linear regime with two resonant modes in the PC nano-cavity. **d** and **e** Field distributions of resonant modes in the PC nano-cavity. **f** and **g** Expanded transmission spectra near the resonant wavelength corresponding to (**d**) and (**e**), respectively. Reprinted with permission from Ref. [31]. © 2005 OSA

- *Transmission spectra in an optical linear regime*:

A transmission spectrum of a PC nano-cavity in a linear regime is measured by a tunable laser with small power, as shown in Fig. 6.15c [31]. Two resonant peaks, mode B and mode A, appear at 1535 nm and 1568.7 nm, respectively. They correspond to resonant modes, as shown in Figs. 6.15d and e, respectively. Figures 6.15f and g show transmission spectra near the resonant wavelengths, respectively.

- *Bi-stable transmission spectra in an optical nonlinear regime:*

Transmission spectra in a nonlinear regime are measured, as shown in Fig. 6.16 [31]. Figure 6.16a shows measured spectra of the output power (P_{OUT}) with the input power (P_{IN}) as a parameter. Each spectrum, broadened in shape for a longer wavelength, presents a sharp drop indicating an optical bi-stable state. Figure 6.16b shows P_{OUT} versus P_{IN} characteristics for upward and downward sweeps in P_{IN} with the wavelength detuning δ from the mode A resonance as a parameter. For δ larger than 60 pm, a hysteresis appears clearly. At $\delta = 180$ pm, the contrast between the upward-beginning and saturating levels is larger than 14 dB. The minimum input power for causing the bi-stable state is about 40 μW, much lower than the conventional nonlinear bi-stable devices such as nonlinear etalons (a few mW), ring-resonators (~ 50 mW), and ultra-small Si photonic micro-ring devices (~ 0.8 mW).

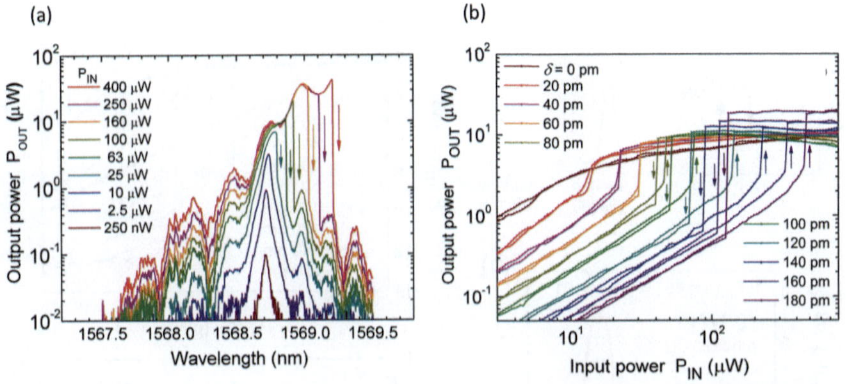

Fig. 6.16 **a** Measured spectra of the output power (P_{OUT}) with an input power (P_{IN}) as a parameter, showing a sharp drop indicating an optical bi-stable state. **b** P_{OUT} versus P_{IN} characteristics for upward and downward sweeps in P_{IN} with the wavelength detuning δ from the mode A resonance as a parameter. The nonlinear state starts from 10 μW and the bi-stable state starts from 40 μW. Reprinted with permission from Ref. [31]. © 2005 OSA

Two-Wavelength All-Optical Bi-Stable Switch

Furthermore, two-wavelength bi-stable operation with mode A and mode B was investigated in more detail using the PC nano-cavity, as shown in Fig. 6.17 [31]. Light beams in mode A and mode B emitted from two tunable lasers (λ_A and λ_B) were incident on the input waveguide. Due to an optical-nonlinearity in the nano-cavity, the incident mode A significantly influences the resonance condition for the mode B, that is, light in mode A as control light can modulate light in mode B as signal light. Figure 6.17a shows a schematic of the hysteresis loop between P_{IN} (A) and P_{OUT} (A), where P_{IN} at P1 excites double P_{OUT} states 1 and 3, while P_{IN} at P2 excites single P_{OUT} state 2 only. Next, two-wavelength bi-stable switch phenomena are investigated, as shown by the right panel in Fig. 6.17a, where signal light in mode B (λ_B) is controlled (ON and OFF) by detuning δ_A from the resonance wavelength of control light in mode A (λ_A). Figure 6.17b shows measured spectra of mode B depending on the values of P_{IN} and, moreover, the order of upward/downward sweeps in P_{IN}. When P_{IN} (A) (P_{IN} of λ_A) changes as 85 μW → 68μ W → 85 μW, corresponding spectra 1, 2, and 3 in mode B are observed. Namely, the spectrum in mode B switches between two different wavelengths (~ 200 pm apart), depicting a clear hysteresis. Figure 6.17c shows measured P_{OUT} (B) (output power in mode B) as a function of P_{IN} (A) (output power in mode A) with λ_B fixed at the peak ($\delta_B = 0$) with λ_A as a parameter. The results clearly indicate bi-stable switching for the signal light (λ_B). When δ_B increases from 0 to 20 and 260 pm with δ_A fixed to 180 pm, bi-stable behaviors of mode B dramatically changes, as shown in Fig. 6.17d. Note that the switching polarity is inversed between the two δ_B values. The result means the first demonstration of an all-optical bi-stable switch using the PC nano-cavity. These demonstrations are continued toward applications to all optical bi-stable switch and memory [32, 33].

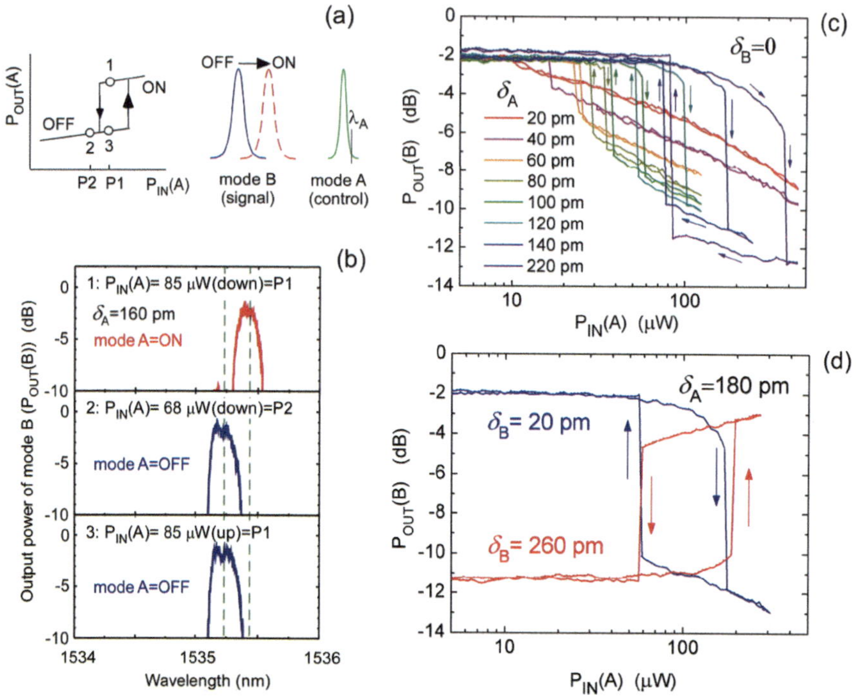

Fig. 6.17 **a** Schematic of all-optical bi-stable switching using two wavelengths. **b** Measured spectra of mode B depending on the values of P_{IN} and, the order of upward/downward sweeps in P_{IN}. **c** Bi-stability of P_{OUT} (B) versus P_{IN} (A) for various δ_A. **d** Bi-stability of P_{OUT} (B) versus P_{IN} (A) for two different δ_B. δ_A is set at 180 pm. Reprinted with permission from Ref. [31]. © 2005 OSA

6.4.3 Bi-stable Switch in InGaAsP/InP BH Laser Nano-cavity

InGaAsP/InP BH Laser Nano-cavity

Bi-stable operation in the PC nano-cavity for all-optical switch/memory was investigated also using PC nano-cavity lasers. Figure 6.18 shows a bird's-eye view of the InGaAsP/InP buried hetero-structure PC (BH-PC) nano-cavity laser and its cross-sectional SEM image (inset) [33]. The BH–PC laser with three InGaAs quantum wells (QWs) with a 1.45 μm wavelength is embedded in an InP PC air-bridge structure. The size of the active region is as small as 4 μm × 0.3 μm × 0.15 μm, thus forming a high-Q nano-cavity. Lasing pump power is two orders smaller than the conventional bi-stable semiconductor lasers. Both ends of the BH–PC laser nano-cavity are connected to barrier defect PC waveguides with $0.98w_0$-widths and input/output waveguides with $1.1w_0$-widths, where w_0 is the width of the laser cavity waveguide. The BH structure has valuable features for bi-stable switch applications. Heat generated within the InGaAsP active region is easily escaped by surrounding InP with

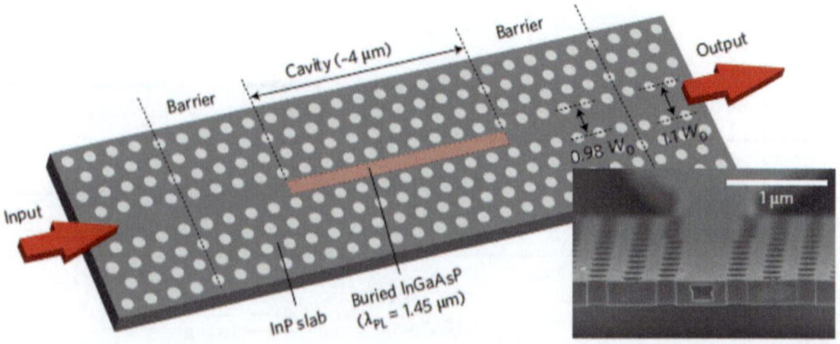

Fig. 6.18 Schematic of the BH–PC nano-cavity laser buried along the missing line-defect waveguide on a triangular-lattice air-bridge substrate. An InGaAsP region is etched to clearly reveal the BH region. Inset shows a cross-sectional SEM image of the BH–PC laser. Reprinted with permission from Ref. [33]. © 2012 Springer Nature

much larger heat conductivity than the InGaAsP cavity. Moreover, carriers in the active region is strongly confined by the InP barrier. Indeed, a carrier relaxation time τ_c in the BH–PC nano-cavity is much longer (~ 7 ns) than in an all-InGaAsP nano-cavity with a non-BH structure (0.24 ns). Furthermore, the cavity has a very high Q (1.7×10^6) with a small V (0.22 μm^3), whereby the BH–PC structure confines both photons and electrons in a sub-μm region, yet allows for heat escaping efficiently. In the transmission spectra for CW light, a bi-stable window is observed, while, in the P_{out} versus P_{in} characteristics, a hysteresis response based on the bi-stable property is also observed, depending on detuning δ from resonance. It is noted that the bi-stable threshold power in the P_{out} versus P_{in} characteristics is as small as nearly—46 dBm (25 nW). These results suggest that the BH–PC nano-cavity laser is extremely promising for application to an optical dynamic RAM, i.e., for optical random access memory (o-RAM) application.

Dynamic One-Bit RAM Operation

The InGaAsP/InP BH–PC laser mentioned above was examined for actual o-RAM operation, as shown in Fig. 6.19 [33]. In the top left panel, an input signal with a continuous-wave bias light, write-pulse, and reset-pulse are shown for writing, storage, and erasing operations. Initially, a wavelength of the bias light is detuned by $\delta = -0.4$ nm, so that the level of the bias light is low. On input of the write-pulse, the resonance in a shorter wavelength occurs, whereby the bias light enters the cavity with a resonance wavelength unchanged. During the time, output level of the bias light remains high. This is an "on" state in memory operation. On input of the reset pulse 1-μs later, the resonance returns to the initial state. This is an "off" state in memory operation. In the top right panel in Fig. 6.19, an output signal with "on" state and "off" state, each with 1-μs holding time and 12 dB contrast is shown. On demonstration, bias power was 500 nW, while writing pulse energy was 13 fJ/bit. In addition, 12 ps of the writing pulse width has found to be enough for a successful o-RAM operation with capability in storage of a 40 Gb/s single-bit pulse. Additionally,

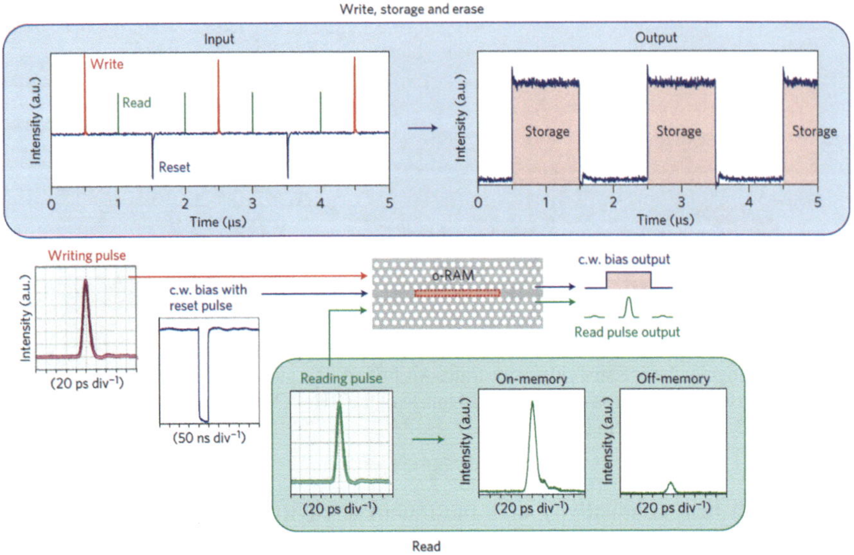

Fig. 6.19 Dynamic one-bit RAM operation. (Top left panel) Input signal with CW bias, write-pulse, read-pulse, and reset pulse. (Top left panel) Output signal with "on" state and "off" state, indicating switching between bi-stable states. (Bottom panel) Demonstrated regeneration of the stored state after injection of a read-pulse after 500-ns delayed write-pulse. Reprinted with permission from Ref. [33]. © 2012 Springer Nature

injection of a read-pulse after 500-ns delayed write-pulse successfully confirmed regeneration of the stored state, as shown in the bottom panel in Fig. 6.19, where 9 dB contrast of the output signal with "On-memory" and "Off-memory" pulses is found. As a whole, optical writing, storage, reading, and erasing operations, namely, random access memory (RAM) have proved to be demonstrated successfully. After demonstration mentioned above, the device was up-graded to an integrated chip for a four-bit RAM. See [33] for detail.

6.4.4 Large-Scale All-Optical DRAM with PC Nano-cavity

Aforementioned optical-nonlinearity-based bi-stable and hysteresis phenomena in the PC high-Q nano-cavities [33] were intensively investigated for large-scale all-optical DRAM operation using both InGaAsP/InP BH-PC laser [8, 32, 33] and Si 2-D PC waveguide [8]. To this end, the following consideration is the most important; arrangement both with a large number of nano-cavities in a compact area and with strong coupling between the nano-cavity and a common bus waveguide for input/output signals as well as small power consumption. In the previous bi-stable memory operations with one-bit or four-bit (not introduced here) o-RAMs using single BH nano-cavity laser [28], nano-cavities are end-coupled to the waveguide in a parallel

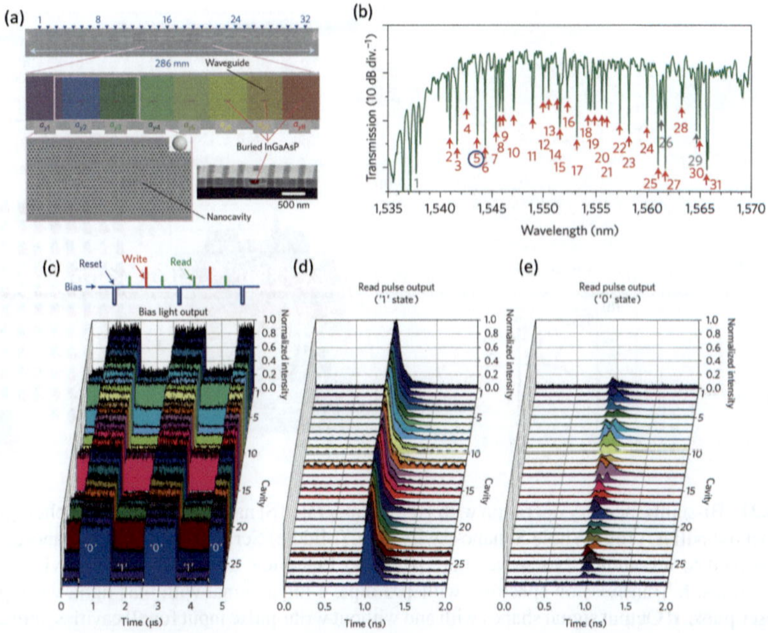

Fig. 6.20 Bi-stable operation with 32-cascaded InGaAsP/InP BH-PC nano-cavities lasers **a** SEM top-view images of BH–PC nano-cavities lasers. (Top) Whole BH–PC laser chip. (Middle) Cascaded BH–PC nano-cavities coupled to a common bus waveguide. (Bottom) Plan-view and cross-sectional views of BH–PC nano-cavity lasers. **b** Transmission spectrum with 32 dips. **c** to **e** Output signals for 28-bit write/readout bit-memory operation: **c** Output signals of the bias light with write-, reset-, and read-pulses. **d** Output signals for the read-pulse at state "1". **e** Output signals for the read-pulse at state "0". Reprinted with permission from Ref. [8]. © 2014 Springer Nature

integration. In contrast, for large-scale integration of o-RAMs, side-coupling between the cavities and bus waveguide to form a serial integration is almost essential, as shown later. Indeed, the latter side-coupling configuration was successfully applied to realizing 25-bit (Fig. 6.20a) and 128-bit (Fig. 6.21a) DRAM operations. They are briefly overviewed as follows:

25-Bit DRAM with InGaAsP/InP BH Laser Nano-cavity

Figure 6.20 shows 32-cascaded integrated InGaAsP/InP BH-PC nano-cavity lasers and their 25-bit optical RAM (o-RAM) operation [8]. In the middle panel in Fig. 6.20a, a series of buried InGaAsP nano-cavities are side-coupled to a common bus line-defect waveguide to implement a large-scale wavelength-addressable o-RAM for a practical photonic network application. A lattice constant in each nano-cavity is changed by 0.5 nm steps between 420 nm and 435.5 nm. The top panel in (a) shows an SEM photograph of a plan-view whole chip with a total length of 286 mm, while bottom panels show magnified SEM images of a top-view and cross-sectional view of nano-cavities. Figure 6.20b shows demonstrated transmission spectra. The spectra involves 31 dips with different cavities, from which 28 cavity modes are

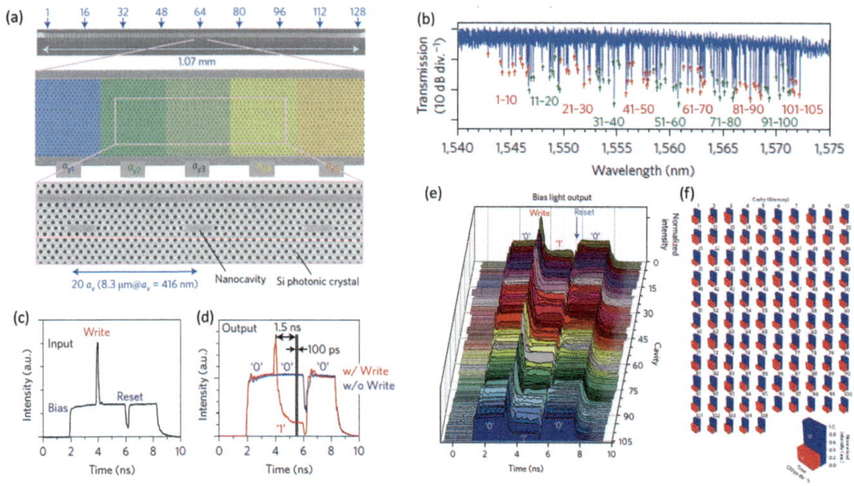

Fig. 6.21 Bi-stable memory operation with 105-bit integrated Si nano-cavities. **a** SEM photographs of Si nano-cavities. (Top) Whole Si nano-cavity chip. (Middle) Serially integrated Si nano-cavities side-coupled to a common bus waveguide. (Bottom) Plan-view and cross-sectional views of Si nano-cavities. **b** Transmission spectrum with 105 dips. **c** Input signal with bias light, write-pulse, and reset-pulse. **d** Output signal shapes with and without write-pulse input for all cavities, indicating the states "1" and "0", respectively. **e** Output signals of the bias light with temporal waveforms for all channels showing momentary memory information, either "1" or "0". **f** Screenshots with (red) and without (blue) the write-pulse for clear visualization of all 105-bit memory states. Reprinted with permission from Ref. [8]. © 2014 Springer Nature

used for o-RAM operation, as shown in Fig. 6.20c–e. Figure 6.20c shows output signals with bias-light (like the top right panel in Fig. 6.19) for 28 channels in the bit-memory operation. An initial state is "0". On input of write-pulse, it is switched to the state "1", which is switched back to the state "0" when the reset-pulse is on. At arbitrary timing, the bit information "0" or "1" can be read by the read pulse. Figure 6.20d shows read-pulse output waveforms for 28 channels at the state "1", whereas Fig. 6.20e shows those at the state "0". Note that "0" and "1" states correspond to high and low intensities for expressing output signals for the bias-light in Fig. 6.20c, while these relations are opposite for read-pulses in Fig. 6.20d and e. On demonstration of o-RAM operation, bias power was 1.6–7.9 µW, while write-/read-pulse energy was 100–200 fJ. As a result, 28-bit memory operation was achieved with total power consumption of 137 µW using InGaAsP/InP BH-PC nano-cavity lasers.

128-Bit DRAM with Si Nano-cavity

When compared with the InGaAsP/InP BH–PC laser, the Si nano-cavity has high superiority in two factors; (i) a loaded high-Q with large coupling between a cavity and waveguide for high-contrast bi-stable switching and (ii) a large spacing between cavity modes (FSR: free spectral range. Refer to Sect. 1.4.3). Furthermore, careful modification of design in the Si PC nano-cavity with missing 3 air-holes around the cavity end has found to lead to the significantly large Q of ~ 1 million; an order of magnitude larger than that in the BH–PC laser. Based on these features, an integrated 105-bit memory operation using Si PC nano-cavities has been demonstrated, as shown in Fig. 6.21 [8]. Figure 6.21a shows SEM photographs of the Si o-RAM device with (top) a whole chip, (middle) several nano-cavity zones with different lattice constants, and (bottom) magnified three nano-cavity zones. Nano-cavities in a serial integration are side-coupled to a common bus waveguide, similar to the BH–PC laser cavities mentioned above. A lattice constant a_y of each cavity varies by 0.125 nm steps and a total length of the Si nano-cavity chip is 1.07 mm. Figure 6.21b shows a demonstrated transmission spectrum of a new o-RAM chip with Si nano-cavities, whereby 115 dips in total and a whole bandwidth of 30 nm are confirmed. An average wavelength spacing of 0.23 nm corresponds to an average WDM channel density of 30 GHz. Figure 6.21c shows input signals with bias light, write-pulse, and reset- pulse for the bi-stable memory operation, while Fig. 6.21d shows output signal shapes with and without write-pulse input for all cavities (1–105), indicating the states "1" and "0", respectively. Figure 6.21e shows that output signals of the bias light with temporal waveforms for all channels reveal momentary memory states, reflecting the time-dependent output intensity in red in Fig. 6.21d. A memory state, initially "0", is switched to "1" on input of a write-pulse, continuing until input of a reset-pulse. In this way, the bi-stable operation is confirmed for all 105 bits in this chip, similar to the result in Fig. 6.20c. For clear visualization of memory states, all data for 105 bits are shown in Fig. 6.21f, where all operating bits are depicted by screenshots with (red) and without (blue) the write-pulse. They are clipped for the fixed 100-ps time window, as shown by the doubled vertical lines in Fig. 6.21d. Resultantly, the clear contrast between the red and blue screenshots verifies that both "1" and "0" memory information for all 105 bits has been recognized successfully. Demonstrated bias power was 79–400 μW, larger than for the case of the BH–PC laser mentioned above. Nevertheless, note that the Si nano-cavity o-RAM chip with high fabrication accuracy enabled a large-scale memory integration. These results present an important step toward future highly accomplished large-scale photonic integration.

6.5 Extreme High-Power Photonic Crystal Surface-Emitting Laser

Thus far, photonic band gap (PBG) properties of the PC to block light selectively into a nano-scale area have been applied to 2-D PC lasers such as PBG defect-mode lasers [34] and 2-D DFB lasers [35]. In particular, a 2-D PC surface-emitting laser (PCSEL), lasing at a singular Γ-point in the 2-D lattice, is capable of single-mode and extremely high-power oscillation with a narrow beam divergence by current injection in a large area [36, 37]. This section overviews the following excellent aspects of the PCSEL:

(1) Room temperature (RT) and continuous wave (CW) emission with an extremely high beam-quality [36] and watt-class power [10].
(2) Proposal of a unique double-hole lattice PCSEL (DH-PCSEL) with tens-of-watt class power [11].
(3) Proposal of a unique theoretical design of PCSEL for 100 W to 1 k W class high-power over a cm-class large area with a single-mode oscillation [12].
(4) Demonstration of outstanding 50 W power at CW operation with a single mode and high beam-quality of 1GW/cm^2/sr, comparable to conventional gas/fiber lasers [14].
(5) Proposal of a unique dually modulated PCSEL (DM-PCSEL) for a new scheme of 3-D LiDAR application [15].

6.5.1 Photonic Crystal Surface-Emitting Laser: PCSEL

Triangular-Lattice PCSEL

The first lasing oscillation of the PCSEL was observed with pulsed excitation at room temperature (RT) by using a triangular-lattice PC slab with InGaAsP-based multiple quantum-wells (MQWs) [37]. Figure 6.22a shows a schematic of the PCSEL structure: WAFER A involves a lasing layer with a MQW active region and an InP cladding region, while WAFER B involves a triangular-lattice PC region on an InP substrate. WAFER A is combined to WAFER B with a wafer-fusion technique in metal organic vaper phase epitaxy (MOVPE) growth. Figure 6.22b shows a calculated band diagram of a 2-D PC triangular-lattice device (upper panel) and an expanded diagram including Γ-points in the Γ-X direction (lower panel). Here, unique behaviors of the light beam at the Γ-point in the 2-D PC slab in general are reviewed in more detail using Fig. 2.7b, as follows: The light beam on the Γ-point, located within the light cone specific to the 2-D PC slab, is radiated out of the slab surface. In addition, at the Γ-point in Fig. 6.22b (light line in vacuum and light cone are not noted here), the group velocity as well as the wave vector in the Γ-X direction are zero. This means that the light beam is vertically emitted under the Bragg diffraction specific to the lattice configuration. A square-lattice PC slab, as shown in Fig. 2.15b, emits light vertically as a consequence of exciting a resonant-wave mode caused by four (0°, ± 90°, 180°) equivalent in-plane Bragg diffractions. On the other hand, a triangular-lattice, as shown in Fig. 6.22a, emits vertical light under six (0°, ± 60°, ± 120°, 180°) equivalent in-plane Bragg diffraction. In Fig. 6.22b, normalized frequencies at the Γ-points (A and B) are found to be nearly 0.35, the result of which is discussed later.

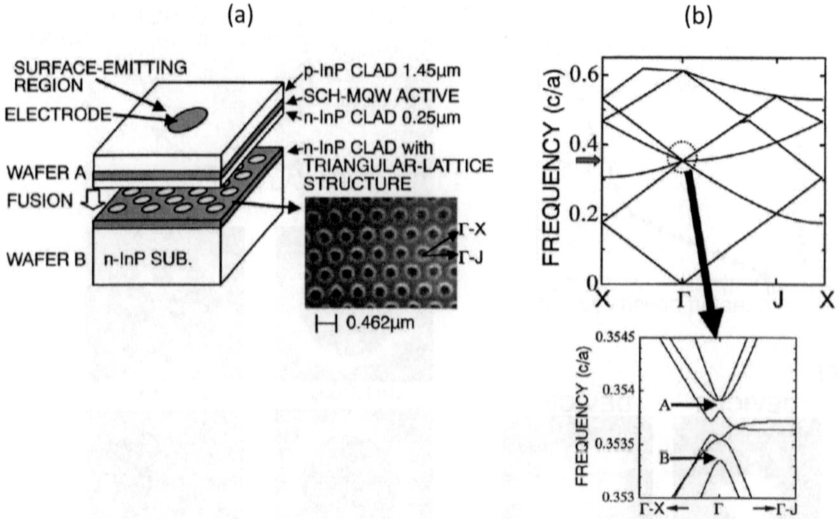

Fig. 6.22 **a** Schematic of the triangular-lattice PCSEL, as shown by the SEM photograph in the inset. **b** Calculated band diagram of a triangular-lattice 2-D PC slab. Reprinted with permission from Ref. [37]. © 1999 AIP

In-Plane Resonant Γ-Point Emission

Characteristics of the aforementioned triangular-lattice PCSEL are shown here.

Figure 6.23a shows *I-L* characteristics of the pulse-excited PCSEL at RT [37]. Here, two samples with different electrode configurations are prepared for investigating detailed in-plane resonant Γ-point laser emission, as shown in Fig. 6.23b. One is DEVICE-I with a circular uniform electrode and the other is DEVICE-II with an asymmetric electrode by masking a part of it. An emission area of the PCSEL was as large as over 500 μm. For an injection current $J = 1.56\, J_{\mathrm{th}}$ (J_{th}: threshold current density of 3.2 kA/cm^2), DEVICE-I exhibited two-mode oscillation in the lasing spectrum, which is found to attribute to the two band edges A and B in Fig. 6.22b. Figure 6.23b shows near-field patterns in DEVICE-I (left) and DEVICE-II (right). In DEVICE-I, uniform oscillation at a 2-D large area is observed for any injection current, suggesting a coherent lasing oscillation with an effective coupling among light propagating in the six equivalent Γ-X directions mentioned above. However, DEVICE-II exhibits anomalous lasing oscillation, suggesting an incoherent lasing oscillation with anomalous coupling from different Γ-X directions depending on the injection current. Figure 6.23c shows far-field patterns in DEVICE-I (left) (very narrow divergence angle of less than 1.8°) and DEVICE-II (right) (asymmetric line-crossed beam profile). In summary, the triangular-lattice PCSEL with the InGaAsP-based SCH-MQW structure has successfully demonstrated the first 2-D coherent (single-mode) lasing oscillation with a very narrow divergence resulting from the coherent light coupling among six equivalent Γ-X directions.

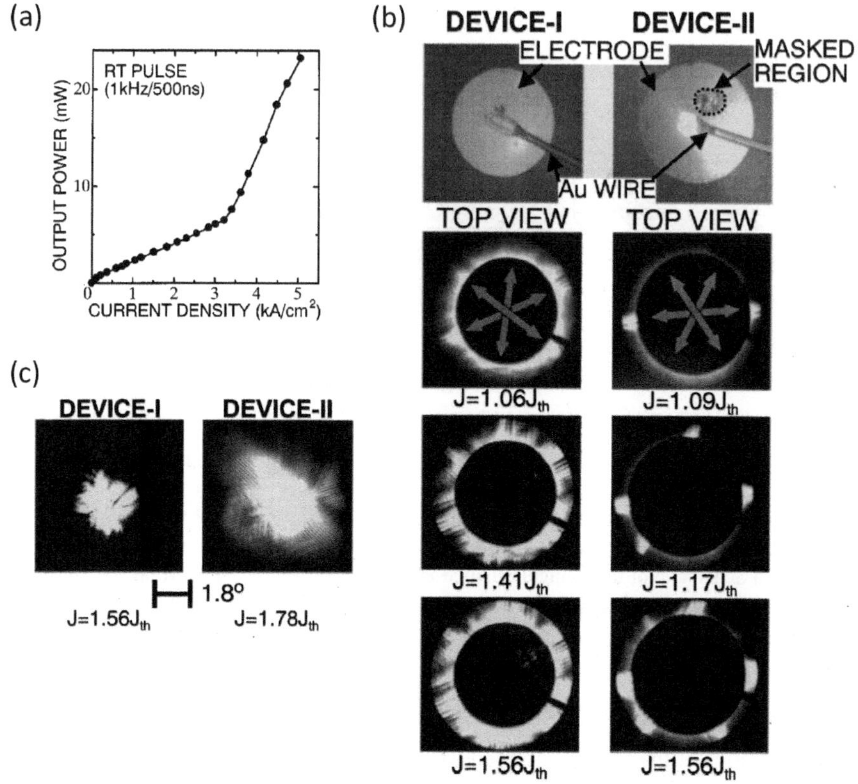

Fig. 6.23 **a** Output power versus current density in the PCSEL (DEVICE-I) with RT and pulsed operation. **b** Near-field patterns in DEVICE-I (left) and DEVICE-II (right). **c** Corresponding far-field patterns. Reprinted with permission from Ref. [37]. © 1999 AIP

6.5.2 Polarization Control by Unit Shape

Single-Mode Lasing with Elliptical Unit Cell

The aforementioned PCSEL based on an InGaAsP SCH-MQW structure with a large lasing area successfully implemented the first 2-D coherent (single-mode) lasing emission with a narrow beam diversion (less than 1.8°) [37]. Yet, the result involved a challenging issue against a complex polarization mode in the lasing oscillation caused by a complex band-edge structure, as shown in Fig. 6.22b. An optical field distribution cannot be determined precisely due to the undesirable degeneracy of an operational condition leading to an unstable state. Consequently, it was pointed out that control of polarization mode is a key issue for operation with single mode, high output power, and narrow beam divergence angle in a large emission area. This guide line was achieved by a new shape of the 2-D PC lattice, that is, an "elliptical-shaped" unit cell instead of a conventional "circular-shaped" unit cell in a square-lattice 2-D PCSEL [9]. In a calculated band diagram in the new PCSEL with an elliptical hole-shape, degeneracy of the band edges, appearing in Fig. 6.22b, disappears and an energy separation occurs at four points in the new band diagram. Interestingly, unified or linear electromagnetic field distributions, that is, simple linear polarizations occur. In addition, it was found that a frequency separation depends on the elliptical ratio of individual unit cells.

These results were experimentally confirmed using an InGaAsP/InP MQW PCSEL with the new elliptical-shaped unit cell mentioned above. For a large lasing emission area with a more than 300-μm diameter, a single-mode and linear polarization-pulsed lasing with a peak power of as high as 20 mW was successfully achieved at RT.

6.5.3 Room-Temperature Continuous-Wave PCSEL

Low Threshold Current with Strong Carrier Confinement

The 2-D PCSEL over a broad lasing area [37] progressed continuously even after successful polarization control by an elliptical unit shape introduced in the previous section [9]. Afterward, the current-injected RT CW operation of the 2-D PCSEL was first realized by improving carrier confinement structure near the active region; by inserting a "sub-cladding layer" above the MQW layer [36]. An $Al_{0.30}$ $Ga_{0.70}$ As sub-cladding layer was inserted between the InGaAs/GaAs three-quantum-wells and the p-AlGaAs cladding layer, whereby unwanted carrier relaxation at the damage in the GaAs/GaAs interface induced by wafer bonding was mitigated. Owing to strong carrier confinement based on this structure improvement, the first current-injected lasing operation over a broad area (over 50 μm) under the RT CW condition was realized with a threshold current of 65 mA and light output power of 4 mW at lasing wavelength of 959.44 nm and a beam divergence angle of as narrow as ~ 1.1°.

6.5.4 Watt-Class RT CW–PCSEL

1.5 W Single-Mode Laser with < 2° Beam Divergence

So far, the 2-D PCSEL has achieved excellent characteristics in polarization-independent coherent (single-mode) lasing emission with low beam diversion over a broad area under a RT CW condition. Yet, typical output power remained in ~ mW level [36]. Afterward, an excellent improvement in the shape of an air-hole in the 2-D PC lattice enabled watt-class high-power lasing oscillation under the RT CW condition [10]: The shape of the unit cell in the new PCSEL is modified from the previous symmetric (circle or ellipse) air-hole to the new asymmetric (right-angled isosceles triangle) air-hole, as shown in Fig. 6.24a, where a plan-view SEM image of the 2-D PC after air-hole processing is revealed. Figure 6.24b indicates a cross-sectional SEM image of the new asymmetric air-holes with narrowed lower regions rather than the upper regions. The GaAs 2-D PC layer on the InGaAs/AlGaAs MQWs, grown by MOCVD for PCSEL structures on the n-GaAs substrate, was nano-fabricated into the asymmetric air-holes by electron beam (EB) lithography and dry etching. Then, the PCSEL is completed after a p-GaAs contact layer is overgrown on the p-AlGaAs cladding layer over the GaAs 2-D PC.

Figure 6.24c shows an output power-current characteristic for CW operation at 20 °C, where maximum output power is 1.5 W at an input current of 2.5 A. Figure 6.24d shows a single-mode spectrum with a peak wavelength of 941.51 nm, while Fig. 6.24e–g show far-field patterns at current of 300 mA, 1200 mA, and 2000 mA, respectively. Beam divergences of < 3° are very narrow up to output power of 1.5 W under RT and CW operations. Measured beam quality was M^2 ~ 1.0 in both x- and y-directions up to a current of 1.0 A, where $M^2 = 1$ corresponds to an ideal Gaussian beam and increases with degradation of the beam quality. The measured M^2 value is more than 10 times smaller than those (typically ~ 20) of conventional VCSELs. It results in reduction of the focus spot size by two orders of magnitude when compared with the conventional VCSEL.

As a unique demonstration of burning a paper with this device, the PCSEL directly irradiated a narrow-divergence beam on a paper located 8.5 cm from it with lens-free configuration. Light output of 0.86 W at a current of 1.7 A resulted in generation of

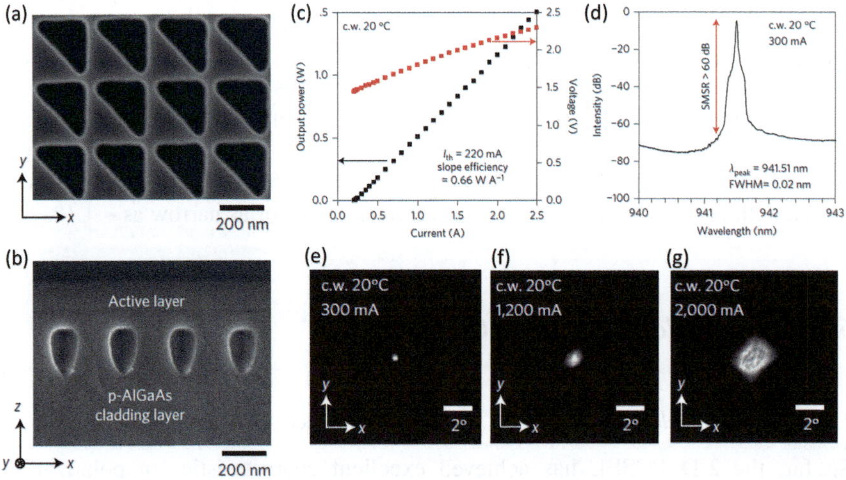

Fig. 6.24 **a** SEM photograph of the plan-view PC with right-angled isosceles triangle air-hole array. **b** SEM photograph of cross-sectional air holes. **c** Light–current characteristics showing maximum 1.5 W at 2.5 A injection current and RT (20 °C)-CW operation. **d** Single-mode spectrum of a peak wavelength of 941.51 nm. **e** to **g** Far-field patterns depending on the current. The beam has FWHM < 3° up to 2.5 A. Reprinted with permission from Ref. [10]. © 2014 Springer Nature

a line of smoke from a burned hole on the paper. The result suggests a large potential for lens-free applications.

Current VCSELs are practically used as various terminal equipment's with low-power operation in applications of optical communications and interconnections. In contrast, the PCSEL currently with watt-class high-power performances enables thousand-times broader coherent-oscillation areas: It is continuously growing aiming at the outstanding high power level of several tens W or more, as follows.

6.5.5 Double-Hole Lattice PCSEL for 10 W-Class Lasing

Double-Hole Lattice PCSEL: DH-PCSEL

Recently, the PCSEL has caused significant change in a 2-D PC lattice configuration from a conventional hole lattice to a new double-hole lattice (DH) to realize single-mode oscillation with high-power even in a large lasing area, as shown schematically in Fig. 6.25 [11]. For increasing a lasing area to more than 500 μm in diameter for more than 10 W-class high output power in the conventional single-hole lattice configuration, generation of higher-order modes other than the single-order mode degrade the advantage of high brightness PCSEL with symmetric narrow-divergence beams.

As shown in Fig. 6.25a, Yoshida, et al. proposed a new concept of the double-hole lattice (DH) PCSEL, i.e., DH-PCSEL, where two sets of 2-D PC hole-lattices with different positions and sizes are superimposed. Positions are shifted by $d \sim 0.25a$ (a: wavelength), i.e., a quarter wavelength both in x- and y-directions, while sizes and positions are both optimized to minimize optical higher-order diffractions, i.e., 180° as well as 180°/90° mixed diffractions. Figure 6.25b shows SEM images of patterned top-view double-hole lattices (upper panel) and their cross-sections after regrown by MOCVD (lower panels, refer to the similar structures in Fig. 6.24b). Here

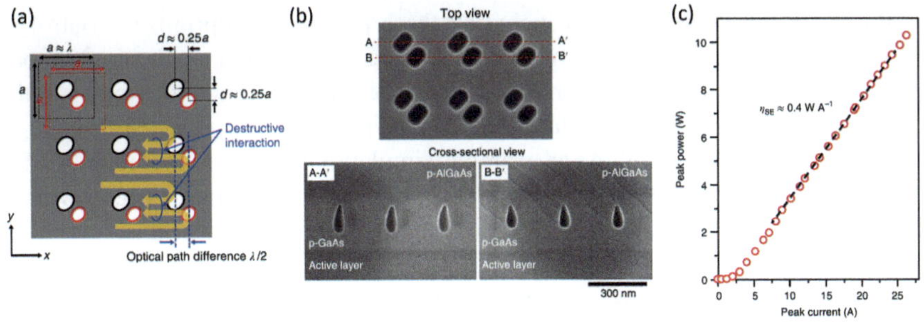

Fig. 6.25 **a** Top-view schematic of double-hole lattice PCSEL. **b** SEM image of patterned top-view double-hole lattices (upper) and their cross-sections after regrown by MOCVD (lower). **c** Output power versus injection current in pulsed operation. Reprinted with permission from Ref. [11]. © 2019 Springer Nature

the top-view sizes are designed as the same but the depths are different between the black and red holes in (a).

Figure 6.25c shows the output power versus peak injection current characteristic for pulsed operation. An emission area was 500 μm in diameter. It is noted that a maximum peak power exceeds ~ 10 W at an injection current of ~ 25 A. From observed near-field and far-field patterns, emission of a single-mode beam with a very narrow divergence angle of less that 0.3° was confirmed. It corresponds to a very high brightness of 300 $MW/cm^2/sr$ at a peak power of 10 W. CW operation has also showed an output power of ~ 7 W with a brightness of ~ 180 $MW/cm^2/sr$.

3 mm-Diameter 50-W Class DH-PCSEL

Although the DH-PCSEL realized a single-mode oscillation with 7–10 W power using an emission area of 500 μm (and also 1 mm later) in diameter, larger emission areas of more than 3–10 mm in diameter were difficult in holding a single-mode operation for higher output power.

Recently, an innovative theory for designing such a large-area DH-PCSEL performing a stable single-mode operation with theoretically 100 W-class output power was reported [12, 14]. Figure 6.26 shows an outline of the DH-PCSEL with such an outstanding performance [14]. The theoretical design is based on the two schematics, as shown in Fig. 6.26a and b. Figure 6.26a shows an in-plain coupling between 2-D PC propagation modes with 180° and 90° diffractions without vertical emission loss, where κ_{1D} and κ_{2D}- mean coupling coefficients for 180° and 90° diffractions. These diffraction modes are controllable by changing a size of the lattice hole, *x*, and separation distance of two holes, *d*, depending on the emission area in the DH-PCSEL [12]. Meanwhile, Fig. 6.26b shows a vertical coupling between in-plain diffractions modes and vertical (up- and down-wards) emission modes via a coupling factor, $i\mu e^{\pm i\theta_{pc}}$, where μ is a coupling coefficient and θ_{pc} is a phase change between in-plain 180°-diffraction modes accompanying the vertical coupling. The parameter, μ, depending on the distance, t_{pclad}, between the 2-D PC layer and the distributed Bragg reflector (DBR) located at the bottom, has an important role of controlling the vertical emission [12, 14].

Based on the PCSEL design including these parameters, optimized control of these two couplings depending of the emission area enables a stable single-mode lasing with extremely high power. Ref. [12] established theoretical treatment on the optimization by deriving close relationship among parameters, κ_{1D}, κ_{2D}-, μ, θ_{pc}, t_{pclad}, *d*, and *x*. Refer to it for detail. As an example, simulation for a 3 mm-diameter PCSEL resulted in more than 100 W output power under ~ 140 A injection current, while a 10 mm-diameter PCSEL revealed 1 kW-class output power, both with single-mode beam divergence angles of < 0.03°.

Although aforementioned new design of the DH-PCSEL for 500 μm to 1 mm emission area successfully brought about peak output power of 70 W under pulsed operation, CW operation for high-power condition with emission areas of more than 3 mm in diameter caused problems of heat generation [13, 38]. To solve this problem, a DH-PCSEL with a size of 3 mm in diameter was fabricated by improving the design of the 2-D PC lattice distribution which compensates thermally generated

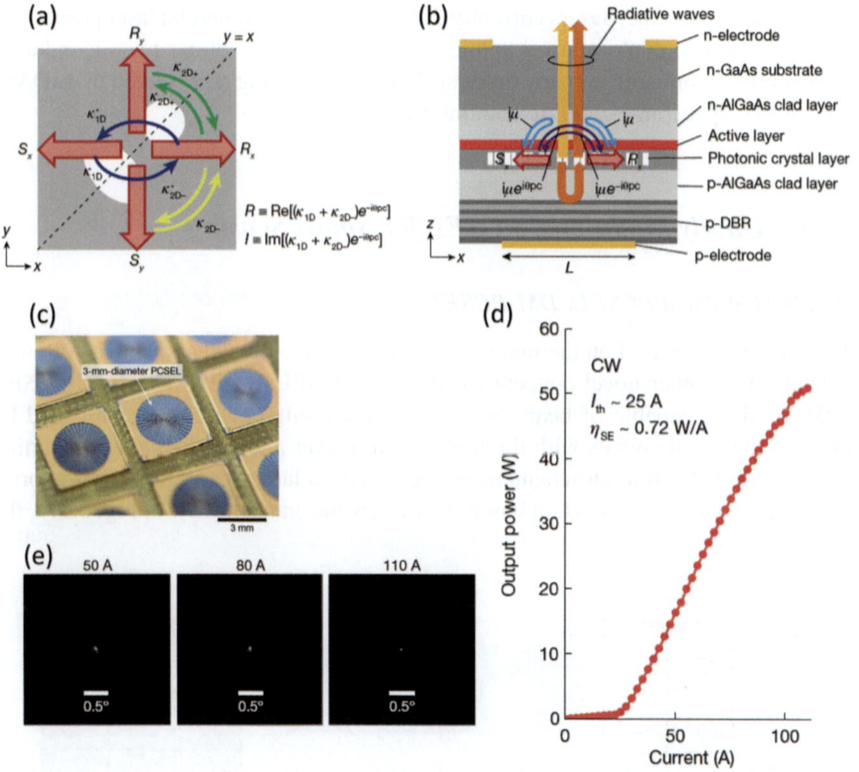

Fig. 6.26 Challenge of a large-area DH-PCSEL for 100W-class output power. (a) Schematic of coupling between in-plain 180° and 90° diffraction modes. (b) Schematic of coupling between in-plain 180°-diffraction mode and vertical emission mode. (c) Optical microscope image of fabricated 3 mm-area PCSEL. (d) Output power versus injection current characteristic showing maximum 50W output power under CW operation. (e) Narrow beam divergence angle of ~ 0.05°. Reproduced from [14] (composite of Figs. 1c and d, 2a and 4b and c), released under a CC BY 4.0 license; https://doi.org/10.1038/s41586-023-06059-8

non-uniform distribution in refractive index of the 2-D PC lattice [14]. Figure 6.26c shows a top-view microscope photograph of the PCSEL before packaging. Owing to an extremely novel design, an unprecedentedly excellent lasing characteristic was established under CW operation, as shown in Fig. 6.26d. Considerably high output power of 50 W was realized at 100 A injection current. In addition, extremely narrow beam divergence angles of ~ 0.05° were observed at injected currents ranging from 50 to 110 A, as shown in Fig. 6.26e. It is noted that the brightness, 1GW/cm^2/sr, corresponding to these results is equivalent to those of main gas-/fiber-lasers in industry.

The DH-PCSEL is ever-growing toward single-mode kW-class output power operation by optimizing design of an aforementioned in-plain and vertical mode coupling in a more than 10 mm-area sample [14]. Realization of such PCSELs will not only catch up with gas-/fiber-lasers currently dominant in the commercial laser processing industry, but also explore novel applications to extremely compact semiconductor laser sources promising in many fields such as smart sensing represented by LiDAR, optical free-space/inter-satellite communications, etc.

6.5.6 Dually Modulated PCSEL for Smart Sensing

Dually Modulated PCSEL: DM-PCSEL

The DH-PCSEL based on the innovative concept as mentioned above was followed recently by another novel concept for the new PCSEL; dually-modulated PCSEL (DM-PCSEL), capable of beam emission in the arbitrary direction controlled by external electrical signals with the high output power and beam quality remaining [15]. Figure 6.27 shows a schematic of the proposed lattice configuration and principle for optical beam emission based on the new mechanism [15]. In Fig. 6.27a, the

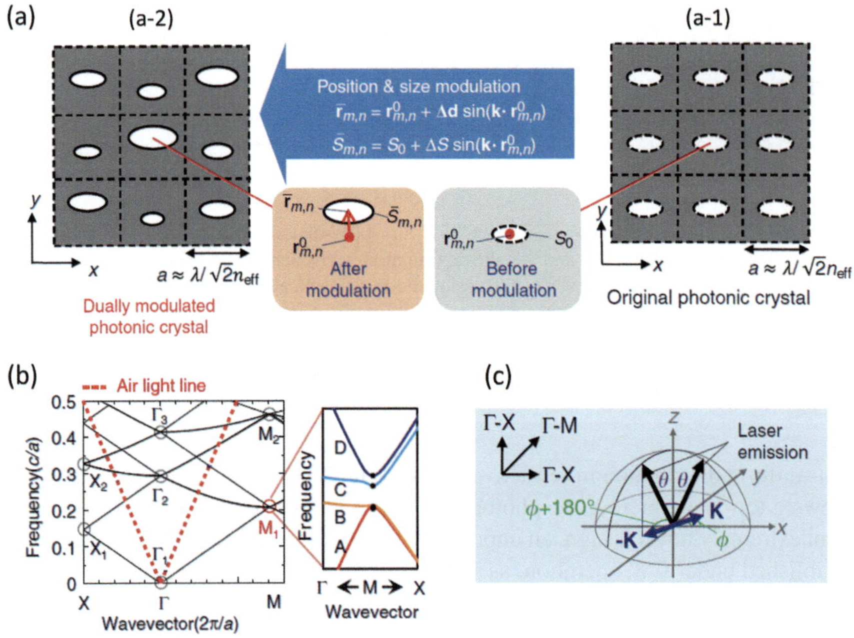

Fig. 6.27 Schematic of proposed lattice configuration and principle of the DM-PCSEL. **a** Lattice configurations before (a-1) and after (a-2) modulation. (b) 2-D PC band diagram of the square lattice including air light lines, and Γ-/M-points. Right panel shows an expanded diagram around M_1 with four resonance points, A, B, C, and D. **b** Band diagram showing Γ-point within the air light line and M-point out of the air light line in the DM-PCSEL. **c** Shift of k vector from M-point to Γ-point enabling laser emission. Adapted from [15] (Fig. 1 with panel e cropped), released under a CC BY 4.0 license; https://doi.org/10.1038/s41467-020-17092-w

panel (a-1) shows a 2-D PC square lattice before the new proposal, while the panel (a-2) shows a new lattice with a hole size and its position changed by $S_{m,n}$ and $r_{m,n}$, respectively. Figure 6.27b shows a band diagram (left) of the 2-D PC square lattice including air light lines in red. The point M_1 in red, showing the resonance state to be considered here, is expanded to the detailed diagram (right) with four resonance points, A, B, C, and D.

Mechanism of the laser emission by the new style is described as follows: In the previous DH-PCSEL, a laser beam is vertically emitted based on the Γ-point resonance emission (Γ-point: located within the air light lines). Meanwhile, before modulation, the DM-PCSEL is based on the M-point resonance (M-point: located out of the air light line), so that no vertical emission occurs at any of four fundamental modes, A, B, C, and D due to cancelation of electric fields around each lattice hole under lasing operation. After modulation, however, the electric fields positively interfere to each other and laser emission occurs in $\pm\,\theta$ directions given by the modulation condition of the air-hole lattice, as shown by the black arrow in Fig. 6.27c.

Figure 6.28a shows a tilted schematic view of the DM-PCSEL with arrayed p-/n-line electrodes and directions indicating 2-D beam scanning [15]. A right panel shows a magnified structure of the pixel in the arrayed DM-PCSEL. Figure 6.28b shows a microscope image of the 3 mm-square DM-PCSEL chip with 10 × 10 crossing p-/n-line electrodes, each of which surrounds a 100 μm-diameter current-injection pixel. Each pixel is separated by etched mesas, thus allowing for independent laser operation. Figure 6.28c shows measured far field patterns emitted from different

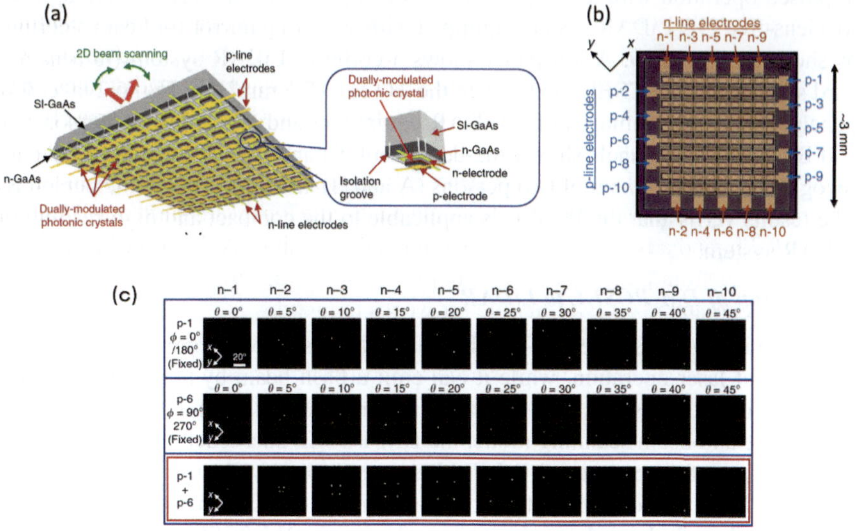

Fig. 6.28 **a** Schematic view of the arrayed DM-PCSEL. **b** Microscope image of the 3 mm-square DM-PCSEL chip with 10 × 10 crossing p-/n-line electrodes. **c** Measured far-field patterns emitted from different pixels by different laser operation. Reproduced from [15] (composite of Figs. 4a, 5b and 6a), released under a CC BY 4.0 license; https://doi.org/10.1038/s41467-020-17092-w

pixels by different laser operation using ten n-line electrodes (n-1 to n-10) and three sets of p-line electrodes (p-1, p-6, and p-1 + p-6). θ is changed from 0° to 45° for fixed $\phi = 0°/180°$ and 90°/270°, as shown in the figures.

6.5.7 Application of PCSEL to LiDAR Sensing Mobility

Section 5.2 reviewed several types of solid state LiDAR systems with Si photonics-based phased arrays for beam scanning and broad-area type semiconductor lasers for external light sources. Due to low-beam qualities and astigmatism of their light sources, external lens systems and their precise adjustment were almost inevitable. On the other hand, PCSELs capable of high output power and narrow beam-divergence angles were recently applied to demonstration of compact and high-resolution LiDAR with no need of complex lens system mentioned above in the conventional edge-emission type laser-based LiDAR [15, 39, 40]. In this section, two types of demonstration of LiDAR using DH-PCSELs and DM-PCSELs are reviewed, as follows.

Application of DH-PCSEL to LiDAR

An early-stage DH-PCSEL, introduced in Sect. 6.5.5, was improved in output power afterward by providing a DBR structure at the bottom of the 2-D PC slab, resulting in an excellent peak power of more than 10W and beam divergence angle of ~ 0.1° at pulsed operation with a 500 μm-diameter sample [39]. The results were applied to a lens-free 2-D LiDAR system equipped with a rotating mirror for beam steering, as shown in Fig. 6.29. A left panel shows a compact LiDAR system (in a name-card size) with the PCSEL mounted in the TO can (5.6 mm$^{\phi}$). In Fig. 6.29a–c, left panels show camera images of a LiDAR equipment and two subjects (persons, A and B), while right panels show time-dependent distance maps, capable of delicate recognition for movement of two persons (A and B) and both arms of the person B. The results verify that the PCSEL is applicable to the compact and high-resolution LiDAR system.

Application of DM-PCSEL to LiDAR

As shown in Fig. 6.28c, the DM-PCSEL can emit a laser beam from a different pixel by different laser operation, while it can emit a flash beam by simultaneous laser operation of all pixels at a time. By combination of these two emission modes, i.e., flash mode and beam scanning mode, the DM-PCSEL succeeded in demonstration of a new type of the compact 3-D LiDAR system [40, 41]. The system uses a time-of-flight (ToF) method for high-resolution sensing. As shown in Fig. 6.30a, the system involves flash-mode and beam-scanning mode DM-PCSELs and a ToF infra-red (IR) camera [40]. Figure 6.30b shows an IR image including white and black objects taken by the ToF camera. Low-reflective black objects are quite dark. Due to this situation, Fig. 6.30c shows distance measurement after flash-mode illumination, where the black object was not measured different from the white object. Then, the

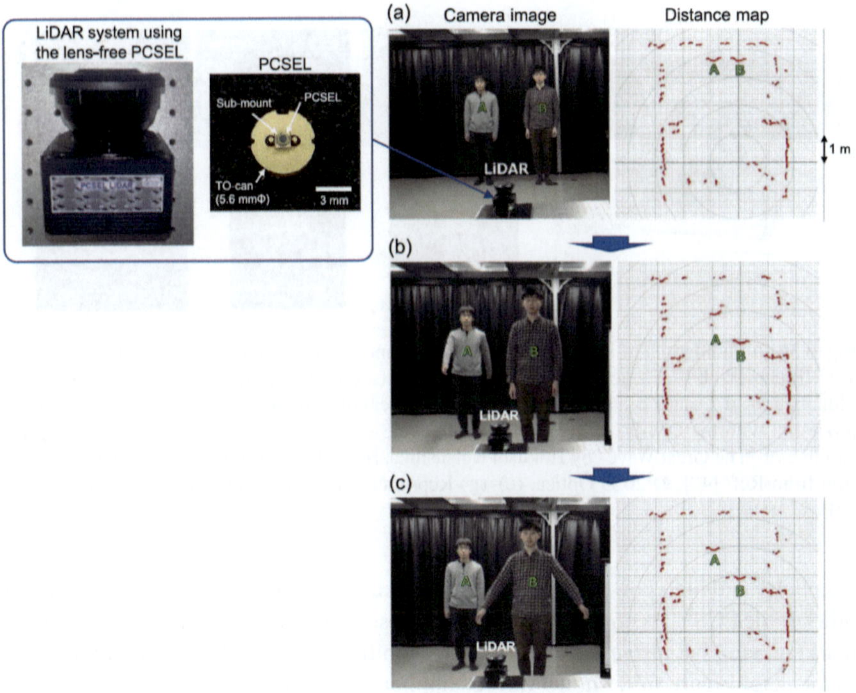

Fig. 6.29 (Left panel) Compact LiDAR system with the DH-PCSEL. **a–c** Camera images (left) and distance maps (right). Reprinted with permission from Ref. [39]. © 2021 IOP Publishing

IR image of the black object was automatically tracked by the beam-scanning mode illumination, resulting in the successful measurement of the position, as shown in Fig. 6.30d.

The beam-scanning mode DM-PCSEL opens another novel application to emission of laser-beam processed fine patterns. Figure 6.30e–g shows fine patterns processed with the DM-PCSEL; a circular-arranged 48-dot pattern, a logo of Kyoto University, and a painting of "The Great Wave" by Hokusai Katsushika, are shown, respectively [38].

Application of PCSEL in Next-Generation Society 5.0

Gas-/fiber-lasers, dominant in the current commercial laser processing fields, have problems of large size, high price, low energy-efficiency, etc. In addition, conventional edge-emission type semiconductor lasers suffer from low beam quality and a wide wavelength spectrum originated from their large beam emission angles, resulting in necessity of precise beam arrangement for sensing application and crucial deficiency of beam brightness for laser processing.

On the contrary, PCSELs have solved many problems for realizing high output power and beam quality in a large emission area. As a result, the brightness is now comparable to fiber lasers, while features of compactness, low price, structural

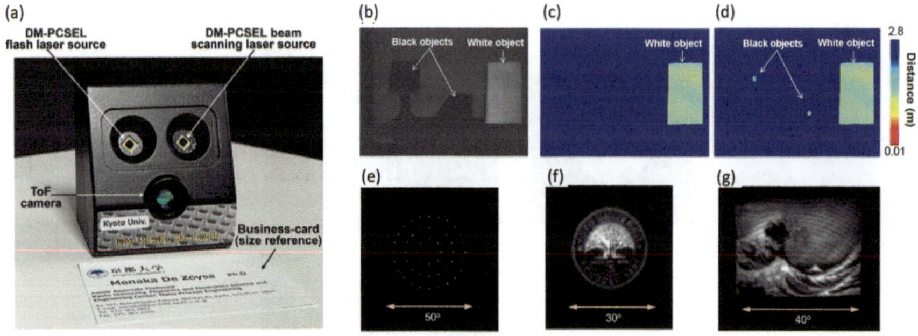

Fig. 6.30 **a** Photograph of LiDAR with flash-mode and beam-scanning mode DM-PCSELs and a ToF–IR camera. **b** IR image of white and black objects. **c** Distance measurement after flash-mode illumination. **d** Distance measurement after beam-scanning mode illumination. **e** to **g** Patterns processed with the DM-PCSEL, circular-arranged 48-dot pattern, logo of Kyoto University, and painting of "The Great Wave" by Hokusai Katsushika, respectively. (a)–(d) Reprinted with permission from Ref. [40]. © 2023 Optica. (e)–(g) Reprinted with permission from Ref. [41]. © 2023 AIP

simplicity and high power efficiency are far exceeding them. Consequently, applicability of the PCSEL is not limited to laser processing including metal cutting, welding, marking, etc., but also extended to smart mobility such as LiDAR and autonomous robots, etc. Figure 6.31 shows various applications of PCSELs for next-generation smart Society 5.0 [39]. Here, Society 5.0 is defined by Cabinet Office, Japan, as "a human-centered society that balances economic advancement with the resolution of social problems by a system that highly integrates cyberspace and physical space [42]. PCSELs are expected to play an active part in many industries such as smart laser processing (for electronics, automobiles, solar cells, etc.), smart mobility (for autonomous robots, farm machines, construction machines, automobiles, etc.), LiDAR for weather, medical applications, entertainment, and so on.

6.6 Summary

Applications of 2-D PC waveguides to various signal processing and lasing devices, overcoming several performance limits that current electronics encounters in the photonic networks, were overviewed here.

Slow-light in the lattice-parameter-modified 2-D PC waveguides was experimentally confirmed by proposed chirped PC-coupled waveguide and lattice-shifted PC waveguide (LSPCW). The LSPCW was integrated with Si wire, heater, metal wire, etc., on a SOI wafer to apply to the tunable dispersion compensator, where 50-ps and more delay were tuned by heating. In the symbol-rate variable DQPSK receiver, a slow-light-based PC micro-ring array with one-bit tunable delay and symbol rates of ~ 15 to 18 Gb/s was successfully demonstrated. Slow-light was applied also to

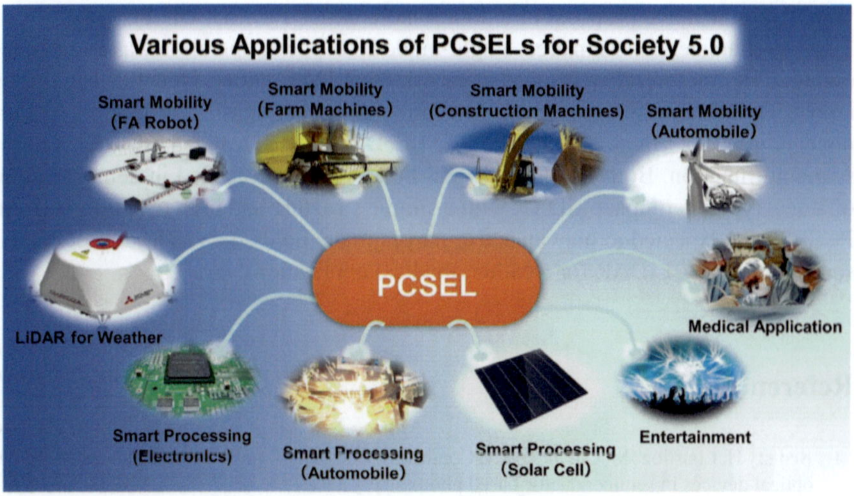

Fig. 6.31 Various applications of PCSELs for next-generation smart Society 5.0. Reprinted with permission from Ref. [39]. © 2021 IOP Publishing

the compact MZI modulator with much reduced arm length of sub-100 μm for phase shifters, resulting in 40 Gb/s RF modulation.

Optical nonlinear coupling between symmetrical MZ PC waveguides on GaAs/AlGaAs substrate and embedded InGaAs QDs achieved a low-power and ultra-fast (20–40 GHz) analog all-optical switch (PC–SMZ). In addition, combination of identical PC–SMZs, forming a compact feedback loop, was proposed to create an ultra-fast flip-flop digital switch (PC–FF). Proof of concept of the PC–FF was confirmed using discrete photonic devices and certified high-speed demonstration corresponding to 10 and 40 Gb/s in the flip-flop arrangement.

Meanwhile, light field strongly confined in the high-Q PC nano-cavity induces optical nonlinearity to provide an optical bi-stable Fabry–Perot resonator effectively. Integration of series of high-Q PC nano-cavities coupled in parallel to a common bus waveguide, using Si PC and InGaAsP/InP BH-PC laser substrates, separately confirmed bi-stable switches/memories. Initial bias-light and write pulse, reset pulse, and read pulse input into the PC waveguide produce bit information in the bias-light output, thus implementing a 28 to 128-bit optical DRAM operation. Besides these bi-stable switches/memories, serial integration of bi-stable elements using high-Q PC nano-cavities will be inevitable for integration of optical logic devices.

Finally, a 2-D PC surface-emitting laser, PCSEL, for single-mode and high output power operation based on the band-edge (Γ-point) resonance was widely reviewed from its birth to the latest results including unprecedented performances. In early stages, various modification in asymmetric in-plane and vertical air-hole shapes brought W-class output power with high beam quality of $M^2 < 1.1$ and a narrow beam divergence angle of $< 2°$ in an emission area of less than 1 mm^{ϕ}. Afterward, the PCSEL was extremely improved in a large emission area of more than

$1mm^{\phi}$ by innovative proposal of a double-hole lattice PCSEL; DH-PCSEL. With a 3 mm-in-diameter emission area, single-mode 50W power operation was established owing to an innovatively advanced design of the 2-D PC lattice. Note that brightness reached 1 GW/cm^2/sr, comparable to commercially core gas-/fiber-lasers. Another novel proposal of a dually modulated PCSEL, DM-PCSEL realized a new laser beam scanning function. Both DH-PCSEL and DM-PCSEL succeeded in demonstration of a new type of compact LiDAR with a simple structure and high resolution. These PCSELs are expected to play an active part in many fields such as laser processing, sensing mobility, LiDAR for weather, medical applications, etc.

References

1. Benisty H, Lourtioz JM, Chelnokov A, Combrie S, Checoury X (2006) Recent advances toward optical devices in semiconductor-based photonic crystals. Proc IEEE 94:997–1023
2. Pendry B (2000) Negative refraction makes a perfect lens. Phys Rev Lett 85:3966–3969
3. Yariv YX, Lee RK, Scherer A (1999) Coupled-resonator optical waveguide: a proposal and analysis. Opt Lett 24:711–713
4. Fan S, Villeneuve PR, Joannopoulos JD, Haus HA (1998) Channel drop tunneling through localized states. Phys Rev Lett 80(5):960–963
5. Olivier S, Benisty H, Weisbuch C, Smith CJM, Krauss TF, Houdre R, Oesterle U (2003) All-photonic crystal add–drop filter exploiting low group-velocity modes. Opt Lett 28:2246–2248
6. Baba H, Nguyen C, Ishikura N, Suzuki K, Shinkawa M, Hayakawa R, Kondo K (2013) Photonic crystal slow light devices fabricated by CMOS-compatible process. IEICE Electron Express 10:1–15
7. Asakawa K, Sugimoto Y, Ikeda N, Tsuya D, Koide Y, Watanabe Y, Ozaki N, Ohkouchi S, Nomura T, Inoue D, Matsui T, Miura A, Fujikawa H, Sato K (2012) Nano-photonics based on semiconductor-photonic crystal/quantum dot and metal-/semiconductor plasmonics. IEICE Trans Electron E95.C:178–187
8. Kuramochil E, Nozaki K,. Shinya K, Takeda K, Sato T, Matsuo S, Taniyama H, Sumikura H, Notomi M (2014) Large-scale integration of wavelength-addressable all-optical memories on a photonic crystal chip. Nat Photon 8:474
9. Noda S, Yokoyama M, Imada M, Chutinan A, Mochizuki M (2001) Polarization mode control of two-dimensional photonic crystal laser by unit cell structure design. Science 293:1123–1125
10. Hirose K, Liang Y, Kurosaka Y, Watanabe A, Sugiyama T, Noda S (2014) Watt-class high-power, high-beam-quality photonic-crystal lasers. Nat Photonics 8:406
11. Yoshida M, Zoysa MD, Ishizaki K, Tanaka Y, Kawasaki M, Hatsuda R, Song B, Gelleta J, Noda S (2019) Double-lattice photonic-crystal resonators enabling high-brightness semiconductor lasers with symmetric narrow-divergence beams. Nat Mater 18:121–128
12. Inoue T, Yoshida M, Gelleta J, Izumi K, Yoshida K, Ishizaki K, Zoysa MD, Noda S (2022) General recipe to realize photonic-crystal surface-emitting lasers with 100-W-to-1-kW single mode operation. Nat Commun 13:3262
13. Zoysa MD et al (2020) Thermal management for CW operation of large-area double-lattice photonic-crystal lasers. JOSA:B 37(12):3882–3887
14. Yoshida M, Katsuno S, Inoue T, Gelleta J, Izumi K, Zoysa MD, Ishizaki K, Noda S (2023) High-brightness scalable continuous-wave single-mode photonic-crystal laser. Nature 618(22):727–732
15. Sakata R, Ishizaki K, Zoysa MD, Fukuhara S, Inoue T, Tanaka Y, Iwata K, Hatsuda R, Yoshida M, Gelleta J, Noda S (2020) Dually modulated photonic crystals enabling high-power high-beam-quality two dimensional beam scanning lasers. Nat Commun 11:3487

16. Baba T, Adachi J, Ishikura N, Hamachi Y (2009) Review: dispersion-controlled slow light in photonic crystal waveguides. Proc Jap Acad Ser B 85:443–453
17. Mori D, Baba T (2005) Wideband and low dispersion slow light by chirped photonic crystal coupled waveguide. Opt Express 13:9398–9408
18. Kubo S, Mori D, Baba T (2007) Low-group-velocity and low-dispersion slow light in photonic crystal waveguides. Opt Lett 32:2981–2983
19. Shinkawa M, Ishikura N, Hama Y, Suzuki K, Baba T (2011) Nonlinear enhancement in photonic crystal slow light waveguides fabricated using CMOS-compatible process. Opt Express 19:22208–22218
20. Dulkeith E, Vlasov YA, Chen X, Panoiu NC, Osgood RM Jr (2006) Self-phase-modulation in submicron silicon-on-insulator photonic wires. Opt Express 14(12):5524–5534
21. Ishikura N, Hayakawa R, Hosoi R, Tamanuki T, Shinkawa M, Baba T (2012) Photonic crystal tunable slow light device integrated with multi-heaters. Appl Phys Lett 100:221110
22. Suzuki K, Nguyen HC, Tamanuki T, Shinobu F, Saito Y, Sakai Y, Baba T (2012) Slow-light-based variable symbol-rate silicon photonics DQPSK receiver. Opt Express 20:4796–4804
23. Nguyen HC, Hashimoto S, Shinkawa M, Baba T (2012) Compact and fast photonic crystal silicon optical modulators. Opt Express 20:22465–22474
24. Asakawa K, Sugimoto Y, Watanabe Y, Ozaki N, Mizutani A, Takata Y, Kitagawa Y, Ishikawa H, Ikeda N, Awazu K, Wang X, Watanabe A, Nakamura S, Ohkouchi S, Inoue K, Kristensen M, Sigmund O, Borel PI, Baets R (2006) Photonic crystal and quantum dot technologies for all-optical switch and logic devices. New J Phys 8:208
25. Nakamura H, Sugimoto Y, Kanamoto K, Ikeda N, Tanaka Y, Nakamura Y, Ohkouchi S, Watanabe Y, Inoue K, Ishikawa H, Asakawa K (2004) Ultra-fast photonic crystal/quantum dot all-optical switch for future photonic networks. Opt Express 12:6606–6614
26. Nakamura S, Watanabe A, Wang X, Ikeda N, Sugimoto Y, Ozaki N, Watanabe Y, Asakawa K (2008) Optical flip-flop based on coupled ultra-small Mach-Zehnder all-optical switches. OFC/NFOEC 2008, San Diego, USA, pp 045–10
27. Bendsøe MP, Sigmund O (2003) Topology optimization—theory, methods and applications. Springer, Berlin
28. Borel PI, Harpøth A, Frandsen LH, Kristensen M, Shi P, Jensen JS, Sigmund O (2004) Topology optimization and fabrication of photonic crystal structures. Opt Express 12:1996–2001
29. Watanabe Y, Ikeda N, Takata Y, Kitagawa Y, Ozaki N, Sugimoto Y, Asakawa K (2008) Topology optimization of wavelength selective Y-junction for 2D photonic crystal waveguides. J Phys D 41:175109
30. Smith SD (1986) Optical bi-stability, photonic logic, and optical computation. Appl Opt 25:1550–1564
31. Notomi M, Shinya A, Mitsugi S, Kira G, Kuramochi E, Tanabe T (2005) Optical bi-stable Switching action of Si high-Q photonic-crystal nano-cavities. Opt Express 13:2678–2687
32. Chen C-H, Matsuo S, Nozaki K, Shinya A, Sato T, Kawaguchi Y, Sumikura H, Notomi M (2011) All-optical memory based on injection-locking bistability in photonic crystal lasers. Opt Express 19:3387–3395
33. Nozaki K, Shinya A, Matsuo S, Suzaki Y, Segawa T, Sato T, Kawaguchi Y, Takahashi R, Notomi M (2012) Ultralow-power all-optical RAM based on nano-cavities. Nat Photonics 6:248–252
34. Painter O, Lee RK, Scherer A, Yariv A, O'Brien JD, Dapkus PD, Kim I (1999) Two-dimensional photonic band-gap defect mode laser. Science 284:1819–1821
35. Meier M, Mekis A, Dodabalapur A, Timko A, Slusher RE, Joannopoulos JD, Nalamasu O (1999) Laser action from two-dimensional distributed feedback in photonic crystals. Appl Phys Lett 74:7–9
36. Ohnishi D, Okano T, Imada M, Noda S (2004) Room temperature continuous wave operation of a surface-emitting two-dimensional photonic crystal diode laser. Opt Express 12:1562–1568
37. Imada M, Noda S, Chutinan A, Tokuda T, Murata M, Sasaki G (1999) Coherent two-dimensional lasing action in surface-emitting laser with triangular-lattice photonic crystal structure. Appl Phys Lett 75:316–318

38. Katsuno S et al (2021) Self-consistent analysis of photonic-crystal surface-emitting lasers under continuous-wave operation. Opt Express 29(16):25118–25132
39. Yoshida M, De Zoysa M, Ishizaki K, Kunishi W, Inoue T, Izumi K, Hatsuda R, Noda S (2021) Photonic-crystal lasers with high-quality narrow-divergence symmetric beams and their application to LiDAR. J Phys Photon 3:022006
40. Zoysa MD, Sakata R, Ishizaki K, Inoue T, Yoshida M, Gelleta J, Mineyama Y, Akahori T, Aoyama S, Noda1 S (2023) Non-mechanical three-dimensional LiDAR system based on flash and beam-scanning dually modulated photonic crystal lasers. Optica 10(2):264–268
41. Sakata R, Ishizaki K, Zoysa MD, Kitamura K, Inoue T, Gelleta J, Noda S (2023) Photonic-crystal surface-emitting lasers with modulated photonic crystals enabling 2D beam scanning and various beam pattern emission. Appl Phys Lett 122:130503
42. https://www8.cao.go.jp/cstp/english/society5_0/
43. Noda S, Inoue T, Yoshida M, Gelleta J, De Zoysa M, Ishizaki K (2023) High-power and high-beam-quality photonic-crystal surface-emitting lasers: a tutorial. Adv Opt Photon 15(4):977–1032

Chapter 7
Applications of Surface Plasmon Waveguides

Abstract This chapter overviews applications of surface plasmon (SP) waveguides to advanced photonic devices/networks. The first topic introduces several configurations of diffraction-limit free compact SP waveguides. Notably, slot-type hybrid metal-insulator-metal (HMIM) and insulator-metal-insulator (HIMI) configurations exhibit SP field distributions mitigating deficient propagation length inherent to the SP waveguides, thus preferable for devices having features with both a size of an electronic device and a speed of a photonic device. The second topic is applications of hybrid SP waveguides to functional devices such as optical mode converter, modulator and detector with high-efficiency, high-speed, and broadband. They have a potential to function in the next-generation optical communication networks. The third topic is applications of the SPs to microwave plasmonics in wireless communications. SP devices positively challenge the high-speed and ultra-wideband demand in short optical links; optical interconnect in the sub-THz photonic world. 500 GHz plasmonic modulators with CMOS-processed Si photonics aggressively challenge the targets.

7.1 Introduction

Si photonics and photonic crystals (PCs) as dielectric photonic devices play an active part in the modern photonic networks, demonstrating their own unique propagation characteristics, as shown previously. Yet, their sizes are limited to an order of a wavelength of light due to an intrinsic law of diffraction, thus forced to be orders of magnitude larger than nano-scale electronic components. Such disagreement in size seriously causes a major obstacle for integrating electronic and photonic components.

Incidentally, surface plasmon (SPs) are electromagnetic (EM) waves coupled to a coherent motion of free electrons in the metal surface, thus propagating along the metal–dielectric interface. Owing to such a unique motion of electron gases on the nano-scale metal surface, SPs present diffraction-limit free properties. Metal stripe and metal slot waveguides with nano-scale widths are typical diffraction-limit free SP waveguides. Figure 7.1 is a schematic diagram showing a standing

K. Asakawa et al., *Nano-photonics for Advanced Networks*,
https://doi.org/10.1007/978-3-031-80637-7_7

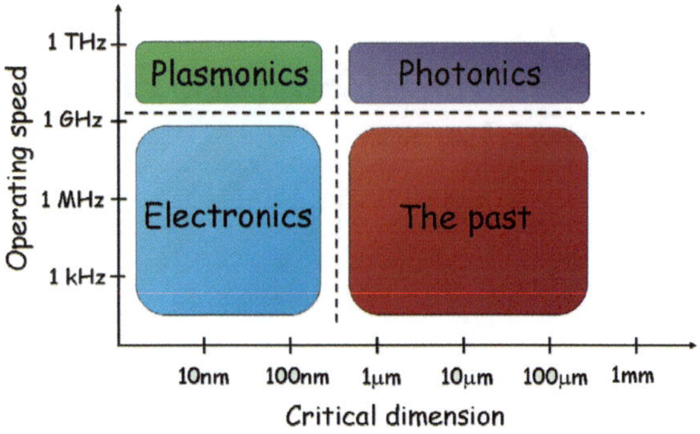

Fig. 7.1 Schematic of symbolic relations among electronic, photonic, and plasmonic chip from viewpoint of operating speeds and critical dimensions. Reprinted with permission from Ref. [1] © 2006 Elsevier

position of the SP chip, when compared with electronic and photonic chips, from viewpoints of operating speeds and critical dimensions [1]. The figure symbolizes that the plasmonic chip has a capability of a unique position in mediating a photonic chip with a similar speed and an electronic chip with a similar size in future integrated circuits. Fusion of these technologies will open the door for the next-generation electronics-photonics chip-scale technologies.

Plasmonic waveguides inherently have a challenge in propagation loss due to the electron motions in the metal, allowing for deficient propagation length for applications to large-scale integrated circuits. Nevertheless, they have been largely improved by combination of a variety of modified waveguide materials and configurations. Figure 7.2 shows schematics of various plasmonic waveguide (PW) proposed recently: Typical examples include (a) metal grooves [2], (b) metal stripes [1], (c)

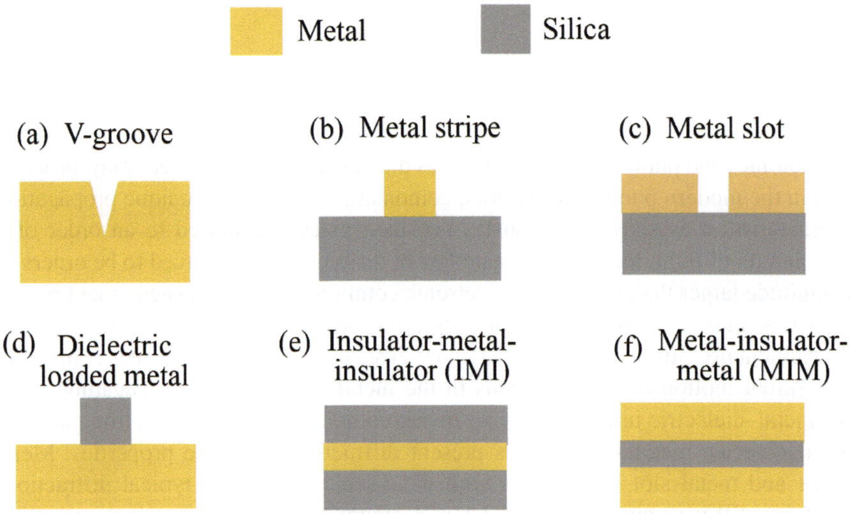

Fig. 7.2 Plasmonic waveguides with V-groove (**a**), metal stripe (**b**), metal slot (**c**), dielectric-loaded metal (**d**), IMI (**e**), and MIM (**f**) structures

metal slots [3], (d) dielectric-loaded metal [4], (e) insulator-metal-insulator (IMI) structures [5] and (f) metal-insulator-metal (MIM) structures [6]. Applicability of these PWs are determined by propagation losses, field confinements, physical dimensions, and other waveguide performances. Some of the simply modified typical PWs are discussed next in Sect. 7.2. Moreover, applications of SP-based several waveguide devices to microwave plasmonics in wireless communications are briefly introduced in Sects. 7.3 and 7.4.

7.2 Development of Plasmonic Waveguides

7.2.1 Metal-Based Plasmonic Waveguide

Metal Stripe Plasmonic Waveguide

A metal stripe PW on a glass substrate as an example in Fig. 7.2b is introduced first. Figure 7.3a shows a simulated field distribution of a "leaky" SP mode located on the metal/air (top) interface, while Fig. 7.3b shows a simulated field distribution of a "bound" SP mode on the metal/glass (bottom) interface. The waveguide is a gold (Au) stripe with a thickness of 55 nm and a width of 3.5 μm on the glass at $\lambda = 800$ nm [7]. Incidentally in Fig. 3.9, excitation of a SP mode along the metal/air interface is explained by the Kretschmann configuration, where light incident into the prism excites the SP on the metal surface via coupling between evanescent fields of light and SP within the metal film, namely, via total internal reflection (TIR) at the metal/prism interface. In the metal stripe PW with a finite waveguide width, however, existence of both a leaky SP mode along the top metal/air interface and a bound SP mode along the bottom metal/glass interface via ATR (attenuated total reflection) is analytically simulated [7]. In Fig. 3.9 again, a bound mode exists out of the light cone, while a leaky mode exists within the light cone, reflecting ATR. Since these SP modes are eigenmodes attached to the top and bottom Au surfaces, they are available for independent propagating-SP modes. Figure 7.3c shows simulated real part (top panel) and imaginary part (bottom panel) of the complex propagation constant ($\beta_{sp} + i\,\alpha_{sp}$) in the leaky SP mode as a function of Au stripe widths, Ws. The simulation was conducted for eight lowest-order leaky SP modes. The geometry of the Au stripe is depicted in the inset in the bottom panel [7]. The result suggests that, first, there is a cut-off waveguide width in the leaky SP mode and second, the Au stripe width of nearly ≥ 5 μm significantly reduces the α_{sp}, i.e., the propagation loss for the lowest-order leaky SP mode. Note that, though there are two SP modes as mentioned above, the leaky SP mode is excited with much larger efficiency and considerably higher field intensity than the bound SP mode. See [7] for behaviors of these modes in more detail.

Figure 7.3d shows an optical micrograph of varying-width Au stripes on a large Au plate for observing a SP mode propagation along the Au stripes. An inset is a scanning electron microscope (SEM) photograph showing a stripe with a width of

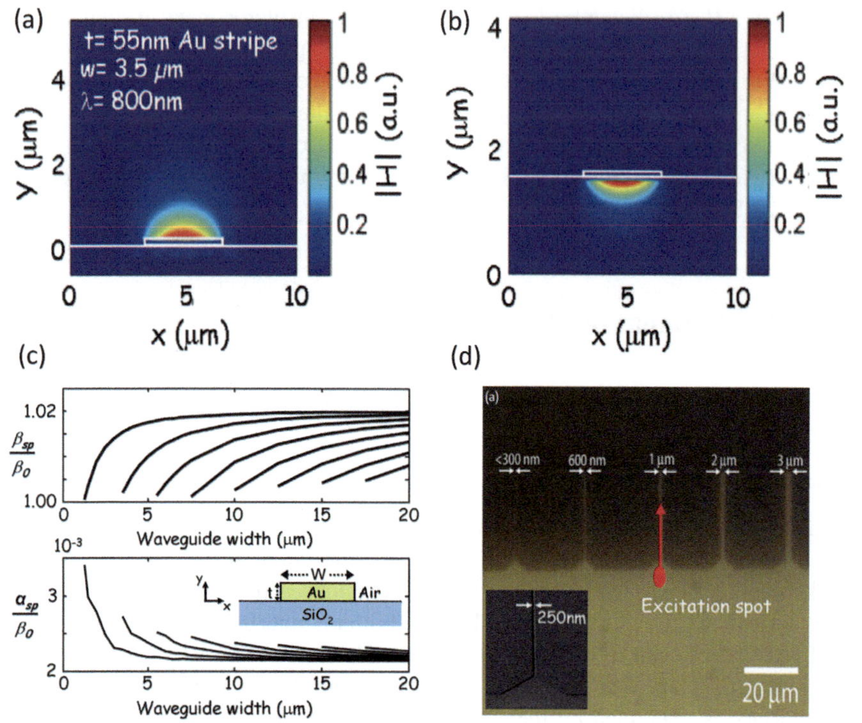

Fig. 7.3 **a**, **b** Simulated field distributions of leaky SP modes at an air/metal (top) interface and bound SP mode at a metal/glass (bottom) interface, respectively. **c** Calculated real part (top panel) and imaginary part (bottom panel) of complex propagation constants ($\beta_{sp} + i\ \alpha_{sp}$) versus Au waveguide width. **d** Optical micrograph of an array of Au stripes attached to a large launch pad. (a–c) Reprinted with permission from [7] © 2005 American Physical Society. (d) Reprinted with permission from [1] © 2006 Elsevier

250 nm. A red arrow shows a spot of a laser launched on the stripe with a width of 1 μm [1]. In addition, PSTM (photon scanning tunneling microscope) images of the SPs propagating along Au stripes with different widths of 3.0-, 1.5-, and 0.5-μm were observed. As a result, the Au stripe with a width of 3.0 μm was found to be applicable to SP waveguides with lengths of over several tens of microns.

Metal Slot Plasmonic Waveguide

3-D plasmonic slot waveguides with widths of sub-wavelength, as classified in Fig. 7.2c, are also capable of supporting special SP modes. Figure 7.4a shows a schematic of a metal slot waveguide with a width of sub-wavelength surrounded by (or on) a dielectric medium [3]. Light propagates along an arrow in the slot. A symmetric slot configuration possesses the fundamental bound mode irrespective of wavelength and waveguide parameters. Meanwhile, a metal (silver)-slot waveguide in Fig. 7.4b is vertically asymmetric, surrounded by air (above) and silica (below).

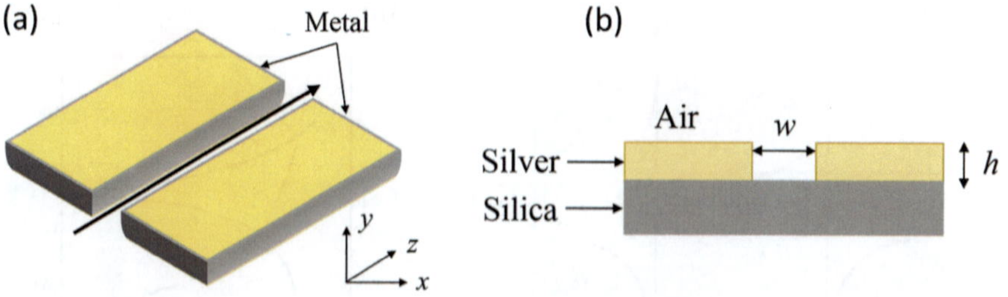

Fig. 7.4 **a** Schematic of SP symmetric metal slot waveguide. **b** Schematic of cross-sectional SP asymmetric metal slot waveguide used for the calculation below

Different from the symmetric slot waveguide, the fundamental propagation mode in the asymmetric waveguide is not always bound mode. In the article [3], a propagation length L_p of the fundamental mode is shown as a function of the slot width of w, where the L_p is defined as the propagation distance needed before the intensity of the SP field decreases to 1/e. See Eq. (3.21) in Sect. 3.3.2 for definition of L_p. The silver film thickness and wavelength are 50 nm and 1.55 μm, respectively. L_p reaches as long as ~ 30 μm. L_p ~ 30 μm is suitable for the SP device in some applications. The lateral HMIM (hybrid metal-insulator-metal) structures are presented for the SP slot-type MIM waveguide devices: Refer to Sect. 7.3 for mode converter and modulator. See Sect. 7.4 also for applications to high-speed and broadband microwave plasmonics.

Toward Long Propagation Length

The SP IMI waveguide, classified in type Fig. 7.2e, has a possibility of much large L_p on the order of centimeter [8–10]. The long-range SP (LRSP), i.e., an SP mode with a long propagation length is reviewed precisely in [8]. As shown in Fig. 7.5a, a practical configuration of the LRSP mode is a metal slab (thin film) waveguide with a thickness of t and relative permittivity of $\varepsilon_{r,2}$, sandwiched by clad dielectric media with relative permittivity of $\varepsilon_{r,1}$ and $\varepsilon_{r,3}$. The metal slab becomes symmetric or asymmetric depending on whether $\varepsilon_{r,1}$ and $\varepsilon_{r,3}$ are equal or not. Red curves show main transverse electric fields (E_ys) of an *asymmetric bound* mode a_b and a *symmetric bound* mode s_b, propagating along $a + z$ axis. Charge densities in the metal for modes a_b and s_b have symmetric (++) and asymmetric (+−) distributions, respectively, as shown in Fig. 7.5a. These are transverse magnetic TM-polarized bound modes ($H_x \neq 0$, $E_x = H_y = H_z = 0$) called *coupled modes*. Figure 7.5b shows calculated β/β_0 (normalized phase) and α/β_0 (normalized attenuation) versus a metal thickness t. α/β_0 for the *symmetric bound* mode s_b is abruptly reduced for t ≤ ~ 30 nm, indicating a long propagation length L_p ~ 300 μm [10]. Another approach for the LRSP mode with a long propagation length has been made using a different hybrid SP IMI waveguide [11]. Propagation lengths L_ps are simulated as a function of metal thicknesses. The possibility of L_p in the centimeter regime for the metal thickness ~ 30 nm is shown. See [11] for more detail. The SP IMI waveguide with a thin meal

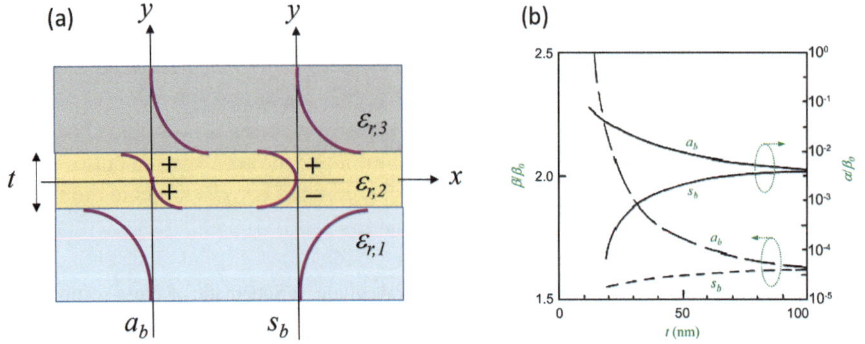

Fig. 7.5 **a** Metal slab (thin film) waveguide with thickness of t and relative permittivity of $\varepsilon_{r,2}$, sandwiched by clad dielectric media with relative permittivity of $\varepsilon_{r,1}$ and $\varepsilon_{r,3}$. **b** Calculated β/β_0 and α/β_0 versus metal thickness t. (b) Reprinted with permission from Ref. [10] © 1981 APS

thickness of typically ~ 30 nm, as mentioned above, has proved to contribute to the long propagation length on the order of centimeter to sub-centimeter.

7.2.2 *Hybrid Plasmonic Waveguide*

Hybrid Metal-Dielectric Structure

As a typical hybrid metal-dielectric structure, classified in type Fig. 7.2d, a hybrid dielectric-loaded plasmonic waveguide (HDLPW) was demonstrated for guiding plasmonic waves for a long distance [4]. Figure 7.6a shows a cross-section of the HDLPW composed of a stacked Si nano-wire and a SiO_2 thin layer on a 100 nm thick silver (Ag) metal layer on a glass substrate. The role of the low-index ($n_2 = 1.44$) SiO_2 layer inserted between the high-index ($n_1 = 3.48$) Si layer and the Ag layer was investigated by varying the SiO_2 thickness t. Figure 7.6b indicates calculation of normalized effective refractive indices N_{eff}/k_0s ($k_0 = \omega/c$) of the SP mode and their propagation length L_{p}s versus SiO_2 thickness of t for Si nanowires with different thicknesses (hs) of 100, 150, and 200 nm. A width of w in the HDLPW is set to 200 nm, allowing only single mode at $\lambda = 1.55$ μm. L_{p} increases to ~ 250 μm when the thickness of t increases to 200 nm. In addition, when t decreases to 50 nm with h remaining 150 nm, N_{eff} and L_p decrease to 1.27 and 64 μm, respectively. Consequently, a fundamental-mode field penetrates deeply into the low-index SiO_2 stripe. In the end, the HDLPW causes large confinement of an optical field into the low-index region (SiO_2 layer), thus exhibiting low loss and strong field confinement into the waveguide in a compact structure. In this work, further simulation on 90°-circular bends, S-shaped waveguide bends, directional couplers, etc., have been implemented [4].

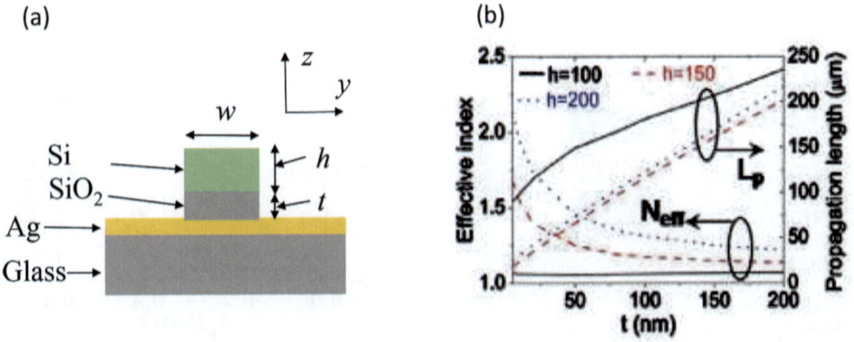

Fig. 7.6 **a** Schematic of a cross-section in the HDLPW. **b** Calculation of N_{eff}/k_0 of SP mode and L_p versus SiO_2 thickness of t for different hs as parameters. (b) Reprinted with the permission from Ref. [4] © 2010 AIP

Hybrid Insulator-Metal-Insulator Structure

As a hybrid insulator-metal-insulator (HIMI) plasmonic waveguides, classified in type Fig. 7.2e, another hybrid-layer-structured SP waveguide was reported, namely, a SiO_2-silver (Ag)-SiO_2 SP waveguide sandwiched by Si layers on a glass substrate, as shown in Fig. 7.7a [5]. Investigation at a wavelength of 1.55-μm includes theoretical characteristics on transmission and loss in 90°-shaped, S-shaped, and U-shaped bend waveguides as well as straight waveguide. It should be noted first that the proposal brought a realistic planar HIMI SP waveguide with two important features: One is an adequate balance between a long-range propagation length L_p (~ 250 μm) and well confinement (~ 38%) of an optical energy near the metal layer. The other is a compatibility with SOI CMOS process fabrication. Several theoretical investigations first derived an effective refractive-index Re (N_{eff}) in the dispersion property of the HIMI SP waveguide by changing the metal layer thickness t_m and derived two modes; asymmetric TMha[1] and symmetric TMhs[2], both strongly depending on the thickness t_m. Difference of the Re (N_{eff}) between TMha[1] and TMhs[2] is small in the larger t_m regime, while it is large in the smaller t_m regime. Accordingly, smaller t_m, 30 nm, was selected for preferential single-mode (TMhs[2]) operation and suppression of lossy asymmetric mode TMha[1]. Next, the Re (N_{eff}) and P.L. (propagation length) versus thickness of t_h in the Si layer was investigated, as shown in Fig. 7.7b. The mode starting as a leaky mode at smaller t_h converts into the dielectric mode at larger t_h with the long P. L. In the middle t_h, the mode is the HIMI SP mode resulted from a combined effect of the metal, spacer and high-index Si layer. Note that the HIMI SP waveguide can confine light strongly in the spacer regions as a result of coupling between the SP mode and dielectric mode in the spacer regions. The result is similar to the case for the HDLPW, as shown in Fig. 7.6. Energy distribution reaches maximum when $t_h = 0.2$ μm. Moreover, transmission characteristics for several HIMI SP bend waveguides were investigated: for 90°-bend, S-shape bend, and U-shaped bend with bend radii of 1.6 μm, 2.1 μm, and 1.7 μm, excellent maximum transmittance of 87%,

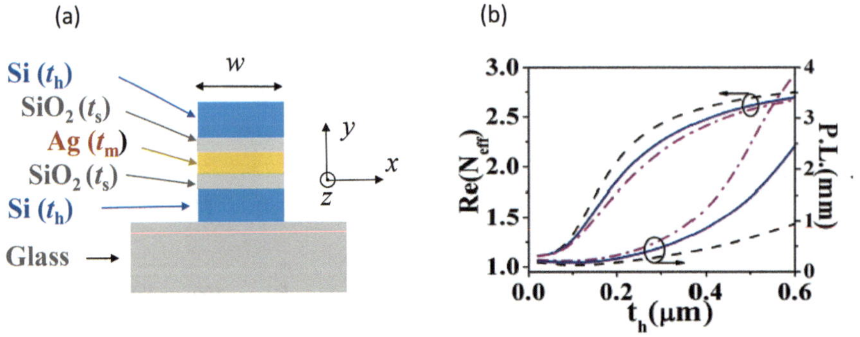

Fig. 7.7 **a** Schematic cross-sectional view of hybrid-layer-structured SP waveguide. **b** Re (N_{eff}) and L_p versus thickness of t_h in the Si layer. (b) Reprinted with permission from Ref. [5] © 2017 IEEE

80%, and 79% were calculated, respectively, all at cross-section areas of 0.47 × 0.25 μm^2. As a whole, results shown here are likely to pave the way for providing efficient SP waveguides.

7.3 High-Speed and Broadband Plasmonic Devices

This section introduces applications of plasmonic waveguides mentioned above to several high-speed and wide bandwidth devices that have attracted ever-growing attention to microwave plasmonics in recent years. Specifically, a high-efficiency plasmonic mode converter, high-speed and broadband plasmonic modulator and detector are overviewed.

7.3.1 High-Efficiency Plasmonic Mode Converter

Structure of Tapering Mode Coupler

Among several SP waveguides listed up in Fig. 7.2, the SP slot-type MIM waveguide was applied to an efficient plasmonic mode converter connected to a Si nano-scale waveguide on silica [12]. Figure 7.8a shows a bird's-eye view of a proposed whole SP waveguide mode converter. A thin Au plate in yellow with a narrow slot shows the SP slot-type MIM waveguide region, while a single-mode Si nano-wire waveguide in dark gray with a taper region largely differs in size and thickness from the SP slot-type MIM region. Note that the Si taper region is laterally separated from the Au thin plate by a small air-gap W_{gap}. Cross-sectional electric field distributions along the broken lines on the metal (i) and Si (ii) are shown in Fig. 7.8b, where panels (i) (left) and (ii) (right) correspond to field distributions on the metal (i)

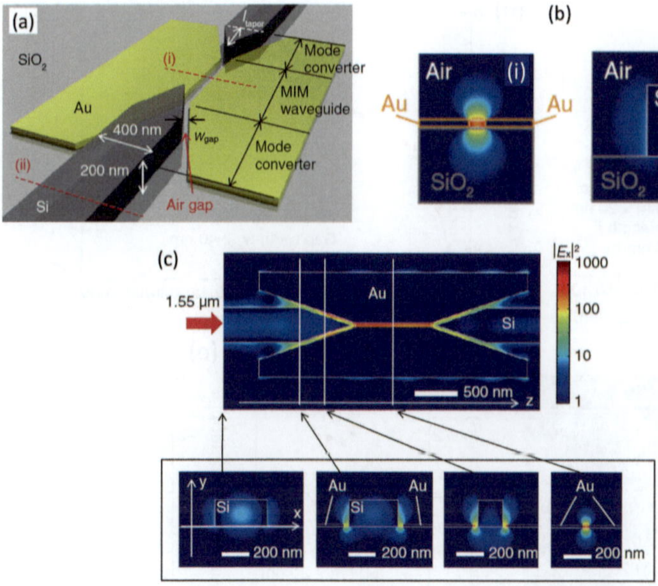

Fig. 7.8 **a** Schematic of the plasmonic mode converter. **b** (i) Field distribution of the propagation mode at the metal slot (along the broken line (i) in (**a**)). (ii) Field distribution of the propagation mode at the Si nano-wire waveguide (along the broken line (ii) in (**a**)). **c** Top-view (upper panel) and cross-sectional views (bottom panel) of field distributions at positions indicated in the upper panel. Reprinted with permission from Ref. [12] © 2018 NTT Technical Review

and Si (ii), respectively. A typical size parameter is 50 nm × 20 nm in the MIM slot core. A conventional 2-D plasmonic mode converter has been formed by a laterally tapered structure with the same height, which performs mode conversion by compression in the lateral direction only. However, the above SP slot-type MIM converter exhibits excellent performance by introducing a small air-gap between Si and metal, as mentioned above. Figure 7.8c shows calculated top-view and cross-sectional electric field distributions in propagating modes of 1.55-μm input light in a Si nano-wire waveguide, mode converter, and slot-type MIM waveguide. Note that the side lobes of the propagating mode in the Si nano-wire penetrate deeply into the metal region, thus indicating significant coupling between the Si nano-wire and slot-type MIM waveguide established against their large difference in height. Device characteristics are shown text.

Characteristics of Mode Converter

Figure 7.9a, b shows a calculated coupling efficiency as a function of air-gap width w_{gap} and taper length l_{taper}, respectively [12]. A width and height of the Si nano-wire are 400 nm and 200 nm, respectively, while those of the slot waveguide are 50 nm and 20 nm, respectively. The coupling efficiency is maximum at $w_{gap} \approx$ 20 nm and $l_{taper} = 600$ nm, so that a very small air-gap is important to achieve an efficient mode conversion. Figure 7.9c shows a SEM photograph of the fabricated

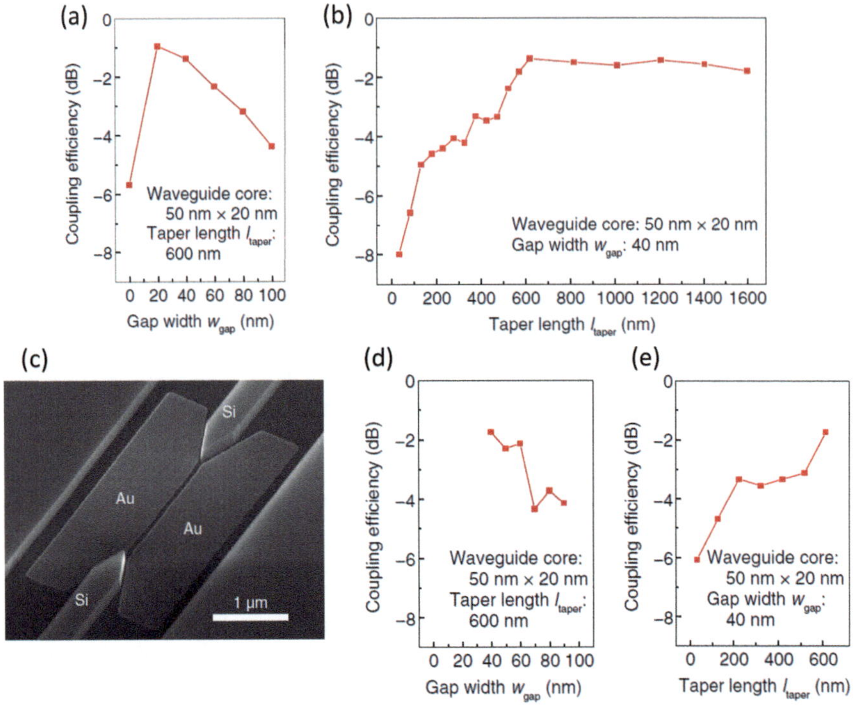

Fig. 7.9 **a**, **b** Calculated coupling efficiencies versus air-gap width w_{gap} and taper length l_{taper}, respectively. **c** SEM photograph of the fabricated sample. **d**, **e** Measured coupling efficiencies versus w_{gap} and $l_{taper.}$, respectively. Reprinted with permission from Ref. [12] © 2018 NTT Technical Review

device. Figure 7.9d, e shows measured coupling efficiencies versus w_{gap} and l_{taper}, respectively. Measured coupling efficiencies of ~ 1.7 dB for $w_{gap} = 40$ nm and $l_{taper} = 600$ nm show good agreement with the calculated value (−1.4 dB). Note that a small and simple taper with length of less than 1 μm is capable of achieving a high-efficiency plasmonic mode converter that connects waveguides of highly different heights. The result is also important for reducing a footprint of the device in the future Si photonic integrated circuit.

7.3.2 *High-Speed and Broadband Plasmonic Modulator*

The SP slot-type MIM waveguide exhibits strong optical field confinement in the narrow sub-wavelength metal-gap regions, as shown above. If the metal-gap is filled with nonlinear media or semiconductor, high-efficiency plasmonic active devices such as modulator and detector are available, as shown by the following demonstrations.

Figure 7.10a shows a schematic of a plasmonic Si photonic MZI (Mach–Zehnder interferometer) modulator on silica [13]. Both arms in the MZI are composed of SP slot-type MIM waveguides to exploit a Pockels effect in DLD164 (nonlinear material) filled in the slot region. The plasmonic modulator operates with a bandwidth beyond 170 GHz, finally aiming at as high as 500 GHz (see Sect. 7.4.3). Input and output ends are connected to Si photonic strip waveguides with grating couplers for coupling with external light. Incident light is divided into two MZI arms by a MMI (multi-mode interferometer) coupler. Photonic mode in the Si nano-wire and plasmonic mode in the metal slot waveguide are converted efficiently via a tapering section. A length and width of the slot-type MIM waveguide are 12.5 μm and 100 nm, respectively. The modulator chip is 287 μm × 270 μm in total. Figure 7.10b shows a schematic of the metal-DLD164 (nonlinear material)-metal slot structure [14, 15]. Gold electrodes (150 nm height) serve as both the MIM slot waveguide and electrical contacts. Refractive-index change of the DLD164, Δn_{eff}, based on the electro-optic effect is known to increase with decrease of the width in the metal slot [14, 15]. Figure 7.10c shows measured frequency response of the SP modulator in a wide range from 75 MHz to 170 GHz, as shown in [13]. No prominent indication of a bandwidth limit is observed in a whole frequency range. Moreover, 2-level and 4-level pulse amplitude modulations (PAM-2 and PAM-4) were implemented. For instance, usage of a 100 GBd non-return-to-zero (NRZ) PAM-2 signal resulted in successful observation of the eye diagram. Including other results with 50–60 GBd PAM-4 multi-level signals, demonstrated results mentioned above are the first operation of the SP modulator at the level of 100 GBd NRZ signal [13].

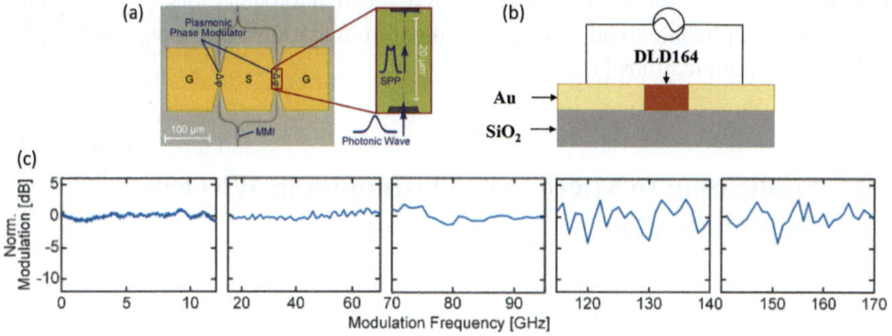

Fig. 7.10 **a** Schematic of the SP slot-type MIM modulator consisting of two narrow gold slots, filled with organic nonlinear material (DLD164). **b** Schematic cross-section of the metal-DLD164 (nonlinear material)-metal slot structure. **c** Measured frequency response of the modulator. (a) and (c) Reprinted with permission from Ref. [13] © 2017 OSA

7.3.3 High-Speed and Broadband Plasmonic Detector

A high-speed SP photodetector was demonstrated using a Si slot-type MIM waveguide with an amorphous Ge (α-Ge) inserted into the SP slot, that is, a metal–semiconductor–metal (MSM) slot for exploiting its photoconductive effect [16]. An incident optical signal in the Si nano-wire waveguides couples to the nano-scale SP MSM slot-type waveguide through the taper region, similar to the SP slot-type MIM modulator in Fig. 7.10. A slot width and length are 160 nm and 3 μm, respectively. The optical field in the nano-scale MSM slot region is strongly confined, while the DC field in the Ge slot is distributed uniformly by flow of DC.

RF response vs bandwidth characteristics of the MSM slot detector were studied with two different methods: At a lower frequency range (100 MHz–20 GHz), a 1310 nm carrier with an intensity-modulated RF signal was used, while at a higher frequency range (1–100 GHz), an approach with beating by two lasers was used. A measured frequency response was found to be almost flat up to 100 GHz. Moreover, an eye diagram for a 72 Gbit/s ON − OFF keying bit sequence and BER of 1.6×10^{-2} was observed. Measured results are limited by the available modulator with a 3-dB bandwidth of 30 GHz at 1300 nm. Nevertheless, the demonstrated MSM slot detector was found to indicate no limitation in speed up to 72 Gbit/s. Conventional Ge/Si and/or III − V/Si photodetectors provide typical characteristics of 100 GHz bandwidths and data rates up to 100 Gbit/s. In contrast, a photoconductive MSM system is preferable in higher-speed applications on account of its small resistance–capacitance product originated from carriers with a short transit time. Since demonstrated SP slot-type MSM detectors are capable of fabrication by CMOS-compatible technologies, they are promising for a cost-effective high-speed photodetectors over 100 GHz bandwidths, applicable to advanced optical communications, microwave photonics, and THz communications [16].

7.4 Application to Microwave Plasmonics in Wireless Communication

This section overviews applications of plasmonic active devices such as high-efficiency plasmonic mode converter, high-speed, and broadband plasmonic modulator and detector, as mentioned above, to microwave plasmonics in wireless communication, specifically to plasmonic nano-antenna, optical interconnect with plasmonic device array, and 500 GHz plasmonic modulator for sub-THz photonics.

7.4.1 Plasmonic Nano-antenna

In the recent wireless communication, electrical signals detected by an antenna were amplified, converted to a baseband with a millimeter-wave (MMW) electronic receiver, electrically amplified again, and converted to an optical signals with a photonic modulator. In contrast, the latest plasmonic nano-antenna capable of directly converting MMW (and even THz wave) signals to optical signals have brought much attention for advanced wireless communication to encode high data-rate (e.g., Tbit/s) wireless signals to an optical carrier [17, 18].

For instance, a plasmonic nano-antenna based on a plasmonic phase modulator (PPM) was proposed [19]. The PPM is connected to a pair of metal antennae, thus significantly contributing to low-cost wireless THz communication. The PPM is an SP slot-type MIM waveguide with a nonlinear material (DLD164) in the slot [14]. RF power incident on the antenna is strongly confined in the MIM slot by applied voltage and resonantly enhanced from the incident field by several orders of magnitude, where the resonant condition of the antenna is designed to be tuned to a carrier frequency of, e.g., 60 GHz. On the other hand, input/output Si nano-wire waveguides are connected to the PPM with the Au antenna and MIM slot. Light incident in the SP slot waveguide is strongly confined to the slot and almost perfectly overlaps with the RF electric field owing to strong interaction with nonlinear material (DLD164) in the slot medium. Thus incident RF field can be efficiently encoded to an optical carrier.

For demonstration of the plasmonic antenna, a CW laser was incident on a grating coupler, while a 60 GHz RF signal fed into the antenna modulated the phase of the PPM [19]. An optical spectrum was measured using an antenna with a width of 75 nm, where transmitted MMW frequencies (f_{RF} of 60 GHz) were clearly observed as sidebands of the optical carrier. Electric field enhancement (FE) factors, defined as a ratio of enhanced electric field within the nano-slot against incident wave field, were measured as a function of the slot width. As a result, FE factors as high as 35,000, 17,000, and 15,000 were derived for samples with slot widths of 40, 75, and 90 nm. The measured FEs coincide with the simulated FEs. Accordingly, the SP modulator antenna has proved to achieve large resonant field enhancement owing to strong nonlinear interaction between optical and electrical fields in the SP slot. The results also suggest a potential for a promising device in the next-generation Tbit/s wireless communication system.

7.4.2 Optical Interconnect with Plasmonic Device Array

As shown in Sect. 7.3.3, the SP MZM with a slot-type MIM waveguide on silica exhibited a modulation bandwidth beyond 170 GHz [13]. An integrated array of such a SP modulator was applied to a high-speed (4×36 Gb/s) optical interconnect [14, 20]. Figure 7.11a, b show compact high-speed optical interconnect transceivers

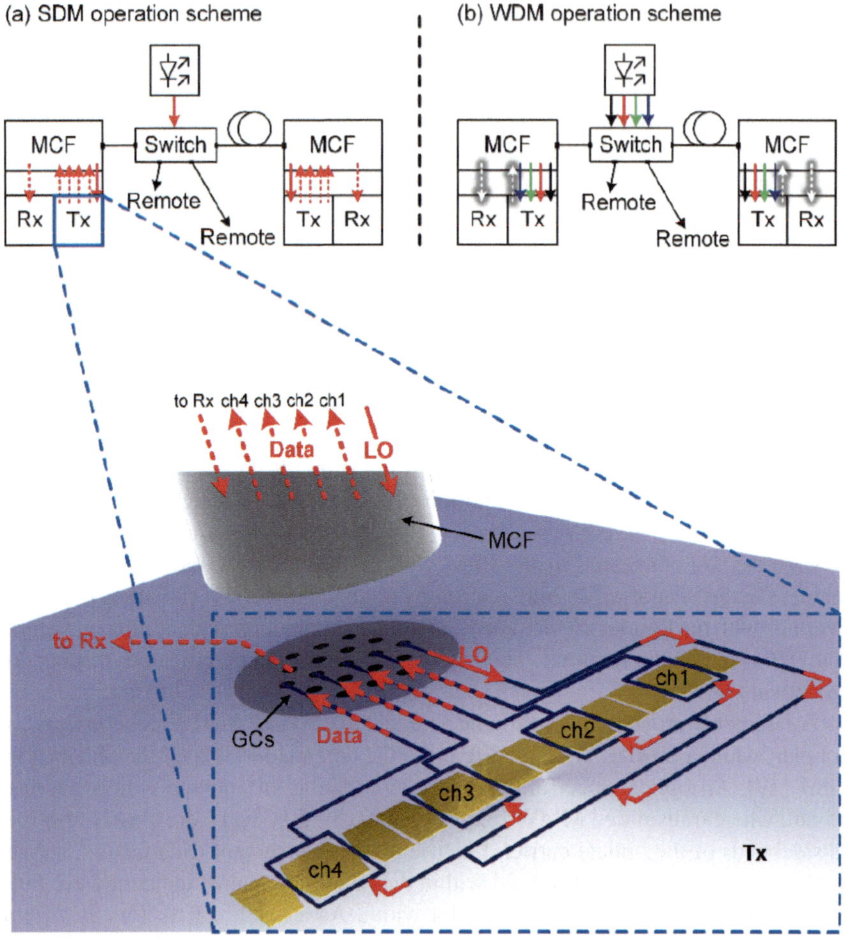

Fig. 7.11 (Bottom panel) Schematic of a compact high-speed optical transceiver with integrated plasmonic MZM and photodiode arrays. (Upper panel) **a** SDM scheme with a compact plasmonic Tx and Ge Rx. **b** WDM scheme using multiplexed/de-multiplexed laser light with different wavelength on the chip or externally through MCFs. Reprinted with permission from Ref. [14] © 2015 OSA

based on two schemes of space-division multiplexing (SDM) and WDM operations, respectively. In the lower panel, a magnified transmitter (Tx) array for the SDM scheme is depicted [14]. Input light emitted from a centralized laser (LO, solid line) is guided through an optical switch and multi-core fibers (MCFs) to integrated transceivers via on-chip GC (grating coupler) arrays. Then, light is further divided into four SP slot-type MZM channels (ch1–ch4) with a phase shifter length of 20 μm and a channel spacing of 50 μm. There, information is encoded. Modulated data are fed back to a central switch through a different MCF path and delivered partly to remote routs and partly to Ge photodiode Rxs (receivers) [15, 21, 22]. In the

WDM scheme, external lasers with four wavelengths (in this example) feed light into interconnect transceivers. At Tx, a modulated array is connected to an on-chip or out-of-chip multiplexer, while, at Rx, a de-multiplexer separates wavelength channels or individually detects them.

In experiments on data modulation, data signals of 36 Gbit/s in binary phase shift keying (BPSK) originated from four CW laser sources were fed to the transceiver. As a result, eye diagrams and constellation diagrams as well as bit error ratios (BERs) of below 4.5×10^{-3} were clearly confirmed for all channels [14]. See [20] for more detail.

7.4.3 500 GHz Plasmonic Modulator for Sub-THz Photonics

THz microwave photonics is attracting much attention to a rapidly growing wireless communication system with unprecedented data rates in the next-generation (5G) era [23, 24]. To enable expected transmission rates (~ hundreds Gb/s), carrier frequencies (above 300 GHz) are now under consideration in the unallocated spectrum range. In the MMW (millimeter-wave) frequencies (60 and 70–95 GHz), an available contiguous bandwidth ($\leq$ 9 GHz) is insufficient for expected high data rates. An alternative THz band (300 GHz–10 THz) is now suffering from free space path loss and water vapor-based atmospheric absorption. Yet, it has several tens of GHz bandwidth with reduced absorption between 200 and 450 GHz, thus suitable for wireless links. On the other hand, photonic technologies for treating THz signals above 300 GHz need converters between optical and THz signals with a bandwidth above 300 GHz, high power handling, and high linearity. So far, uni-traveling-carrier photodiodes (UTC-PDs) have been developed for optical-to-THz conversion [25]. However, an efficient THz-to-optical conversion has not been realized due to a lack of modulators with a sufficiently wide (over 300 GHz) electro-optic bandwidth, high power handling, and very high linearity [26].

To solve this problem, a valuable 500 GHz SP MZM (Mach–Zehnder modulator) with a compact size of a 25-μm length and high linearity was developed recently [23]. Figure 7.12a shows a remote antenna unit (RAU), where an improved version of the SP MZM, once developed previously as shown in Fig. 7.10, is schematically depicted in the middle panel (a1). An antenna receives THz wireless signals, which is converted to an optical region to be assigned over a radio-over-fiber (RoF) link. A lower right panel (a2) shows an image of the SP slot-type MIM phase modulator waveguide with a 120-nm width (w), 20-μm length (L), and nonlinear organic material (HD-BB-OH/YLD124) filled in the slot. Refer to [14] for detail. Figure 7.12b shows frequency responses of the SP MZM as a function of input electrical signals from 75 MHz up to 500 GHz onto an optical carrier at 1547.5 nm. Precisely, responses from frequency ranges in 15–70 GHz (in red) and 200–500 GHz (in black) were obtained with different setups using the same SP MZM, while those from frequency ranges in 75 MHz–15 GHz, 70–95 GHz, and 115–170 GHz (all gray) were obtained with a similar SP MZM device in the different setup [27].

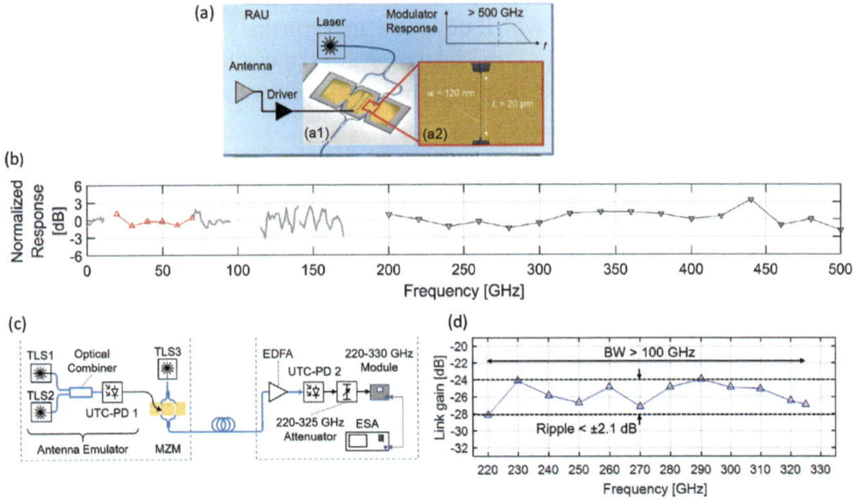

Fig. 7.12 The image consists of four parts, (a) to (d). (a) Schematic of 500 GHz SP MZM in remote antenna unit (RAU) enabling sub-THz microwave photonics. (a1) Micrograph of the SP MZM with a plasmonic phase modulator. (a2) Magnified phase modulator with 120 nm-wide and 20 μm-long slot waveguide. (b) Measured frequency responses of the SP MZM versus input electrical signals. (c) Experimental setup for testing capability of the SP MZM to implement an analog RoF link via operation of direct THz-to-optical-to-THz conversion. (d) Measured flat responses through the RoF link transmission. Reproduced from [23] (composite of Figs. 1b, 2a and 3b and c), released under a CC BY 4.0 license; https://doi.org/10.1063/1.5086868

For testing capability of the SP MZM mentioned above to demonstrate a RoF link, operation of direct conversion was investigated in a THz-to-optical-to-THz range with a bandwidth of over > 100 GHz (220–325 GHz range). Figure 7.12c shows an experimental setup for connecting a remote antenna unit to a central office (both omitted in the panel) via the RoF link. To this end, two different wavelengths emitted from lasers (TLS1 and TLS2) are converted by an optical combiner to a beat frequency, which is incident on the UTC-PD 1. The output sub-THz signal from the UTC-PD 1 directly modulates an optical carrier by the SP MZM. Thus, modulated (sub-THz) optical signals are generated and propagate through the RoF link. The received optical signals are amplified and detected with the UTC-PD 2. Figure 7.12d shows flat responses through the RoF link transmission with a bandwidth of > 100 GHz and a ripple of < ± 2.1 dB. The results indicate that the SP MZM is capable of direct THz-to-optical conversion showing a flat response.

It should be noted that plasmonic devices introduced in this section possess a potential for advanced tools applicable to microwave and sub-THz photonics saving high performance needed for 5G wireless, IoT, remote sensing, etc.

7.5 Summary

Applications of SPs to advanced photonic networks were overviewed based on the fundamental properties of the SPs introduced in Chap. 3. First, typical diffraction-limit free compact SP waveguide configurations were surveyed; metal V-groove, metal stripe, metal slot and hybrid dielectric-metal (HIMI and HMIM) configurations. A deficient propagation length inherent to the SP waveguides in general was largely mitigated by sophisticated constructions of plasmonic hybrid IMI and slot-type MIM waveguides. For instance, CMOS-compatible planar plasmonic IMI (SiO_2-Ag-SiO_2) waveguides theoretically verified a good balance showing a propagation length (~ 250 μm) long enough for practical application and a field confinement (38%) near the metal layer thick enough for reservation of a hybrid SP mode.

Next, SP slot-type MIM waveguides were applied to active devices; high-efficiency mode converters, and high-speed modulators/photodetectors. SP mode converters connecting tapered Si-wire waveguides to the slot-type MIM waveguides on silica showed a high coupling efficiency with a taper length of 600 nm and a slot air-gap width of 20 nm. A high-speed MZM with slot-type MIM arms presented an operation bandwidth beyond 170 GHz owing to nonlinear materials (DLD164) in the slot region. It also implemented PAMs using 50 GBd–100 GBd PAM-2 and PAM-4 signals with NRZ and square-root raised cosine pulse shapes. A high-speed SP photodetector, exploiting a photoconductive effect in amorphous Ge within the MIM slot, showed flat frequency response up to 100 GHz. These results have a potential to function in high-speed and over 100 GHz broadband optical communication networks.

Finally, SP functional devices were applied to microwave ~ THz plasmonics for next-generation wireless communications. A compact high-speed plasmonic nano-antenna with a plasmonic phase modulator involving nonlinear materials in the slot directly encoded a wireless millimeter-wave (60 GHz) signal onto an optical carrier, thus verifying a solution for practical Tbit/s wireless-fiber links. Chip-to-chip interconnects using SP MZM transmitter arrays with 20-μm-length phase shifters realized an interconnect operation at 36 Gbit/s in BPSK as compact and high-speed SP transceivers. Note that a 500 GHz SP MZM implemented a flat frequency response up to 500 GHz using a compact (25-μm-length) MZM. It also demonstrated an analog RoF link up to 325 GHz. These SP MZMs were verified to have potential for high-performance sub-THz photonics.

References

1. Zia R, Schuller JA, Chandran A, Brongersma ML (2006) Plasmonics: the next chip-scale technology. Mater Today 9:20–27
2. Bozhevolnyi SI, Volkov VS, Devaux E, Laluet JY, Ebbesen TW (2005) Channel plasmon-polariton guiding by subwavelength metal grooves. Phys Rev Lett 95:046802

3. Veronis G, Fan S (2007) Modes of subwavelength plasmonic slot waveguides. J Lightwave Technol 25:2511–2521
4. Chu H-S, Li E-P, Bai P, Hegde R (2010) Optical performance of single-mode hybrid dielectric-loaded plasmonic waveguide-based components. Appl Phys Lett 96:221103
5. Sharma P, Kumar VD (2017) Hybrid insulator metal insulator planar plasmonic waveguide-based components. IEEE Photonics Technol Lett 29:1360–1363
6. Alam MZ, Aitchison JS, Mojahedi M (2014) A marriage of convenience: hybridization of surface plasmon and dielectric waveguide modes. Laser Photon Rev 8:394–408
7. Zia R, Selker MD, Brongersma ML (2005) Leaky and bound modes of surface plasmon waveguides. Phys Rev B 71:165431
8. Berini P (2009) Long-range surface plasmon polaritons. Adv Opt Photonics 1:484–588
9. Berini P (2006) Figures of merit for surface plasmon waveguides. Opt Express 14:13030–13042
10. Sarid D (1981) Long-range surface-plasma waves on very thin metal films. Phys Rev Lett 47:1927–1930
11. Zia R, Selker MD, Catrysse PB, Brongersma ML (2004) Geometries and materials for subwavelength surface plasmon modes. J Opt Soc Am A 21:2442
12. Ono M, Taniyama H, Kuramochi E, Nozaki K, Notomi M (2018) Toward application of plasmonic waveguides to optical devices. NTT Tech Rev 16:14–19
13. Hoessbacher C, Josten A, Baeuerle B, Fedoryshyn Y, Hettrich H, Salamin Y, Heni W, Haffner C, Kaiser C, Schmid R, Elder DL, Hillerkuss D, Moller M, Dalton LR, Leuthold J (2017) Plasmonic modulator with >170 GHz bandwidth demonstrated at 100 GBd NRZ. Opt Express 25:1762–1768
14. Heni W, Hoessbacher C, Haffner C, Fedoryshyn Y, Baeuerle B, Josten A, Hillerkuss D, Salamin Y, Bonjour R, Melikyan A, Kohl M, Elder DL, Dalton LR, Hafner C, Leuthold J (2015) High speed plasmonic modulator array enabling dense optical interconnect solutions. Opt Express 23:29746–29757
15. Haffner C, Heni W, Fedoryshyn Y, Niegemann J, Melikyan A, Elder DL, Baeuerle B, Salamin Y, Josten A, Koch U, Hoessbacher C, Ducry F, Juchli L, Emboras A, Hillerkuss D, Kohl M, Dalton LR, Hafner C, Leuthold J (2015) All-plasmonic Mach-Zehnder modulator enabling optical high-speed communication at the microscale. Nat Photonics 9:525–528
16. Salamin Y, Ma P, Baeuerle B, Emboras A, Fedoryshyn Y, Heni W, Cheng B, Josten A, Leuthold J (2018) 100 GHz plasmonic photodetector. ACS Photonics 5:3291−3297. https://pubs.acs.org/doi/10.1021/acsphotonics.8b00525
17. Wells J (2009) Faster than fiber: the future of multi-G/s wireless. IEEE Microwave Mag 10:104–112
18. Koenig S, Lopez-Diaz D, Antes J, Boes F, Henneberger R, Leuther A, Tessmann A, Schmogrow R, Hillerkuss D, Palmer R, Zwick T, Koos C, Freude W, Ambacher O, Leuthold J, Kallfass I (2013) Wireless sub-THz communication system with high data rate. Nat Photonics 7:977–981
19. Salamin Y, Heni W, Haffner C, Fedoryshyn Y, Hoessbacher C, Bonjour R, Zahner M, Hillerkuss D, Leuchtmann P, Elder DL, Dalton LR, Hafner C, Leuthold J (2015) Direct conversion of free space millimeter waves to optical domain by plasmonic modulator antenna. Nano Lett 15:8342−8346. https://pubs.acs.org/doi/10.1021/acs.nanolett.5b04025
20. Hoessbacher C, Salamin Y, Fedoryshyn Y, Heni W, Baeuerle B, Josten A, Haffner C, Zahner M, Chen H, Elder DL, Wehrli S, Hillerkuss D, Van Thourhout D, Van Campenhout J, Dalton LR, Hafner C, Leuthold J (2017) Optical interconnect solution with plasmonic modulator and Ge photodetector array. IEEE Photonics Technol Lett 29(21):1760–1763
21. Leuthold J, Hoessbacher C, Muehlbrandt S, Melikyan A, Kohl M, Koos C, Freude W, Dolores-Calzadilla V, Smit M, Suarez I, Martínez-Pastor J, Fitrakis EP, Tomkos I (2013) Plasmonic communications: Light on a wire. Opt Photonics News 24:28–35
22. Heni W, Haffner C, Baeuerle B, Fedoryshyn Y, Josten A, Hillerkuss D, Niegemann J, Melikyan A, Kohl M, Elder DL, Dalton LR, Hafner C, Leuthold J (2016) 108 Gbit/s plasmonic Mach-Zehnder modulator with >70-GHz electrical bandwidth. J Lightwave Technol 34:393–400
23. Burla M, Hoessbacher C, Heni W, Haffner C, Fedoryshyn Y, Werner D, Watanabe T, Massler H, Elder DL, Dalton LR, Leuthold J (2019) 500 GHz plasmonic Mach-Zehnder modulator enabling sub-THz microwave photonics. APL Photonics 4:056106

24. Seeds AJ, Shams H, Fice MJ et al (2015) Tera-Hertz photonics for wireless communications. J Lightwave Technol 33:579–587
25. Jia S, Pang X, Ozolins O et al (2018) 0.4 THz photonic-wireless link with 106 Gb/s single channel bitrate. J Lightwave Technol 36:610–616
26. Wang C, Zhang M, Chen X et al (2018) Integrated lithium niobate electro-optic modulators operating at CMOS-compatible voltages. Nature 562(7725):101–104
27. Hoessbacher C, Josten A, Baeuerle B et al (2017) Plasmonic modulator with >170 GHz bandwidth demonstrated at 100 GBd NRZ. Opt Express 25:1762–1768

Index

A
Abrupt bend, 35
Absorption, 61
Absorption coefficient, 50
Absorption cross section, 61
Absorption saturation, 162
Absorption spectra, 63
Activation function, 130
Add-drop filters, 152
Add-drop MRR filter, 17
Add/Drop Multiplexer (ADM), 16, 39
Add/drop multiplexing, 76
AI accelerators, 112, 129
Air-bridge, 26
Air-bridge structure, 163
Air-bridge type, 36
Air-bridge type PC, 36
Air-hole, 26
Air-hole shift, 155
Air-hole triangular lattice, 30
All-optical switch, 152
All-optical switch/memory, 174
All-Photonic Network (APN), 72
Alternative path, 97
Amplitude control line, 117
ANN architecture, 130
Anomalous-dispersive medium, 8
Arrayed DM-PCSEL, 189
Arrayed Waveguide Grating (AWG), 17, 72
Arrayed waveguides, 153
Artificial Intelligence (AI), 112
Artificial Neural Network (ANN), 112, 129
Artificial neurons, 130
Asymmetric air-holes, 184
Asymmetric bound mode, 201
Asymmetric electrode, 181
Asymmetric planar dielectric slab waveguide, 5
Asymmetric waveguide, 201

B
Back and forth motion, 156
Back-end SOI CMOS photonics, 85
Backward propagation, 132
Bacterial Foraging Optimization (BFO), 140
Ball Grid Array (BGA), 84
Band-edge emission, 43
Band-edge (G-point) emission, 153
Bar port, 14
Beam combiner, 12
Beam quality, 184
Beam scanning mode, 190
Beam splitter, 12
Beam steering, 113
Beat frequency, 124
Bend curvature radius, 35
Bend waveguide, 35, 44
Beneš, 96
Beyond CMOS, 129
BH-PC laser nano-cavity, 174
Bias light, 175
Bias power, 175
Binary-class data, 139
Bi-sable threshold power, 175
Bi-stable optical memory, 152
Bi-stable switches/memories, 153
Bit Error Rates (BERs), 90
Bloch folding line, 32
Blocking characteristics, 97
Boolean logics, 140

K. Asakawa et al., *Nano-photonics for Advanced Networks*,
https://doi.org/10.1007/978-3-031-80637-7

Bound mode, 53
Bound SP mode, 199
BOX layer, 86
Bragg diffraction, 180
Brightness, 187
Brillouin zone, 28
Broadband plasmonic modulator, 206
Broad lasing area, 183
Bulk plasmon, 48
Bulk plasmon frequency, 63
Buried Oxide (BOX) layer, 75
Buried waveguide, 9
Butt-coupling, 21

C
Cache memory based on SRAM, 86
Cantilever coupling, 74
Carrier light wave, 8
Carrier relaxation time, 175
Carrier wave velocity, 42
Cascading stages, 97
CDC–ROADM system, 95
Centralized laser, 210
Channel waveguide, 12
Chip-bonding, 78
Chip bonding on Si, 78
Circular air-holes, 168
Circular-shaped unit cell, 183
Circular uniform electrode, 181
Clad region, 10
Classification accuracy, 144
Classification and regression, 139
Classification of the two circles, 137
CMOS-compatible fabrication, 73
CMOS process, 73
Coherent electron-photon interaction, 40
Coherent lasing oscillation, 181
Coherent LiDAR, 122
Coherent oscillation, 47
Coherent oscillatory distribution, 52
Collective wave, 47
Collision rate, 49
Colorless, Directionless, and Contention-less (CDC) functions, 94
Colorless function, 95
Commercial laser processing fields, 191
Commercial laser processing industry, 188
Common bus waveguide, 176
Compact high-speed optical interconnect transceivers, 209
Compact MZ modulator, 159
Compact 3-D LiDAR system, 190
Complex band-edge structure, 182
Complex conjugate, 130
Complex dielectric function, 52
Complex polarization mode, 182
Complex-valued encodes, 134
Complex-valued neural network, 134
Complex-valued operations, 134
Complex-valued Optical Neural Chip (ONC), 134
Complex wave vector, 56
Conduction electron, 49
Contention-less function, 95
Control light, 173
Convolutional Neural Network (CNN), 142
Convolution operation, 142
Core network, 93
Core region, 9
Corning LX MMF, 90
Correct identification, 132
Correlation matrix, 132
Coupled common bus waveguide, 152
Coupled modes, 14, 201
Coupled mode theory, 13
Coupling factor, 186
Coupling length, 14
CP wavelength, 164
Critical angle, 2
Crossbar, 96
Crossbar-type switch matrix, 105
Cross coupling, 19
Cross-coupling coefficient, 16
Cross port, 14
Cross section, 61
Crystalline Si (c-Si) layer, 86
Cut-off frequency, 5
Cut-off thicknesses, 5
Cylindrical lens, 12

D
Data-center, 93
Data-processing speed, 112
DBR structure, 190
Deep learning, 112, 129
Defect-based eigen-mode, 32
Defect mode, 31, 32
De-multiplexer, 82, 211
Dense metallic electron, 49
Detuning wavelength, 164
Diagonal matrix, 134
Dielectric constant, 26
Dielectric function, 48, 49
Dielectric light line, 55

Dielectric-loaded metal, 199
Dielectric prism, 57
Differential Quadrature Phases-Shift Keying (DQPSK), 159
Diffraction grating, 59
Diffraction-limit free properties, 197
Diffractive cell, 142
Diffused waveguide, 10
Digital CMOS image sensor, 124
Digital coherent technology, 72
Digital flip-flop switching, 152
Digital PLL (phase locked loop), 88
Dimer nano-antenna, 65
Dipole antennas, 65
Dipole moment, 49
Directional Couplers (DC), 10, 36, 202
Directionless function, 95
Discharge in a fluorescent tube, 47
Dispersion compensation, 153
Dispersion curve, 8, 29
Dispersion relation, 28
Dispersion relation of plasmon, 51
Distributed Bragg Reflector (DBR), 79
Doppler frequency (fd) shift, 121
Doppler shift, 124, 128
Double-core region, 20
Double hetero-structure cavity, 41
Double-Hole lattice (DH), 185
Double-lattice configuration, 153
Double micro-rings, 102
Double nano-particles, 63
Double particles, 64
Drop port, 16
Drude-model, 48
Dual LSPCW–MZM, 161
Dually-Modulated PCSEL (DM-PCSEL), 188

E

Earth's upper atmosphere, 47
Edge-coupling, 74
Effective electron mass, 49
Effective-medium theory, 75
Effective phase matching, 22
Effective refractive index, 4
Eigen-frequency, 29
Eigen mode, 14, 26
Eigenvalue equation, 7, 29
Electrical and thermal phenomena, 48
Electric field penetration depth, 56
Electric field wave equation, 53
Electromagnetic-wave approach, 1
Electron-beam lithography, 36
Electron flux density, 49
Electron gas, 47, 197
Electron gas density, 47
Electronic charge, 48
Electro-optic effect, 207
Electro-optic Pockels effect, 77
Elementary excitation, 48
Elliptical hole-shape, 183
Elliptical-shaped unit, 183
Embedded semiconductor quantum dots (QDs), 152
Embedding four InAs-QD layers, 163
End-coupled, 176
End mirror, 15
Entangled shape, 137
Envelope, 8
Envelope velocity, 42
EO effect (Pockels effect), 104
Epitaxial growth, 78
Epitaxial SiGe PD, 87
Equation of motion, 49
Erasing operations, 175
Evanescent field distribution, 57
Evanescent wave, 59
Excellent beam quality, 44
Exciton, 48
Exclusive algorithm, 132
Exclusive OR (XOR), 139
External laser array, 82
External phase shifter, 132
External splitter, 97
Extinction, 61
Extinction coefficient, 50
Extinction cross section, 61
Extinction Ratio (ER), 77
Extinction spectra, 63
Extraordinary refraction, 38
Extraordinary refraction model, 39, 40
Extreme anisotropy, 40
Eye-opened outputs, 168
Eye pattern, 90

F

Fab, 81
Fabless, 81
Fabry–Perot cavity, 79
Fabry–Perot Resonator (FPR), 10, 14, 170
Face-centered rectangular-lattice, 41
Failure recovery applications, 99
Fan-Shaped Far Field Pattern (FFP), 127
Far-field patterns, 181

Feedback loop, 165
Fiber-to-chip light coupler, 11, 21
Field confinements, 199
Field-Programmable Gate Array (FPGA), 88
50 Gbps optical link, 84
Fingertip-size Si photonic transceiver, 88
Finite difference method, 133
Finite-Difference Time-Domain (FDTD), 36, 156
Finite-Difference Time-Domain (FDTD) method, 36
5G mobile-network, 91
1st-order Bragg diffraction, 75
Flash mode, 190
Flat-band property, 156
Flip-flop MZI digital switch, 152
FMCW-LiDAR, 119
Focal Plane Array (FPA), 124
45-nm CMOS-compatible process, 103
45 nm Si photonic die, 86
Forward propagation, 132
Foundry, 81
Fourier transform, 142
Free electron, 47
Free electron gas theory, 48
Free electron model, 50
Free-propagation region, 17
Free-propagation slab waveguides, 142
Free-space communication, 111
Free space path loss, 211
Free Spectral Range (FSR), 15, 170
Frequency-Modulated Continuous Wave (FMCW), 119
Fresnel reflection, 22
Front/back loop mirrors, 80
Front-end SOI CMOS photonics, 85
Front mirror, 15
Full PBG, 30
Fully-etched GC, 22
Fully-integrated LiDAR system, 117

G

GaAs substrate, 36
Gallium Arsenide (GaAs), 78
Γ-X direction, 180
Gas-/fiber-lasers, 187, 191
Gaussian-like envelope, 41
Genetic algorithm, 146
Ge photodiode Rxs (receivers), 210
Gold nanoparticle, 62
Goose–Hänchen shift, 3
Gradient descent, 132
Grating antenna, 122
Grating array, 114
Grating Coupler (GC), 11, 22, 74, 209
Group velocity, 7, 42
Group Velocity Dispersion (GVD), 42
Group-Velocity-Dispersion (GVD) control, 153
Guided light wave, 1
Guided mode, 3
Gyro (gimbal), 111

H

Hadamard product, 142
Handwriting recognition task, 136
Handwritten digit, 145
Harmonic mode, 29
Harmonic time-dependence, 49
Heterodyne receiver, 124
Heterogeneously die-bonded, 116
Heterogeneously integrated narrow-linewidth laser, 79
Heterogeneous Si photonics sensing, 115
Hidden layer, 130
High-Bandwidth Memories (HBMs), 85
High-Δ (refractive index ratio), 105
High fabrication accuracy, 179
High index-contrast, 28, 151
High Index-Contrast Grating (HCG), 78
High-index wave-guiding core region, 9
High on-off contrast, 99
High port-count, 103
High-power oscillation, 179
High-Q nano-cavity, 40
High-Q nano-cavity resonator, 171
High-speed plasmonic modulator, 204
Homodyne configuration, 119
Hot-spot, 63
Hybrid Dielectric-Loaded Plasmonic Waveguide (HDLPW), 202
Hybrid III-V laser, 78
Hybrid Insulator-Metal-Insulator (HIMI), 203
Hybrid Insulator-Metal-Insulator (HIMI) plasmonic waveguides, 203
Hybrid-integrated III–V/Si platform, 115
Hybrid-integrated silicon OPA, 114
Hysteresis loop, 171

I

Image sensing matrix, 124
InAs-QDs, 163

Incoherent lasing oscillation, 181
Index-chirped DC, 156
Index-chirping, 42
In-direct bandgap, 78
Inference task, 132
Infra-red frequency region, 55
Innovative Optical and Wireless Network (IOWN), 72
In-plain 180°-diffraction, 186
In-plane Bragg diffractions, 180
In-plane-coupling spectra, 78
In-plane first-order Bragg diffraction, 43
In-plane resonant Γ-point laser emission, 181
Input vector, 136
Insulator-Metal-Insulator (IMS) structures, 199
Integrated Diffractive Neural Network (IDNN), 142
Integrated diffractive ONN, 142
Integrated narrow-linewidth laser, 80
Integrated optical beam steering arrays, 112
Integrated optical DRAM, 152
Integrated photonics, 73
Integrated solid state Si LiDAR, 112
Integrated two-dimensional (2-D) OPAs, 113
Interference pattern, 12
Interferometric waveguide, 10
Interleaved p and n junctions, 87
Inter-leaver block (INT), 101
Intermediate Frequency (IF), 120
Internet of Things (IoT), 71
Interposer, 85
Inverse taper coupler, 11, 22, 74
Ionosphere, 47
Iris, 139
Iris classification, 141
Isolated atom, 49

K

Kinetic theory on gass, 48
Kramers-Kronig relation, 163
Kretschmann configuration, 57, 199

L

Land Grid Array (LGA) interposer, 103
Large angular dispersion, 127
Large beam steering range, 127
Large group index, 127
Large/infinity group index, 153
Large optical-field confinement, 165
Large optical nonlinearity, 165
Large-scale all-optical DRAM, 176
Large-scale memory integration, 179
Large scale 2-D Si OPAs, 115
Laser active section, 80
Lateral (ψ) direction, 112
Lattice mismatch, 79
Lattice-shifted air-holes, 159
Lattice-Shifted PC Waveguide (LSPCW), 127, 157
Leaky mode, 2
Leaky region, 40
Leaky SP mode, 199
Lens-free 2-D LiDAR, 190
Light-blocking system, 99
Light cone, 33
Light Detection and Ranging (LiDAR), 111, 188
Light localization, 25
Light-matter interaction, 64, 65, 152
Light wave equation, 51
$LiNbO_3$ waveguide, 10
Linear EO effect, 77
Linearly chirped frequency modulation, 120
Linear matrix operator A, 140
Line-defect 2-D PC waveguide, 152
Line-defect mode, 32
Line-defect PC waveguide, 151
Loading strip, 9
Localization-induced delay, 42
Localized SP resonance, 60
Localized SP Resonance (LSPR) frequency, 63
Localized Surface Plasmon (LSP), 60
Local Oscillator (LO), 120
Local Oscillator (LO) light, 124
Logical inputs, 136
Longitudinal (θ) direction, 112
Longitudinal wave, 47
Long-range propagation length, 203
Long-Range SP (LRSP), 201
Lorentz model, 49
Low cost per data lane, 82
Low dispersion, 154
Low-dispersion slow-light regions, 158
Low group-Velocity Dispersion (LVLD), 157
Low-index wave-guiding clad region, 10
Low-noise photodetector, 64
Low/zero group velocity, 153

M

Machine-learning, 112

Machine-Learning (ML) model, 135
Mach–Zehnder Interferometer (MZI), 10, 77
Mach–Zehnder interferometric waveguide, 11
Magnetic permeability, 29
Mapping functions G(x), 139
Margin of the support vectors, 139
Matrix-Vector Multiplication (MVM), 142
Maximum angular dispersion, 39
Maxwell's equations, 5, 48
Mean Squared Error (MSE), 140
Mesh-based network topology, 94
Metal-air surface, 58
Metal/air (top) interface, 199
Metal–dielectric interface, 52, 197
Metal/glass (bottom) interface, 206
Metal grooves, 198
Metal–Insulator–Metal (MIM) structures, 199
Metallic ion, 48
Metal nano-antenna optics, 63
Metal nano-particle, 63
Metal nano-particle optics, 60
Metal optics, 48
Metal/prism interface., 199
Metal–Semiconductor–Metal (MSM) slot, 208
Metal (silver)-slot waveguide, 200
Metal slots, 199
Metal stripe PW, 199
Metal stripes, 198
Metal surface, 52
Micro-Ring Resonator (MRR), 11, 77, 102
Micro-Ring Resonator (MRR) crossbar array, 146
Microwave plasmonics, 204
Mie scattering, 61
Millimeter-Wave (MMW) frequencies, 211
MIM slot core, 205
Missing air-hole line-defect waveguide, 159, 163
Missing air-rods, 40
Missing line defect, 31
MMI trees, 112
Mode equation, 3
Molecular Beam Epitaxy (MBE), 163
Monolithically integrated OPA transceiver, 117
Monopole, 64, 65
M-point resonance, 189
Multi-cast connection, 97
Multi-cast optical switch, 93
Multi-cast Si photonic switch, 97
Multicast Switch (MCS), 96
Multi-Core Fibers (MCFs), 210
Multi-degree ROADM, 94
Multi-layer dielectric film., 26
Multi-layer IDNN, 142
Multimode interferometer, 10
Multi-Mode Interferometer (MMI) coupler, 99, 207
Multi-Mode Interferometer (MMI)-rib waveguides, 92
Multiple rotation matrix, 134
Multiple wavelength channel, 95
Multiplexer, 82
Multiply-reflected land transmitted ight, 15
Multi-port switches, 97
Multi-ring network, 94
MUX (de-mux block), 101
MZI mesh, 139

N

Nano gap, 65
Nano-monopole, 65
Nano-photonic waveguide devices, 72
Nano-rod, 28
Nano-rod square lattice, 30
Nano-scale cavity, 151
Nanoscale killer photonic device, 73
Nano-scale periodical media, 151
Narrow beam divergence, 179
Narrow sub-wavelength metal-gap regions, 206
Near-field patterns, 181
Near infra-red region, 10
Network of physical objects, 91
Neuron-model, 136
New Si ONN chip, 146
Next-generation smart Society 5.0, 192
90°-circular bends, 202
Non-blocking, 97
Non-circular shapes, 168
Non-dispersive medium, 8
Non-duplicate polarization-diversity, 105
Non-Gaussian far field pattern, 115
Nonlinear activation, 130
Nonlinear activation function, 130
Nonlinear datasets, 136, 140
Nonlinear decision boundaries, 136
Nonlinear etalons, 172
Nonlinear frequency converter, 65
Nonlinearity enhancement device, 157
Nonlinear mapping-based classification, 138

Nonlinear material (DLD164), 209
Nonlinear organic material (HD-BB-OH/YLD124), 211
Nonlinear PC devices, 152
Nonlinear projection operator G(x), 140
Nonlinear refractive index, 170
Nonlinear XOR gate, 136
Non-PC ridge waveguide, 164
Non-Return-to-Zero (NRZ), 161
Non-Return-to-Zero (NRZ) PAM-2 signal, 207
Non-uniform distribution, 187
Normal-dispersive medium, 8
No vertical emission, 189
N/p contact metallization, 116
N-plane first-order Bragg diffraction, 43
NRZ transmitter die, 87
Nucleus of charge, 48

O

Off-chip laser source, 115
Off-memory, 176
OM3 fiber, 90
On-chip back propagation, 146
On-chip backpropagation algorithm, 146
On-chip detector, 82
On-chip GC (grating coupler) arrays, 210
On-chip semiconductor laser, 78
1-D PC, 26
1-D input sequence, 143
100 W-class output power, 186
105-bit memory operation, 179
On-memory, 176
ON–OFF keying, 208
On-Si laser, 79
Operation latency, 112
Optical absorption of light, 50
Optical amplifier or attenuator, 130
Optical birefringence, 18
Optical bi-stable devices, 170
Optical buffer, 152
Optical buffer memories, 153
Optical carrier, 209
Optical coherent receiver, 159
Optical delay lines, 152
Optical Discrete Fourier Transform (ODFT), 142
Optical DRAM operations, 152
Optical extinction, 60
Optical flip-flop, 165
Optical free-space/inter-satellite communications, 188
Optical interference unit, 130
Optical Inverse Discrete Fourier Transform (OIDFT), 142
Optical I/O core, 88
Optical linear regime, 172
Optical link, 82
Optical matrix multiplication, 130
Optical nano-antenna, 63
Optical Network Unit (ONU), 92
Optical Neural Network (ONN), 129
Optical non-linear regime, 172
Optical Phased Array (OPA), 112
Optical pin, 89
Optical Random Access Memory (O-RAM), 175
Optical resonator, 10
Optical switching devices, 65
Optical tag, 65
Optical telecom network, 72
Optimum hyperplane, 139
Ordinary refraction, 38
Oscillation of electron, 49
Otto configuration, 58
Out-of-chip multiplexer, 211
Output layer, 130
Overpass access waveguides, 105
Oxide cladding type, 34

P

Packaging ceramic interposer, 105
Packaging techniques, 84
Parallel channel waveguide, 14
Passive Optical Network (PON), 91
Pass port, 16
Path-dependent insertion loss, 105
Path-Independent-Loss (PILOSS), 96
PBG defect-mode lasers, 179
PC Coupled Waveguide (PCCW), 155
PC guided mode coupling, 152
PC nano-cavity, 170
PC-specific structures, 151
PC Surface-Emitting Laser (PCSEL), 153
Periodic air-rod lattice, 170
Periodic lattice component, 28
Periodic medium, 29
Periodic PC lattice, 31
Periodic structure, 25
Phase control line, 117
Phased array, 17
Phased array antenna, 114
Phase velocity, 4, 42
Phonon, 48

Photoconductive effect, 208
Photonic band, 29
Photonic Band Gap (PBG), 25, 179
Photonic Crystal (PC), 25
Photonic crystal slab, 34
Photonic-electronic convergence technology, 72
Photonic machine learning model, 138
Photonic packet router, 152
Photon Scanning Tunneling Microscope (PSTM), 200
Physical dimensions, 199
Pillar array, 31
PILOSS topology, 103
P-i-n and p-n junctions, 77
Plain-wave expansion method., 40
Planar Lightwave Circuit (PLC) fiber connector, 105
Planar Light Wave Circuits (PLCs), 72
Plasma, 47
Plasma chamber, 47
Plasma-dispersion, 76
Plasma-dispersion effect, 77
Plasma frequency, 48
Plasmon, 47
Plasmon frequency, 51
Plasmonic mode converter, 204
Plasmonic nano-antenna, 208
Plasmonic Phase Modulator (PPM), 209
Plasmonic waveguides, 198, 204
Plasmon Polariton (PP), 48
P-n phase shifter, 78
Point cloud 3-D image, 128
Point-defect mode, 32
Point-to-point connection, 97
Polariton, 48
Polarization beam rotator, 11
Polarization Beam Splitter (PBS), 11, 92
Polarization diversity system, 18
Polarization-independent bandpass filter, 92
Polarization-independent MZI modulator, 19
Polarization insensitivity, 99
Polarization Rotator (PR), 92
Polarization Splitter-Rotator (PSR), 105
Positive ion, 47
Positive slope, 155
Power consumption, 129
Power imbalance, 104
Power splitter, 14
Practical (de-) multiplexers, 39
Pre-amplified tunable laser, 114
Processor module, 84
Programmable Nano-Photonic Processor (PNP), 132
Projection-based Photonic Classifier (PPC), 138
Propagation length, 55
Propagation loss, 198
Propagation velocity, 8
PR with double-core structure, 20
PR with mode-evolution, 20
Pseudo flip-flop behavior, 166
Pseudo-random binary sequence signals, 161
PS with crossed waveguide, 20
PS with MMI waveguide, 19
PS with sub-wavelength grating, 19
Pulse amplitude modulations, 207
Pulse-excited PCSEL, 181

Q

Quadrants, 136
Quadratic mapping function, 139
Q-value in the cavity, 40

R

Race-track Micro-Ring Array (MRA)., 160
Race track resonator, 102
Radar, 111
Radiating grating cell, 117
Radiation mode, 33
Radiative mode, 53
Radio-over-Fiber (RoF) link, 211
Randomly distributed ionized gas, 47
Rayleigh scattering, 65
Ray-optics approach, 1
Reactive-ion-beam etching, 36
Read-pulse, 176
Real-valued matrix, 131
Real-valued representations, 134
Received Beam (RX), 120
Receiver, 84
Receiver FPA, 124
Reciprocal lattice, 28
Recognition, 143
Re-configurable Optical Add/drop Multiplexers (ROADMs), 72, 94
Reconfiguring optical path, 97
Rectangular diagonal matrix, 130
Rectangular waveguide, 10
Reduced BZ (the first BZ), 29
Refractive index spectrum, 164
Refreshment clock pulses, 166
Relaxation time, 49

Remote Antenna Unit (RAU), 211
Remote sensing, 112
Resonant-wave mode, 180
Resonator mirror, 170
Resonator mode, 15
RF electric field, 209
Rib waveguide, 10
Ridge waveguide, 9
Ring-resonators, 172
RISC-V processor, 86
Robust light-blocking system, 99
Round-trip, 16
Round-trip phase variation, 15

S
Saturable absorber, 132
Saturable absorption (SP), 162
Scattering, 61
Scattering cross section, 61
Selective chemical etching, 163
Selective wet-etching, 36
Self-coupling coefficient, 16
Self-images, 12
Self-organized bias sputtering, 38
Self-Phase Modulation (SPM), 154, 158
Semiconductor cladding type, 35
Semiconductor Optical Amplifier (SOA), 105
Shallow-etched GC, 22
Short optical links, 82
Si chip modules, 100
Si core, 73
Side-coupling, 177
Si DQPSK receiver, 159
Sigmoid function, 130
Signal light, 173
Silicon-On-Insulator (SOI), 73
Si main bus waveguides, 117
SiN/Si double-layer platform, 105
Si nano-cavity o-RAM chip, 179
Single beat frequency, 124
Single isolated atom, 48
Single LSPCW–MZM, 161
Single micro-ring, 102
Single-mode, 179
Single-Mode Fiber (SMF), 21
Single-mode region, 7
Single neuron, 136
Single-pass, 16
Single Polarization Grating Coupler (SPGC), 101
Single-ring network, 94
Singular beat frequency, 120
Singular Γ-point, 179
SiO_2 cladding, 73
SiO_2-silver (Ag)-SiO_2 SP waveguide, 203
Si ONN, 112
Si photonic ANN accelerator, 112
Si photonic fiber-to-chip coupler, 74
Si photonic Multi-Chip-Module (MCM), 84
Si photonic multi-port switches, 93
Si photonic MZI modulator, 207
Si photonic nano-wire modulator, 76
Si photonic nano-wire switch, 76
Si photonic platform, 73, 81
Si photonic switch, 93
Si photonic transceiver market, 81
Si slot-type MIM waveguide, 208
Six equivalent G-X directions, 181
60°-bend, 36
Skin depth, 50
Slab waveguide, 2, 9
Slot-type MIM waveguide, 205
Slow light, 42
Slow-light effect, 42
Slow-light grating beam scanner, 126
Slow-light PC, 152
Slow-light zone, 155
Small group velocity, 127
Smart laser processing, 192
Smart mobility, 192
Smart sensing, 188
SOI CMOS-process fabrication, 203
SOI-CMOS processing, 10
Solid state LiDAR systems, 190
Space-Division Multiplexing (SDM), 210
Specific full PBG, 30
Specific Photonic Band (PB) diagram, 43
Spectral absorption peak, 60
Spectral scattering peak, 60
SP frequency, 53
Sphere nano-particle, 62, 63
Spherical noble metal particles, 60
Spherical particle, 61
Split-and-select connection, 97
Splitters and selectors, 97
SP MZM (Mach–Zehnder modulator), 211
Spoked ring-modulator, 87
Spot-Size-Converters (SSCs), 158
SP photodetector, 208
SP resonance frequency, 63
SP slot-type MIM waveguide, 207
SP slot-type MSM detectors, 208
SP slot-type MZM channels, 210
SP wavelength, 164

SP wave vector, 57
Square lattice, 28
Square-lattice PC slab, 43, 180
S-shape bend, 203
S-shaped waveguide bends, 202
Stable single-mode operation, 186
Stacked Si nano-wire, 202
Standing wave, 3
Starfish-shaped dispersion surface, 40
Static/moving target, 111
Steering angle, 122
Stop band, 32
Storage operations, 175
Straight waveguide, 36
Strip-loaded waveguide, 9
Strong nonlinear interaction, 209
Sub-cladding layer, 183
Sub-THz photonics, 212
Sub-wavelength GC, 23
Sub-Wavelength Grating (SWG), 75
Sub-wavelength nanostructure, 76
Sub-wavelength-pitch periodic waveguide, 75
Sub-wavelength region, 76
Successful bi-stable operation, 166
Super-prism effect, 38
Support-Vector Machine (SVM), 138
Surface-coupling, 74
Surface-/edge-emitting lasers, 152
Surface-emitting lasing, 44
Surface-normal direction, 44
Surface Plasmon (SP), 47, 52, 197
Surface Plasmon Polariton (SPP), 52
SWG coupler, 75
Switch-and-Select (S&S), 96
Switch cells, 97
Switch fabrics, 93
Switching polarity, 173
Switch-off control pulse (reset pulse), 165
Switch-off CP, 163
Switch-on Control Pulse (CP), 163
Switch-on control pulse (set pulse), 165
Switch topology, 96
Symmetrical MZI analog switch, 152
Symmetric bound mode, 201
Symmetric Directional Coupler (DC), 155
Symmetric slab waveguide, 5
Symmetric slot waveguide, 201
System-on-Chip (SoC), 86

T
Taper coupler, 21
Tapered Si waveguide, 20
Target image, 142
Tbit/s wireless communication system, 209
Test image, 142
Thermal equilibrium, 49
Thermal phase tuner, 80
Thermo-Optics (TO), 93
Thinner slab terrace, 92
III-V compound semiconductor, 73
III-V laser, 78
III-V Optoelectronic Integrated Circuits (OEICs), 73
III-Vs on Ge-on-Si (Ge/Si), 79
III–V/Si hybrid MOS phase shifter, 146
III–V/Si phase shifter array, 116
Third-order optical nonlinearity, 162
3-D imaging LiDAR, 124
3-D LiDAR, 123
3-D PC, 26
3-D plasmonic slot waveguides, 200
3-D resin polymer waveguide, 89
3-D TSV (through silicon via), 85
3mm-square DM-PCSEL chip, 189
THz band, 211
THz microwave photonics, 211
THz-to-optical conversion, 211
THz-to-optical-to-THz range, 212
THz wireless signals, 211
Tightly bound electron, 48
Time-dependent ONL phase shift, 168
Time-differential MZI, 163
Time-differential phase modulation, 165
Time Domain Multiplexing (TDM), 72
Time-of-Flight (ToF) method, 190
Titanium (Ti) diffused core, 10
TO attenuator, 118
TO-based MZI switch elements, 101
TO-based phase shifter, 99, 132
TO phase shifter, 118
Topology Optimization (TO), 168
Total Internal Reflection (TIR), 1, 58, 199
TPA sub-system, 100
TPA switch unit, 97
Traffic explosions, 71
Training data, 132
Training task, 111
Transmission capacity, 72
Transmitted Beam (TX), 119
Transmitter, 82
Transmitter FPA, 124
Transmitter (Tx) array, 210
Transponder Aggregator (TPA), 95
Transpose matrix operation, 146

Transverse multi-modes, 153
Transverse wave, 48
Triangular lattice, 28
Triangular-lattice LSPCW, 157
Tri-angular lattice PC slab, 180
T-shaped 2-D PC waveguide branch, 36
Tunable delay devices, 153
Tunable delay/dispersion device, 152
Tunable dispersion compensator., 159
Tunable lasers, 173
28-bit memory operation, 178
Two beat frequencies, 124
Two-bit logic gate, 136
2-D DFB lasers, 179
2-D grating coupler, 19
2-D LiDAR, 123
2-D PC, 26
2-D PC Y-junction and bend, 168
2-D PC high-Q nano-cavities, 152, 170
2-D PC Surface-Emitting Laser (PCSEL), 179
2-D photonic crystal slab, 32
2-D three-digit images, 143
Two-Photon Absorption (TwPA), 154, 158, 171
Two-wavelength bi-stable switch, 173

U
Ultra-Violet (UV) region, 55
Unallocated spectrum range, 211
Uni-cast (multi-port) optical switches, 93
Undesirable degeneracy, 182
Uni-directional VGC, 87
Unitary matrix switches, 132
Unitary transformation, 130
Uni-Traveling-Carrier Photodiodes (UTC-PDs), 211
Upward and downward sweeps, 172
U-shaped bend, 203

V
Valence electron, 48
Variable Optical Attenuators (VOAs), 159
Vertical-Cavity Surface-Emitting Laser (VCSEL), 78
Vertical coupling, 186
Vertical-coupling spectra, 78
Vertical emission loss, 186
Vertical Grating Coupler (VGC), 86
Very-low group velocity, 55
Visible-light frequency region, 55
Von Neumann computing approaches, 112
Vowel recognition, 132

W
Wafer-bonding, 78
Wafer-fusion technique, 180
Water vapor-based atmospheric absorption, 211
Waveguide arms, 11
Waveguide crossings, 105
Waveguide grating array, 116
Waveguide-to-die interconnect, 84
Wavelength Division Multiplexing (WDM), 16, 72
Wavelength-selective device, 16
Wavelength-selective MRR, 105
Wavelength Selective Switch (WSS), 95
Wavelength-selective Y-junction waveguides, 168
Wavelength tuning, 122
Wavelength-tuning ring resonators, 80
Wave vector, 28
Wave vector conservation, 59
WDM operations, 210
Weak bound valence electron, 48
Weight matrix, 136
Weight parameters, 132
Weight vector, 136
Wireless communication, 209
Wireless signals, 209
Write-pulse, 175, 176
Writing operations, 175

Y
Young's interferometer, 10
Y-splitter, 36

Z
Zero-change Si photonic platform, 86

GPSR Compliance

The European Union's (EU) General Product Safety Regulation (GPSR) is a set of rules that requires consumer products to be safe and our obligations to ensure this.

If you have any concerns about our products, you can contact us on ProductSafety@springernature.com

In case Publisher is established outside the EU, the EU authorized representative is:

Springer Nature Customer Service Center GmbH
Europaplatz 3
69115 Heidelberg, Germany

Batch number: 10755227

Printed by Printforce, the Netherlands